Advances in Experimental Medicine and Biology

Proteomics, Metabolomics, Interactomics and Systems Biology

Volume 1510

Advances in Experimental Medicine and Biology provides a platform for scientific contributions in the main disciplines of the biomedicine and the life sciences. This series publishes thematic volumes on contemporary research in the areas of microbiology, immunology, neurosciences, biochemistry, biomedical engineering, genetics, physiology, and cancer research. Covering emerging topics and techniques in basic and clinical science, it brings together clinicians and researchers from various fields.

Advances in Experimental Medicine and Biology has been publishing exceptional works in the field for over 40 years, and is indexed in SCOPUS, Medline (PubMed), EMBASE, Reaxys, EMBiology, the Chemical Abstracts Service (CAS), and Pathway Studio.

2023 CiteScore: 5.9

This series of volumes focuses on concepts, techniques and recent advances in the field of proteomics, interactomics, metabolomics and systems biology. Recent advances in various 'omics' technologies enable quantitative monitoring of myriad various biological molecules in a high-throughput manner, and allow determination of their variation between different biological states on a genomic scale. Now that the sequencing of various genomes, from prokaryotes to humans, has provided the list and linear sequence of proteins and RNA that build living organisms, defining the complete set of interactions that sustain life constitutes one of the key challenges of the postgenomic era. This series is intended to cover experimental approaches for defining protein-protein, protein-RNA, protein-DNA and protein-lipid interactions; as well as theoretical approaches dealing with data analysis, integration and modeling and ethical issues.

José-Luis Capelo-Martínez
Editor

Emerging Sample Treatments in Proteomics

Second Edition

Springer

Editor
José-Luis Capelo-Martínez
BIOSCOPE Research Group, LAQV-REQUIMTE, Department of Chemistry, NOVA School of Science and Technology, Universidade NOVA de Lisboa
OMICS and Analytical Development Group
2829–516 Caparica, Portugal

ISSN 0065-2598 ISSN 2214-8019 (electronic)
Advances in Experimental Medicine and Biology
ISSN 2730-6216 ISSN 2730-6224 (electronic)
Proteomics, Metabolomics, Interactomics and Systems Biology
ISBN 978-3-032-21637-3 ISBN 978-3-032-21638-0 (eBook)
https://doi.org/10.1007/978-3-032-21638-0

New Edition

This Springer imprint is published by the registered company Springer Nature Switzerland AG
The registered company address is: Gewerbestrasse 11, 6330 Cham, Switzerland

Preface

Proteomics has evolved into a central pillar of modern life sciences, bridging molecular biology, analytical chemistry, biotechnology, and medicine. As the functional layer of biological systems, the proteome reflects not only genetic information but also environmental influences, disease states, and dynamic regulatory processes. Advances in mass spectrometry, separation sciences, and bioinformatics have propelled proteomics from descriptive protein cataloguing to a powerful, quantitative, and systems-level discipline.

A critical determinant of success in proteomic workflows is *sample preparation*. Regardless of instrument performance or computational sophistication, the quality, depth, and biological relevance of proteomic data remain fundamentally dependent on how samples are collected, processed, enriched, and prepared for analysis. This book is therefore devoted to *state-of-the-art strategies in proteomic and peptidomic sample treatment*, while also highlighting their integration with complementary omics approaches and real-world applications.

The chapters assembled in this volume reflect the diversity and maturity of contemporary proteomics. They span clinical and environmental matrices, fundamental and applied research, and both established and emerging analytical challenges. Early chapters address liquid chromatography–mass spectrometry (LC–MS)-based metabolomics and proteomics of urine, emphasizing multi-omics integration, as well as metaproteomics applied to water biotechnology, where complex microbial communities demand tailored methodological solutions. Advanced enrichment strategies based on (nano)materials are explored in the context of quantitative phosphoproteomics, underscoring the importance of selective isolation of low-abundance and posttranslationally modified proteins.

Further chapters focus on specialized and biologically critical protein classes, including histone variants and their posttranslational modifications, extracellular vesicles, and protein–ligand interactions. These contributions illustrate how targeted sample preparation and analytical design can reveal regulatory mechanisms that are otherwise inaccessible. The inclusion of case studies from the petroleum industry, particularly addressing biocorrosion, highlights the expanding role of proteomics beyond biomedical research and into industrial and environmental biotechnology. The volume concludes with a dedicated chapter on peptidomics, bringing attention to endogenous peptides and the analytical and preparative challenges that distinguish them from conventional proteomics.

Taken together, this book provides a comprehensive overview of *current methodologies, conceptual frameworks, and practical considerations* in proteomic and peptidomic research. It is intended for researchers, advanced students, and professionals in proteomics, metabolomics, bioanalytical chemistry, biotechnology, and related fields who seek both methodological guidance and insight into emerging applications.

We hope that this volume will serve not only as a reference but also as a source of inspiration, fostering further innovation in proteomic sample preparation and analytical strategies, and contributing to the continued growth and impact of proteomics across scientific disciplines.

The editor and authors would like to express their sincere gratitude to the editorial board of *Springer Nature* for providing the opportunity to publish this book and for their professional support throughout the editorial process. We extend our special thanks to *Professor Daniel Martins-de-Sousa* for his guidance, encouragement, and continuous support, which were essential to the successful completion of this volume.

Caparica, Portugal José-Luis Capelo-Martínez

Contents

LC-MS-Based Metabolomics and Proteomics Sample Preparation of Urine: Methods and Multi-Omics Integration

Muath Khairi Mousa, Nelson C. Soares, and Hugo M. Santos

Abstract Urine is an ideal biological fluid due to the highly metabolomic and proteomic information it provides and its easy collection in large amounts. Urinary biomarkers reported for different types of diseases included the urological tract and systemic diseases. Although the most common approach in omics is single-omics studies, combining multi-omics biomarkers such as metabolomics, proteomics, transcriptomics, and genomics can improve diagnostic accuracy and provide deeper insights into disease mechanisms than single biomarkers. The gold standard technique for bioanalysis is liquid chromatography coupled to mass spectrometry (LC-MS/MS) due to its high sensitivity, specificity, and selectivity for the analysis of metabolites and proteins in complex biological samples. One of the most important aspects in metabolomics and proteomics is sample extraction and preparation before the analysis. Different types of metabolites and protein extraction methods can be used effectively for urine samples, including protein precipitation, liquid–liquid extraction, and solid-phase extraction. However, in the multi-omics approach integrating metabolomics and

M. K. Mousa
BIOSCOPE Research Group, LAQV-REQUIMTE, Department of Chemistry, NOVA School of Science and Technology, Universidade NOVA de Lisboa, Caparica, Portugal

PROTEOMASS Scientific Society, Caparica, Portugal

Research Institute of Sciences and Engineering (RISE), University of Sharjah, Sharjah, United Arab Emirates

N. C. Soares
College of Medicine, Mohammed Bin Rashid University of Medicine and Health Sciences (MBRU), Dubai Health, Dubai, United Arab Emirates

Center for Applied Translational Genomics (CATG) – Mohammed Bin Rashid University of Medicine and Health Sciences (MBRU), Dubai Health, Dubai, United Arab Emirates

H. M. Santos (✉)
BIOSCOPE Research Group, LAQV-REQUIMTE, Department of Chemistry, NOVA School of Science and Technology, Universidade NOVA de Lisboa, Caparica, Portugal

PROTEOMASS Scientific Society, Caparica, Portugal
e-mail: hms14862@fct.unl.pt

J.-L. Capelo-Martínez (ed.), *Emerging Sample Treatments in Proteomics*, Advances in Experimental Medicine and Biology 1510,
https://doi.org/10.1007/978-3-032-21638-0_1

proteomics, sample preparation could be either individual for each or simultaneous for both from a single sample. In this chapter, we discuss aspects of LC-MS-based metabolomics and proteomics sample preparation, as well as their integration for a multi-omics approach. In clinical practice, the reported sample preparation methods for bladder cancer metabolomics and proteomics were also discussed.

Keywords Urinary biomarkers · Metabolomics · Proteomics · Multi-omics · Sample preparation · LC-MS · Bladder cancer

1 Introduction

Urine is an ideal sample source due to it being large, easily accessible, and highly informative. Approximately 4500 metabolites have been identified in urine in relation to approximately 600 human conditions, reflecting numerous physiological and pathological processes occurring within the body [1]. Recently, extensive research has been conducted in metabolomics to predict and diagnose various diseases, including, but not limited to, cancers, kidney diseases, and diabetes [2]. However, in urine proteomics, proteins mainly originate from glomerular plasma filtration and from those secreted by the kidney and urinary tract [3]. Therefore, the proteins in urine are considered as potential biomarkers for specific urogenital diseases and even for other diseases such as cardiovascular disease or appendicitis [4].

A single omics approach, whether metabolomics or proteomics alone, focuses on a single type of biomolecule, often offering a limited perspective and a lack of reliability for robust biomarker discovery. In contrast, integrating multi-omics provides a more comprehensive understanding of biological functions and the molecular mechanisms underlying diseases by integrating gene expression data with protein and metabolic biomarkers across layers. Combining multi-omics biomarkers can enhance diagnostic accuracy and offer deeper insights into disease mechanisms by capturing complementary molecular information compared to using a single biomarker. This approach can also facilitate the identification of novel biomarkers, helping in disease diagnostics and treatment [5].

Liquid chromatography coupled to mass spectrometry (LC-MS/MS) has gained exceptional importance for the analysis of metabolites and proteins in complex biological samples such as urine, plasma, and serum, due to its sensitivity, selectivity, and rapidity, as well as its ability to analyze thousands of metabolites or proteins simultaneously [6]. These features make LC-MS the preferred choice for quantifying and identifying disease biomarkers, such as those in bladder cancer [7]. Before injection, metabolites or proteins should be extracted from biological matrices, including urine, plasma, and cells, into a suitable solvent, such as water, acetonitrile, or methanol. Several methods can be used either for the extraction of metabolites, proteins or both simultaneously. Protein precipitation, liquid–liquid extraction, and solid-phase extraction are among the most common sample preparation methods used in metabolomics and proteomics studies. The selection of the sample preparation method depends on the chemical properties of target analytes and the analytical objectives [8].

In this chapter, we discuss LC-MS-based metabolomics and proteomics sample preparation of urine samples. Furthermore, the importance of integrating multi-omics, metabolomics, and proteomics was discussed, with a case study on bladder cancer.

2 LC-MS-Based Metabolomics Urine Sample Preparation

Prior to liquid chromatography–mass spectrometry (LC-MS) injection, the urine sample should be well prepared to ensure suitability for analysis. Therefore, sample extraction should be applied to extract the samples' metabolites into an appropriate form. The extraction methods are crucial in ensuring the accuracy of metabolic profiles. In this section, we will present common methods for urine metabolic extraction prior to LC-MS analysis. The LC-MS-based metabolic workflow for urine samples, including protein precipitation, liquid–liquid extraction, and solid-phase extraction, is illustrated in Fig. 1.

2.1 Protein Precipitation (PPT)

Protein precipitation is a widely used method for sample preparation in LC-MS-based metabolomics, particularly for biofluids such as plasma, serum, and urine. This method is gaining popularity for its efficiency, speed, cost-effectiveness, and

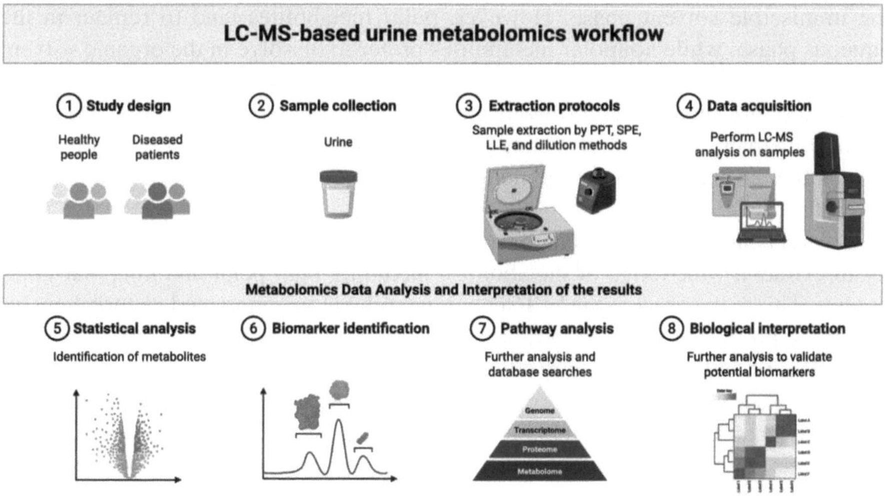

Fig. 1 Illustrates the metabolic workflow for urine samples and LC-MS-based metabolomics

simplicity. In this approach, protein is precipitated to eliminate interference from metabolites, thereby increasing the sensitivity, reproducibility, and accuracy of LC-MS measurements. In PPT, an organic solvent such as methanol, acetonitrile, or a mixture of both with different ratios is added to the sample, leading to protein denaturation and aggregation, resulting in their precipitation followed by centrifugation to precipitate the proteins while keeping the supernatant rich with metabolites [9] . Among these solvents, methanol is the preferred one due to its broad specificity and high accuracy in metabolite extraction [10, 11].

Despite its advantages, PPT has some drawbacks. One of the main challenges in PPT is that it is primarily affected by the matrix effect, which impacts the precision and accuracy of the analysis and, therefore, compromises the quality of the results. However, some techniques, such as optimizing the solvent-to-sample ratio, could help minimize the matrix effect [12]. The other concern about PPT is its throughput limitations, as centrifugation in the traditional PPT method can be time-consuming and challenging to automate. Recently, a specific cartridge can be inserted into the centrifuge tube containing the sample. Then, a manual press or positive pressure can be applied to allow the supernatant to pass through the filter. Therefore, this method can be used instead of centrifugation [13].

2.2 Liquid–Liquid Extraction (LLE)

Liquid–liquid extraction (LLE) is a separation process to extract metabolites from complex biological biofluids into an organic solvent before LC-MS analysis. In LLE, a nonpolar organic solvent is added to an aqueous medium, which is immiscible with this kind of solvent, resulting in the formation of two distinct liquid layers. Since the nature of urine is primarily aqueous, metabolites are distributed into the immiscible solvent phase. However, polar metabolites tend to remain in the aqueous phase, while nonpolar metabolites prefer to dissolve in the organic solvent [14, 15]. The solvent choice significantly affects the metabolite extraction efficiency and commonly includes chloroform, butyl acetate, methyl t-butyl ether (MTBE), dichloromethane, and methyl acetate [16]. For some cases and to cover a wide range of metabolites extraction, a combination of solvents can be used, such as methanol/chloroform/water [17].

Several benefits can be obtained from LLE, the most important one being its ability to extract a wide range of metabolites, including both polar and nonpolar compounds. This feature makes the LLE technique suitable for untargeted, comprehensive metabolic profiling [14]. Furthermore, LLE can be considered as relatively simple, cost-effective and doesn't require advanced instrumentation [14]. Another advantage of LLE is its ability to handle large sample sets, which can be integrated with automated systems such as 96-well plates, thereby increasing processing efficiency [18].

Despite the mentioned advantages, LLE faces several challenges. Matrix effect is one of the most common issues that interfere with LC-MS analysis, leading to ion suppression and decreased analytical accuracy [19]. The potential for emulsion formation upon mixing aqueous and immiscible organic solvents is another disadvantage of LLE, as it reduces extraction efficiency [18]. Additionally, variability in solvent handling can significantly impact the metabolite extraction recovery, making it unreliable for quantitative analysis [20].

2.3 Solid-Phase Extraction (SPE)

Solid-phase extraction (SPE) is an effective technique for extracting metabolites from urine, alongside PPT and LLE. However, the SPE principle is similar to LLE in that samples are eluted through a solid sorbent that retains the target analytes while washing away unwanted compounds such as salts. The retained compounds are then extracted with a suitable solvent for the analyte of interest [21].

Several key roles in SPE, including selecting the right sorbent and eluent solvent, are crucial for extraction efficiency and selectivity. Different SPE sorbents can be used, such as polymeric sorbents like divinylbenzene copolymers, which work well for nonpolar compounds due to their hydrophobicity. Another widely used public sorbent is HLB sorbent, which has both hydrophobic and hydrophilic monomers and is effective for a broad range of metabolites. In contrast, silica-based sorbents such as SCX and WCX are less commonly used sorbents but are helpful for targeted extraction of amino-containing compounds [22].

The eluent solvent's composition, volume, and pH are important factors for optimal extraction efficiency. For example, matching solvent polarity to metabolite polarity, adjusting the optimum volume to avoid incomplete elution or dilution, and tuning pH for ionization improve efficiency, which are vital conditions for optimal extraction. SPE offers multiple advantages, including enhanced sensitivity and specificity in LC-MS detection, the ability to extract a wide range of metabolites, automation, and high reproducibility. However, this technique has challenges, including matrix effects from proteins and lipids, time-consuming parameter optimization, and high cartridge and system costs, especially for large-scale studies [23, 24].

2.4 Dilute and Shoot

The "dilute and shoot" technique is typically suitable for urine due to its low protein content and aqueous nature. This method can be considered reliable due to its simplicity, rapid processing time, and minimal metabolite loss. Despite its advantages, this method is still not recommended because it lacks a cleanup step. The high salt and protein content in the urine sample can interfere with electrospray ionization, leading to ion suppression and reduced sensitivity. Additionally, such unwanted compounds can be retained in the column, leading to low separation efficiency and high back pressure, which reduces the column's lifespan and performance [25].

Table 1 Summary of advantages and disadvantages of LC-MS-based metabolites extraction methods from urine samples

Method	Advantages	Disadvantages
Protein Precipitation (PPT)	Efficient, fast, cost-effective, and simple; Broad metabolite specificity; Increases sensitivity, reproducibility, and accuracy of LC-MS measurements	Susceptible to matrix effects (impacts precision and accuracy); Throughput limitations due to time-consuming centrifugation; Challenging to automate
Liquid–Liquid Extraction (LLE)	Extracts wide range of metabolites (both polar and nonpolar); Suitable for untargeted, comprehensive metabolic profiling; Relatively simple and cost-effective; Can handle large sample sets; Compatible with automated systems (96-well plates)	Matrix effect causes ion suppression and decreased analytical accuracy; Potential for emulsion formation reduces extraction efficiency; Variability in solvent handling impacts metabolite recovery; Unreliable for quantitative analysis
Solid- Phase Extraction (SPE)	Enhanced sensitivity and specificity in LC-MS detection; Extracts wide range of metabolites with different sorbents (polymeric, HLB, silica-based); High reproducibility; Automation capability; Removes unwanted compounds (salts, proteins)	Matrix effects from proteins and lipids; Time-consuming parameter optimization (sorbent selection, eluent composition, volume, pH); High cartridge and system costs, especially for large-scale studies
Dilute and Shoot	Simple and rapid processing; Minimal metabolite loss; Suitable for urine due to low protein content and aqueous nature	Lacks cleanup step; High salt and protein content causes ion suppression and reduced sensitivity; Unwanted compounds retained in column reduce separation efficiency; Leads to high back pressure and reduced column lifespan

As a concluding section, due to the diversity and broad dynamic range of metabolites in urine, no single extraction method can extract all metabolite classes. Therefore, the choice of extraction method should be based on the sample type along with the targeted analytes. Table 1 summarizes the advantages and disadvantages of the extraction methods.

3 LC-MS-Based Proteomics Urine Sample Preparation

Sample preparation before analysis is a critical step in proteomics for urine samples. The chosen extraction technique should be simple to ensure reproducibility and minimal protein loss or modifications. However, normal human urine contains diluted proteins and high salt levels, which interfere with analysis. Therefore, in the initial steps, it's essential to concentrate the proteins while removing salts, commonly via several methods such as protein precipitation, centrifugal filtration, ultracentrifugation, and dialysis, followed by lyophilization. In this section, the sample preparation for LC-MS-based proteomics in urine samples and the underlying principle will be discussed.

3.1 Protein Precipitation

Urine protein precipitation using organic solvents is a widely used method in urinary proteomics. In this method, the organic solvent reduces urine permittivity, breaking the hydration shell on protein surfaces and effectively gathering and precipitating [26]. The most commonly used solvents for protein precipitation in urine proteomics are acetone, trichloroacetic acid (TCA), acetonitrile, and methanol. This method is significantly affected by the solvents used; therefore, solvent selection should be based on the targeted protein groups [27]. However, no single solvent can be used for a comprehensive proteomic study; thus, multiple solvents can be used to cover a broader range of proteins [28]. In a multisolvent study, Saito et al. reported that using a methanol/chloroform (4:1, v:v) mixture results in approximately 80% protein recovery and a higher number of identified proteins than with the solvents tested in the study: acetonitrile, acetone, and ethanol [29]. Once proteins aggregate upon solvent addition, they can be precipitated by centrifugation. The pellets were collected for further reduction, digestion, and treatment for LC-MS analysis [30].

3.2 Centrifugal Filtration

This method is also known as Filter-Aided Sample Preparation (FASP). In this technique, urine is spun at 12,000 × g using a molecular weight cutoff (MWCO) centrifugal column at 4 °C until approximately 1/30 of the initial volume remains [31]. MWCO is a filter membrane used to separate proteins and peptides in biological samples based on their molecular weight [32]. Different material types can be used to make the membrane, including polyethersulfone (PES), polyacrylonitrile (PAN), regenerated Cellulose (RC), and cellulose acetate (CA) with a range of molecular weight cutoff from 3 to 30 kDa [33].

Ultrafiltration has become a widely used technique with broad applications in urine proteomics for protein enrichment and desalting [34]. In this method, salts and undesired small molecules are removed from the proteins. The FASP method offers advantages such as speed and simplicity. However, the use of 96-well MWCO filter plates enables high-throughput sample preparation, a crucial step for large-scale studies [34]. The protein is then reduced, alkylated, and digested with trypsin on the filter. The resulting peptides are recovered by centrifugation through the membrane, thereby avoiding the need for in-solution digestion [35]. However, the in-gel and in-solution digestion methods were typically used to convert proteins extracted from biological material into peptides suitable for mass spectrometry (MS)-based proteome analysis. The in-gel method usually uses sodium dodecyl sulfate (SDS). In contrast, the in-solution method is detergent-free, which may result in an incompletely solubilized proteome and impede digestion by interfering substances. Tryptic digestion with FASP can easily remove SDS and replace it with a urea-containing buffer, which enhances digestion efficiency [36].

Jesus et al. have developed a new FASP-based ultrafast ultrasonic method for shotgun proteomics and label-free protein quantification in urine that uses nitrocellulose membranes for protein separation and in-membrane trypsin digestion, accelerating overnight processing to 4 min with a sonoreactor ultrasonic device. The entire process from separation to identification takes just 3 h [37].

3.3 Ultracentrifugation

The ultracentrifugation method is a simple, straightforward way to isolate the protein from urine samples. In this technique, proteins are extracted by centrifugation at 200,000 × g and 4 °C for about 1–2 h. The proteins, typical vesicle-associated (EV) proteins, will be precipitated as a pellet, collected, and digested for LC-MS analysis [38]. EV isolation by ultracentrifugation co-isolates the contaminant Tamm-Horsfall protein (THP), which can mask lower-abundance proteins in typical mass spectrometry (MS) protocols. To address this challenge, THP polymer networks can be disrupted using dithiothreitol (DTT) and alkaline washing to improve EV isolation and reduce protein contaminants [39]. However, the time-consuming nature and the need for specialized equipment remain drawbacks to using this method [40]. Proteins from urinary EVs can be digested and analyzed by LC-MS [41].

3.4 Dialysis–Lyophilization

Another commonly used method is to combine dialysis with lyophilization. In this method, urine samples were dialyzed against large volumes of distilled water to remove salts. Then, the dialyzed urine is concentrated by lyophilization. The lyophilized urine samples were then resuspended in the solubilizing buffer. Proteins in the dialyzed urine were precipitated using TCA, then collected for protein digestion prior to LC-MS analysis [42].

In conclusion, the urine proteomics sample treatment methods each have their advantages and disadvantages. As illustrated above, centrifugal filtration was found to be the optimal method for rapid, consistent recovery of urinary protein. This method separates proteins based on molecular size, is independent of the chemical variability of urine from different individuals, and is also a convenient way to desalt and remove the confounding effects of small organic molecules present in urine. Though organic solvent precipitation is cheap, rapid, and readily available, it suffers from replicate variability in protein yield. It may therefore not be a method of choice for samples with low protein content. Dialysis and lyophilization yield the highest protein content but suffer from procedural length and complexity, as well as residual salt contamination, which would require an additional cleanup step. Finally, the ultracentrifugation method focuses on the protein in EVs and requires specialized

Table 2 Summary of the pros and cons of urinary proteomics extraction methods

Method	Advantages	Disadvantages
Protein Precipitation	Highe number of identified proteins; Cheap, rapid, and readily available	No single solvent covers comprehensive proteomic study; Solvent selection depends on targeted protein groups; Replicate variability in protein yield; Not recommended for low protein content samples
Centrifugal Filtration (FASP)	Fast and simple technique; Enables high-throughput sample preparation using 96-well MWCO filter plates (crucial for large-scale studies); Effective protein enrichment and desalting; Removes salts and small molecules; On-filter reduction, alkylation, and digestion; Avoids in-solution digestion; Can remove SDS and enhance digestion efficiency; Ultrasonic-assisted method reduces processing from overnight to 4 min	Multiple MWCO filters available (3–30 kDa) requiring optimization; Different membrane materials (PES, PAN, RC, CA) may affect results; Potential protein loss during filtration
Ultracentrifugation	Simple and straightforward; Isolates extracellular vesicle (EV) proteins; Can be optimized with DTT and alkaline washing to reduce Tamm–Horsfall protein (THP) contamination	Time-consuming (1–2 h at 200,000 × g); Requires specialized equipment; Co-isolates contaminant THP which masks lower-abundance proteins
Dialysis–Lyophilization	Highest protein content yield; Effective salt removal through dialysis; Concentrates proteins via lyophilization	Procedurally lengthy and complex; Residual salt contamination requiring additional cleanup step; Most time-consuming method

equipment. Therefore, the preferred method may differ depending on the nature of the downstream application and the amount of protein sample required for proteomic analysis [42, 43]. Table 2 summarizes the advantages and disadvantages of the extraction methods for urinary proteomics analysis.

4 Integration of Multi-Omics (Proteomics and Metabolomics) of Urine Samples

Despite the promising features of urine biomarkers from metabolomics or proteomics studies, the biological interpretation of single-omic data is very challenging due to the complexity of urine samples. Therefore, integrative multi-omics approaches, including genetic, epigenetic, transcriptomic, proteomic, and metabolomic, have become essential for better understanding the biological system.

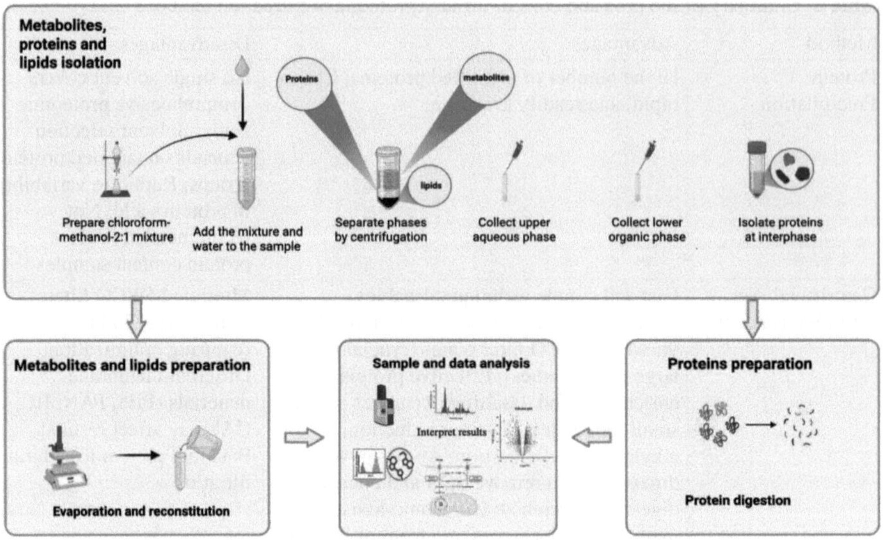

Fig. 2 Illustrates the simultaneous sample preparation steps for multi-omics

Moreover, recent advancements in multi-omics data acquisition and analysis are providing new opportunities for the discovery of novel biomarkers [44].

Sample preparation for metabolomics and proteomics can be done either separately for each feature or simultaneously for both. While preparing samples individually is most common, preparing them simultaneously from a single sample offers benefits such as lower variability, more consistent data, time and cost savings and requires fewer sample quantities [45]. However, the most sample types that can be prepared simultaneously are tissue, cell culture, and plasma samples [46]. For the approach of individually preparing samples, the preparation methods mentioned in the previous sections were used. On the other hand, to date, only one simultaneous extraction method has been reported for urine samples [47]. In this method**,** the metabolite, protein, and lipid extraction (MPLEx) is based on biphasic chloroform–methanol–water phase separation. In these phases, the upper aqueous phase contains hydrophilic metabolites, the lower organic phase contains lipids and hydrophobic compounds, and the proteins precipitate at the interphase between the two layers. The metabolites, proteins, and lipids extraction, preparation, and analysis steps are illustrated in Fig. 2.

However, this method is limited by its complexity, time-consuming nature, and impracticality for implementation with either small sample amounts or high-throughput analyses [48]. Therefore, a novel, quick, and straightforward method that maintains protein and metabolite concentrations is essential.

5 Bladder Cancer Urinary Metabolomics and Proteomics

In this section, we will discuss the integration of urinary metabolomics and proteomics studies and compare them with single-omics studies for bladder cancer disease, highlighting the sample preparation used.

5.1 Bladder Cancer Overview

Bladder cancer (BC) ranks among the top ten most common cancers worldwide, with approximately 550,000 new cases diagnosed annually [49]. Due to its high morbidity and mortality rates, BC places a significant burden on healthcare systems [50]. The dominant histologic type found in over 90% of bladder cancer patients in Europe and the United States is urothelial carcinoma (transitional cell carcinoma). The less common types include non-urothelial carcinomas such as squamous cell carcinoma, adenocarcinoma, and small cell carcinoma [51]. BC can be classified by tumor invasiveness into muscle-invasive bladder cancers (MIBCs) and non-muscle-invasive bladder cancers (NMIBCs) [52]. About 70% of newly diagnosed BC cases are NMIBCs, while the remaining 30% are MIBCs. NMIBCs are further subdivided into noninvasive papillary carcinoma (Ta), invasion into the lamina propria (T1), and carcinoma in situ (CIS), a flat but high-grade lesion. MIBC is characterized by invasion into the muscularis propria (T2) or beyond (T3–T4) [53]. Figure 3 illustrates the stages of bladder cancer and their relationship to the progressively invaded layers of the bladder wall. NMIBC is mainly treated with transurethral resection of the bladder tumor (TURBT) via cystoscopy. Two TURBTs are usually needed: the first identifies the tumor's appearance, size, and location and allows tissue removal

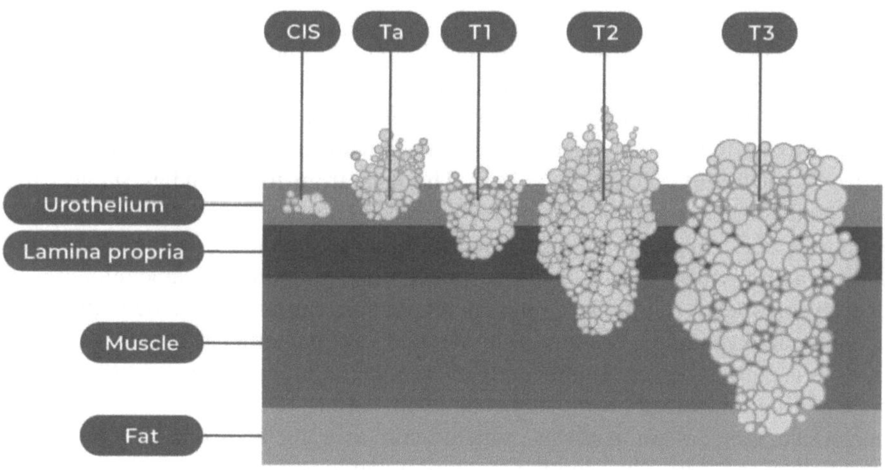

Fig. 3 MBIC and NMBIC pathological staging of bladder cancer progression [57]

for histopathological analysis, while the second is recommended 2–6 weeks after the initial procedure to improve pathological accuracy [54]. The standard treatment for MIBC is radical cystectomy combined with neoadjuvant chemotherapy [55]. Bladder cancer has high rates of progression, recurrence, and cause-specific mortality, which require an intensive surveillance schedule [56].

A further classification of each stage by histological grade is either high-grade or low-grade, based on the extent of cytologic and architectural atypia: high-grade tumors exhibit significant cytologic atypia and architectural disruption, while low-grade tumors maintain organized features. The main risk factors for bladder cancer include gender, advanced age, personal or family history, occupational exposures such as polycyclic aromatic hydrocarbons (PAHs), and smoking. Among these, cigarette smoking is the most confirmed pathogenic factor, with about 50% of bladder cancer patients being smokers [50, 52, 55]. Clinical symptoms of bladder cancer include pain or burning and frequent urination, pelvic and back pain, and hematuria [58, 59].

Different methods can be used for the diagnosis and monitoring of BC. Cystoscopy is the gold standard for diagnosing suspicious bladder lesions. The invasiveness, discomfort, pain, and infectious complications are the main challenges of the procedure [60–62]. As an alternative, cytology can be used. However, the poor sensitivity of cytology urged the need for an improved noninvasive methodology [63]. The development of noninvasive biomarkers with high sensitivity and specificity for early detection and monitoring of bladder cancer is crucial. Both metabolomics and proteomics studies offer a potential method for early detection of bladder cancer [64]. The following provides more details on potential molecular biomarkers of bladder cancer, including metabolomics, proteomics studies, and their integration.

5.2 Metabolomics in Bladder Cancer

Many research articles have studied metabolomics profiling of urine samples to identify potential biomarkers for early diagnosis of different BC subtypes. In studies from 2014 to 2023, more than 270 metabolites were identified as altered in urine samples collected from bladder cancer patients with various types and stages compared to the healthy community [65]. The data from these studies suggest that increased concentrations of metabolites included L-isoleucine, L-carnitine, oleamide, lactic acid, valine, tryptophan, taurine, fructose, glutamine, aspartic acid, uridine, uracil, thymine, ribose, and phosphocholine, palmitamide, arachidonic acid, and glycoursodeoxycholic acid, and decreased concentrations included citric acid, adenine, deoxycytidine, 5-aminolevulinic acid, and pantothenic acid. A variety of metabolic pathways were significantly altered, including glucose metabolism pathway, pyrimidine metabolism, branched-chain amino acid (BCAA) metabolism, nucleotide metabolism pathway, arachidonic acid metabolism, and fatty acid metabolism [65, 66]. The sample preparation methods described in the studies are "dilute and shoot" [67–69] and protein precipitation using acetone, acetonitrile, and methanol [70–72]. Among the sample preparation methods mentioned earlier for metabolic extraction from urine, these two methods were selected in most studies for their speed, simplicity, and broad metabolic extraction coverage.

5.3 Proteomics in Bladder Cancer

Altered protein expression profiles in bladder cancer patients have been extensively studied and reported in various research studies due to their importance in understanding disease mechanisms. Referring to studies of bladder cancer patients compared with a healthy cohort using urine samples and analyzed by LC-MS, spanning 2010 to 2024, a large number of proteins were reported as differentially expressed. Among those proteins, the top-performing biomarkers from these studies in terms of AUC, sensitivity, and specificity are the APOL1 + ITIH3 panel, A1AT, PLK2, HSPE1, and the CO3 + LDHB panel [73–78]. Other potential protein biomarkers were also reported in various studies, including apolipoproteins such as apolipoprotein A-II (APOA2), apolipoprotein B (APOB), and apolipoprotein C-I (APOC1) [79, 80]. Fibrinogen chains, including fibrinogen β and γ chains, were reported to be upregulated in urine from BC patients [7]. Other increased expression proteins in BC patients compared to the control are SAA4 (Serum amyloid A-4), afamin (AFM), HSP beta-1 (HSP27), CD5L (CD5 antigen-like protein), prelamin-A/C (LMNA), cofilin-1 (CFL1), 14-3-3 protein sigma (SFN), 14-3-3 protein zeta/delta (YWHAZ), 14-3-3 protein gamma (YWHAG), heterogeneous nuclear ribonucleoproteins A2/B1 (HNRNPA2B1) and alpha-2 macroglobulin (A2M). Cadherin-related family member 2 (CDHR2), proEGF (Pro-epidermal growth factor), and ITIH2 (Inter-alpha-trypsin inhibitor heavy chain H2) are proteins that were decreased significantly in the urine samples of BC patients [80, 81]. In a recent study, 141 proteins were upregulated in the urine samples of bladder cancer patients compared to non-tumor controls, including SPR, AK1, CD2AP, ADGRF1, GMPS, and C8A [82].

Protein alterations significantly affect pathways involved in cell signaling, metabolism, extracellular remodeling, and immune regulation. Among the most consistently dysregulated pathways are those controlling cell proliferation and survival, including the ERK/MAPK, Ras-Rac-PAK1, and cytokine–TNF signaling cascades, IL-12 signaling, and the inflammatory response pathway [7, 73, 83, 84].

Urine proteins were extracted and concentrated using FASP, protein precipitation, or both. In FASP, urine is loaded onto molecular-weight-cutoff membranes; the most common cutoffs are 3 and 10 kDa [80, 82]. In a study, an additional step was used: washing the extracted proteins on the FASP membrane with 20% acetonitrile, followed by water to remove residual metabolites and salts [79]. For protein precipitation, solvents reported in the literature include methanol/chloroform + EDTA, methanol/chloroform, ethanol, acetone, and acetonitrile [29, 80]. To our knowledge, dialysis–lyophilization is mainly described in general urine proteomics method papers, but not in BC biomarker discovery cohorts, whereas ultracentrifugation has been primarily used to isolate urinary extracellular vesicle components [39, 42]. However, the trypsin digestion protocols were used for urine proteins, either directly on FASP or In-solution [73, 74].

5.4 Multi-Omics Proteomics in Bladder Cancer

Despite numerous studies on bladder cancer metabolomics and proteomics, no biomarkers simultaneously offer high diagnostic accuracy, sensitivity, and specificity for diagnosing bladder cancer and its staging, and none are used in clinical practice [85]. Therefore, an integrated multi-omics approach, including metabolomics and proteomics, is essential. The multi-omics approach for bladder cancer has been reported in only one study. The results showed that both urine metabolites and urine proteins could distinguish the BC group from the healthy group. The results suggest that combining metabolites and proteins could achieve higher accuracy. Similarly, BC clinical stages (high- vs. low-grade BC) could be better distinguished using a panel of metabolites and proteins than either metabolites or proteins alone [86]. The integration of results from the analysis of urinary LC-MS-based metabolites and proteins supports some reported pathways while suggesting new altered pathways. Energy metabolism reprogramming (the Warburg Effect) is one of the pathways validated using metabolomic and proteomic results. While metabolomics clearly demonstrates both increased glycolysis (lactic acid, fructose) and downregulation of the TCA cycle (citric acid), proteomics provides direct enzymatic evidence through elevated LDHB, the key enzyme driving lactate production from pyruvate [7, 65, 66, 73–77].

Multi-omics results linked amino-acid metabolism to nucleotide biosynthesis, whereas metabolomics data showed elevated levels of branched-chain amino acids (L-isoleucine, valine) and glutamine—key nitrogen donors required for de novo nucleotide synthesis. At the same time, reductions in metabolites such as deoxycytidine and adenine reflect increased nucleotide utilization to sustain the high proliferative rate of tumor cells. This linkage was supported by proteomics results showing increased expression of GMP synthase (GMPS), a central enzyme in purine synthesis that leads to nucleotide synthesis [65, 66, 82].

On the other hand, lipid metabolism and transport represent a fully convergent multi-omics pathway in bladder cancer. Metabolomics reveals elevated lipid metabolites (oleamide, palmitamide, phosphocholine), indicating upregulated de novo lipogenesis. Proteomics demonstrates simultaneous elevation of apolipoprotein transporters (APOL1, APOA2, APOB, APOC1), indicating enhanced lipid trafficking. This coordinated signature reveals that cancer cells synthesize lipids for membrane biogenesis to support rapid cell division [65, 79, 80].

For sample preparation in bladder cancer urine multi-omics (metabolomics and proteomics), each method was prepared separately for each approach. The results were then combined for both metabolomics and proteomics. The most common LC-MS-based metabolomics urine preparation techniques are "dilute and shoot" and protein precipitation, while FASP and protein precipitation are used for proteomics studies. However, using a single sample for both metabolomics and proteomics analysis is recommended to minimize biological and technical variability, although it's challenging because optimal extraction conditions differ for metabolites and proteins [87]. Therefore, researchers must put more effort into optimizing the extraction methods for metabolites and proteins from a single sample to reduce biological and technical variability and extraction workloads, while maintaining extraction efficiency for both metabolites and proteins.

In conclusion, integrating urine-based metabolic and proteomic profiling provides significantly greater accuracy and specificity for early detection of bladder cancer than any single-omics approach alone, enabling more reliable, comprehensive, and clinically actionable biomarker panels for improved diagnostic and prognostic assessment.

6 Conclusion

In conclusion, urine has an advantage among biofluid samples due to it being large, easily accessible, and highly informative. The analysis of metabolomics and proteomics provides highly valuable insights into the biological system. Multi-omics is a cutting-edge approach that combines data from multiple biomolecular levels, such as proteins and metabolite markers, to provide a holistic view of how living systems function and interact. However, LC-MS, considered the "gold standard" in bioanalysis, offers superior versatility by enabling the simultaneous analysis of thousands of compounds, ranging from small metabolites to complex peptides and proteins, across complex biological matrices. LC-MS is widely used for metabolomics and proteomics profiling according to its high analytical specificity, accuracy, and sensitivity.

In metabolomics and proteomics, sample preparation is essential for obtaining reliable results. Several methods can be used to extract metabolites or proteins from urine for LC-MS analysis. Each sample treatment method has its advantages and disadvantages; therefore, there is no best method, but all depend on the targeted metabolite or protein groups. Some techniques can be used to extract both metabolites and proteins from a single sample, while others are used only for metabolites or proteins. Bladder cancer has been extensively studied in urinary metabolomics and proteomics individually, but the multi-omics approach was reported in only one article. Thus, we recommend more studies using a multi-omics approach across different disease types, as well as the development of new, reliable extraction methods that simultaneously extract metabolites and proteins from a single sample, enabling lower variability, more consistent data, and time and cost savings, while requiring fewer sample quantities.

Acknowledgments This work was supported by national funds from FCT—Fundação para a Ciência e a Tecnologia, I.P., under the scope of the Project UID/50006/2023 of the Associate Laboratory for Green Chemistry—LAQV REQUIMTE. Details here: https://laqv.requimte.pt/a/74-funding. H.M.S. acknowledges the Associate Laboratory for Green Chemistry—LAQV REQUIMTE (LA/P/0008/2020; DOI: https://doi.org/10.54499/LA/P/0008/2020), funded by FCT I.P. for his research contract. PROTEOMASS Scientific Society is acknowledged by the funding provided to the Laboratory for Biological Mass Spectrometry—*Isabel Moura* (#PM001/2024, #PM001/2019, and #PM003/2016).

References

1. Miller IJ, Peters SR, Overmyer KA, Paulson BR, Westphall MS, Coon JJ (2019) Real-time health monitoring through urine metabolomics. npj Digit Med 2(1):109. https://doi.org/10.1038/s41746-019-0185-y
2. Yu J et al (2024) Using metabolomics and proteomics to identify the potential urine biomarkers for prediction and diagnosis of gestational diabetes. EBioMedicine 101:105008. https://doi.org/10.1016/j.ebiom.2024.105008
3. Shao C, Wang Y, Gao Y (2011) Applications of urinary proteomics in biomarker discovery. Sci China Life Sci 54(5):409–417. https://doi.org/10.1007/s11427-011-4162-1
4. Rodríguez-Suárez E, Siwy J, Zürbig P, Mischak H (2014) Urine as a source for clinical proteome analysis: from discovery to clinical application. Biochim Biophys Acta 1844(5):884–898. https://doi.org/10.1016/j.bbapap.2013.06.016
5. Song J et al (2025) Multi-omics approaches for biomarker discovery and precision diagnosis of prediabetes. Front Endocrinol 16. https://doi.org/10.3389/fendo.2025.1520436
6. Pascali PJ, Giorgetti A, Barone R, Pelletti G, Fais P (2023) Valproic acid determination by liquid chromatography coupled to mass spectrometry (LC–MS/MS) in whole blood for forensic purposes. Drug Test Anal 15(1):128–133. https://doi.org/10.1002/dta.3362
7. Carvalho LB et al (2023) Pathway-guided monitoring of the disease course in bladder cancer with longitudinal urine proteomics. Commun Med 3(1):8
8. Sitnikov GD, Monnin SC, Vuckovic D (2016) Systematic assessment of seven solvent and solid-phase extraction methods for metabolomics analysis of human plasma by LC-MS. Sci Rep 6(1):38885. https://doi.org/10.1038/srep38885
9. Chan W, Zhao Y, Zhang J (2019) Evaluating the performance of sample preparation methods for ultra-performance liquid chromatography/mass spectrometry based serum metabonomics. Rapid Commun Mass Spectrom 33(6):561–568. https://doi.org/10.1002/rcm.8381
10. Bi X et al (2022) Proteomic and metabolomic profiling of urine uncovers immune responses in patients with COVID-19. Cell Rep 38(3):110271. https://doi.org/10.1016/j.celrep.2021.110271
11. Lepoittevin M, Blancart-Remaury Q, Kerforne T, Pellerin L, Hauet T, Thuillier R (2023) Comparison between 5 extractions methods in either plasma or serum to determine the optimal extraction and matrix combination for human metabolomics. Cell Mol Biol Lett 28(1):43. https://doi.org/10.1186/s11658-023-00452-x
12. Côté C, Bergeron A, Mess J-N, Furtado M, Garofolo F (2009) Matrix effect elimination during LC–MS/MS bioanalytical method development. Bioanalysis 1(7):1243–1257. https://doi.org/10.4155/bio.09.117
13. He X, Feng M, Xu X, Zhao H, Qian X (2024) A cartridge structure facilitates clinical mass spectrometry pretreatment and can be integrated into a high-throughput instrument. Sep Sci Plus 7(9). https://doi.org/10.1002/sscp.202400011
14. Huang M, Zhou T (2022) Comprehensive pseudotargeted metabolomics analysis based on two-phase liquid extraction-UHPLC-MS/MS for the investigation of depressive rats. J Sep Sci 45(15):2977–2986. https://doi.org/10.1002/jssc.202200255
15. Snyder WN, Khezam M, Mesaros AC, Worth A, Blair AI (2013) Untargeted metabolomics from biological sources using ultraperformance liquid chromatography-high resolution mass spectrometry (UPLC-HRMS). J Vis Exp 75. https://doi.org/10.3791/50433
16. Darougheh P, Jazan E, Rezayat RM, Jafari TM (2022) Corona discharge ionization ion mobility spectrometry for ultra-trace determination of methamphetamine extracted from urine and plasma samples by dispersive liquid–liquid microextraction. Anal Sci 38(12):1533–1540. https://doi.org/10.1007/s44211-022-00188-2
17. Dutta A, Shetty P, Bhat S, Ramachandra Y, Hegde S (2012) A mass spectrometric study for comparative analysis and evaluation of metabolite recovery from plasma by various solvent systems. J Biomol Tech 23(4):128. https://doi.org/10.7171/jbt.12-2304-001
18. Xue Y-J, Pursley J, Arnold EM (2004) A simple 96-well liquid–liquid extraction with a mixture of acetonitrile and methyl t-butyl ether for the determination of a drug in human plasma

by high-performance liquid chromatography with tandem mass spectrometry. J Pharm Biomed Anal 34(2):369–378. https://doi.org/10.1016/S0731-7085(03)00520-X
19. Deventer K, Pozo OJ, Eenoo VP, Delbeke FT (2009) Qualitative detection of diuretics and acidic metabolites of other doping agents in human urine by high-performance liquid chromatography–tandem mass spectrometry. J Chromatogr A 1216(31):5819–5827. https://doi.org/10.1016/j.chroma.2009.06.003
20. Lee YD, Kind T, Yoon Y-R, Fiehn O, Liu K-H (2014) Comparative evaluation of extraction methods for simultaneous mass-spectrometric analysis of complex lipids and primary metabolites from human blood plasma. Anal Bioanal Chem 406(28):7275–7286. https://doi.org/10.1007/s00216-014-8124-x
21. Chen L et al (2009) On-line coupling of solid-phase extraction to liquid chromatography-a review. J Chromatogr Sci 47(8):614–623. https://doi.org/10.1093/chromsci/47.8.614
22. Badawy IEM, El-Nouby MAM, Kimani KP, Lim WL, Rabea IE (2022) A review of the modern principles and applications of solid-phase extraction techniques in chromatographic analysis. Anal Sci 38(12):1457–1487. https://doi.org/10.1007/s44211-022-00190-8
23. Ferreirós N, Iriarte G, Alonso R, Jiménez R (2007) Development of a solid phase extraction procedure for HPLC–DAD determination of several angiotensin II receptor antagonists in human urine using mixture design. Talanta 73(4):748–756
24. Jin J, Song S-J, Peng Z-J, Guo F, Li P-H (2021) Determination of polybrominated diphenyl ethers and hydroxylated analogues in serum using high performance liquid chromatography-inductively coupled plasma mass spectrometry with online solid phase extraction. Chin J Anal Chem 49(4):520–527. https://doi.org/10.1016/S1872-2040(21)60091-9
25. Shen Y, Luo X, Guan Q, Cheng L (2024) A dilute and shoot method for urinary free cortisol analysis by LC-MS/MS, (in eng). J Chromatogr B Analyt Technol Biomed Life Sci 1239:124127. https://doi.org/10.1016/j.jchromb.2024.124127
26. Roberts JC (2014) Therapeutic protein aggregation: mechanisms, design, and control. Trends Biotechnol 32(7):372–380. https://doi.org/10.1016/j.tibtech.2014.05.005
27. Court M, Garin J, Masselon DC (2015) Urine sample preparation and fractionation for global proteome profiling by LC-MS. In: Methods in molecular biology, pp 175–186
28. Thongboonkerd V, Chutipongtanate S, Kanlaya R (2006) Systematic evaluation of sample preparation methods for gel-based human urinary proteomics: quantity, quality, and variability. J Proteome Res 5(1):183–191. https://doi.org/10.1021/pr0502525
29. Saito S et al (2019) The optimized workflow for sample preparation in LC-MS/MS-based urine proteomics. Methods Protoc 2(2):46. https://doi.org/10.3390/mps2020046
30. Simpson DM, Beynon RJ (2010) Acetone precipitation of proteins and the modification of peptides. J Proteome Res 9(1):444–450. https://doi.org/10.1021/pr900806x
31. Sánchez-Juanes F, González-Buitrago MJ (2019) Sample treatment for urine proteomics. In: Advances in experimental medicine and biology, pp 125–135
32. Fernández A, Zhu Y, Fitzgerald JR, Riera AF (2014) Membrane fractionation of a β-lactoglobulin tryptic digest: effect of the membrane characteristics. J Chem Technol Biotechnol 89(4):508–515. https://doi.org/10.1002/jctb.4144
33. Saxena A, Tripathi PB, Kumar M, Shahi KV (2009) Membrane-based techniques for the separation and purification of proteins: an overview. Adv Colloid Interf Sci 145(1–2):1–22. https://doi.org/10.1016/j.cis.2008.07.004
34. Loftheim H, Nguyen DT, Malerød H, Lundanes E, Åsberg A, Reubsaet L (2010) 2-D hydrophilic interaction liquid chromatography-RP separation in urinary proteomics—minimizing variability through improved downstream workflow compatibility. J Sep Sci 33(6–7):864–872. https://doi.org/10.1002/jssc.200900554
35. Manza LL, Stamer LS, Ham LA-J, Codreanu GS, Liebler CD (2005) Sample preparation and digestion for proteomic analyses using spin filters. Proteomics 5(7):1742–1745. https://doi.org/10.1002/pmic.200401063
36. Wiśniewski RJ, Zougman A, Nagaraj N, Mann M (2009) Universal sample preparation method for proteome analysis. Nat Methods 6(5):359–362. https://doi.org/10.1038/nmeth.1322

37. Jesus RJ, Santos MH, López-Fernández H, Lodeiro C, Arruda ZAM, Capelo JL (2018) Ultrasonic-based membrane aided sample preparation of urine proteomes. Talanta 178:864–869. https://doi.org/10.1016/j.talanta.2017.09.078
38. Eirin A et al (2016) Comparative proteomic analysis of extracellular vesicles isolated from porcine adipose tissue-derived mesenchymal stem/stromal cells. Sci Rep 6(1):36120. https://doi.org/10.1038/srep36120
39. Liu X et al (2015) Intraluminal proteome and peptidome of human urinary extracellular vesicles. Proteomics Clin Appl 9(5–6):568–573. https://doi.org/10.1002/prca.201400085
40. Bijnsdorp VI et al (2017) Feasibility of urinary extracellular vesicle proteome profiling using a robust and simple, clinically applicable isolation method. J Extracell Vesicles 6(1):1313091. https://doi.org/10.1080/20013078.2017.1313091
41. Dai Y, Zhang Y, Qiu F, Li Y-Y, Qiu Z-Y (2010) Analysis of differentially expressed proteome in urinary exosome from non-small cell lung cancer patients. Chin J Anal Chem 38(3):325–331. https://doi.org/10.1016/S1872-2040(09)60030-X
42. Oh J et al (2004) Establishment of a near-standard two-dimensional human urine proteomic map. Proteomics 4(11):3485–3497. https://doi.org/10.1002/pmic.200401018
43. Sigdel KT, Lau K, Schilling J, Sarwal M (2008) Optimizing protein recovery for urinary proteomics, a tool to monitor renal transplantation. Clin Transpl 22(5):617–623. https://doi.org/10.1111/j.1399-0012.2008.00833.x
44. Chen Z, Kim J (2016) Urinary proteomics and metabolomics studies to monitor bladder health and urological diseases. BMC Urol 16(1). https://doi.org/10.1186/s12894-016-0129-7
45. Sapcariu CS, Kanashova T, Weindl D, Ghelfi J, Dittmar G, Hiller K (2014) Simultaneous extraction of proteins and metabolites from cells in culture. MethodsX 1:74–80. https://doi.org/10.1016/j.mex.2014.07.002
46. Gegner MH et al (2024) A single-sample workflow for joint metabolomic and proteomic analysis of clinical specimens. Clin Proteomics 21(1). https://doi.org/10.1186/s12014-024-09501-9
47. Nakayasu ES et al (2016) MPLEx: a robust and universal protocol for single-sample integrative proteomic, metabolomic, and lipidomic analyses. MSystems 1(3):10–1128
48. Faugere J et al (2023) Development of a multi-omics extraction method for ecotoxicology: investigation of the reproductive cycle of Gammarus fossarum. Talanta 253:123806. https://doi.org/10.1016/j.talanta.2022.123806
49. Richters A, Aben KK, Kiemeney LA (2020) The global burden of urinary bladder cancer: an update. World J Urol 38(8):1895–1904
50. van Hoogstraten LM, Vrieling A, van der Heijden AG, Kogevinas M, Richters A, Kiemeney LA (2023) Global trends in the epidemiology of bladder cancer: challenges for public health and clinical practice. Nat Rev Clin Oncol 20(5):287–304
51. Klaile Y, Schlack K, Boegemann M, Steinestel J, Schrader JA, Krabbe L-M (2016) Variant histology in bladder cancer: how it should change the management in non-muscle invasive and muscle invasive disease? Transl Androl Urol 5(5):692–701. https://doi.org/10.21037/tau.2016.06.13
52. Flaig TW et al (2024) Bladder cancer, version 3.2024 featured updates to the NCCN guidelines. JNCCN J Natl Compr Cancer Netw 22(4):216–225
53. Montes ARMJ et al (2025) Differential urinary proteomics and absolute protein quantification reveal biomarkers of muscle-invasive and non-muscle-invasive bladder cancer. J Integr OMICS 15(1):5–11. https://doi.org/10.5584/jiomics.v15i1.245
54. Ahmadi N, Shafee H, Moudi E (2024) Prediction of recurrence risk in patients with non-muscle-invasive bladder cancer. Asian J Urol 11(4):625–632. https://doi.org/10.1016/j.ajur.2023.04.007
55. Zhu C-Z, Ting H-N, Ng K-H, Ong T-A (2019) A review on the accuracy of bladder cancer detection methods. J Cancer 10(17):4038–4044. https://doi.org/10.7150/jca.28989
56. Chamie K et al (2013) Recurrence of high-risk bladder cancer: a population-based analysis. Cancer 119(17):3219–3227. https://doi.org/10.1002/cncr.28147
57. B. C. A. Network (2025) Bladder cancer types, stages & grades. (accessed 19 Nov 2025)

58. Fang W, Yang Z-Y, Chen T-Y, Shen X-F, Zhang C (2020) Ethnicity and survival in bladder cancer: a population-based study based on the SEER database. J Transl Med 18(1). https://doi.org/10.1186/s12967-020-02308-w
59. Jyothi NP, Sreemahalaksmi G, Prathyusha SG, Swapna M (2023) Carcinoma of bladder: a rare case report. World J Biol Pharm Health Sci 13(1):272–276. https://doi.org/10.30574/wjbphs.2023.13.1.0017
60. Crocetto F et al (2022) Liquid biopsy in bladder cancer: state of the art and future perspectives. Crit Rev Oncol Hematol 170:103577
61. Guldhammer CS et al (2023) Cystoscopy accuracy in detecting bladder tumors: a prospective video-confirmed study. Cancer 16(1):160
62. Zlatev DV, Altobelli E, Liao JC (2015) Advances in imaging technologies in the evaluation of high-grade bladder cancer. Urol Clin North Am 42(2):147
63. Chen F, Simsir A, Cheng L (2025) Molecular testing in urinary cytology specimens: current status and future directions. In: Urologic oncology: seminars and original investigations. Elsevier
64. Issaq HJ et al (2008) Detection of bladder cancer in human urine by metabolomic profiling using high performance liquid chromatography/mass spectrometry. J Urol 179(6):2422–2426
65. Pereira F et al (2024) Unmasking the metabolite signature of bladder cancer: a systematic review. Int J Mol Sci 25(6):3347. https://doi.org/10.3390/ijms25063347
66. García-Perdomo AH, Dávila-Raigoza MA, Korkes F (2024) Metabolomics for the diagnosis of bladder cancer: a systematic review. Asian J Urol 11(2):221–241. https://doi.org/10.1016/j.ajur.2022.11.005
67. Shen C et al (2015) Developing urinary metabolomic signatures as early bladder cancer diagnostic markers. OMICS 19(1):1–11. https://doi.org/10.1089/omi.2014.0116
68. Li J, Cheng B, Xie H, Zhan C, Li S, Bai P (2022) Bladder cancer biomarker screening based on non-targeted urine metabolomics. Int Urol Nephrol 54(1):23–29. https://link.springer.com/article/10.1007/s11255-021-03080-6
69. Jin X, Yun JS, Jeong P, Kim YI, Kim W-J, Park S (2014) Diagnosis of bladder cancer and prediction of survival by urinary metabolomics. Oncotarget 5(6):1635–1645. https://doi.org/10.18632/oncotarget.1744
70. Nizioł J et al (2023) Untargeted urinary metabolomics for bladder cancer biomarker screening with ultrahigh-resolution mass spectrometry. Sci Rep 13(1). https://doi.org/10.1038/s41598-023-36874-y
71. Wittmann MB et al (2014) Bladder cancer biomarker discovery using global metabolomic profiling of urine. PLoS One 9(12):e115870. https://doi.org/10.1371/journal.pone.0115870
72. Liu X et al (2018) Investigation of the urinary metabolic variations and the application in bladder cancer biomarker discovery. Int J Cancer 143(2):408–418. https://doi.org/10.1002/ijc.31323
73. Zhou X et al (2024) Exploration of potential biomarkers for early bladder cancer based on urine proteomics. Front Oncol 14. https://doi.org/10.3389/fonc.2024.1309842
74. Tsai C-H et al (2018) Systematic verification of bladder cancer-associated tissue protein biomarker candidates in clinical urine specimens. Oncotarget 9(56):30731–30747. https://doi.org/10.18632/oncotarget.24578
75. Guo J et al (2016) A comprehensive investigation toward the indicative proteins of bladder cancer in urine: from surveying cell secretomes to verifying urine proteins. J Proteome Res 15(7):2164–2177. https://doi.org/10.1021/acs.jproteome.6b00106
76. Kumar P et al (2015) Highly sensitive and specific novel biomarkers for the diagnosis of transitional bladder carcinoma. Oncotarget 6(15):13539–13549. https://doi.org/10.18632/oncotarget.3841
77. Tan L-B, Chen K-T, Yuan Y-C, Liao P-C, Guo H-R (2010) Identification of urine PLK2 as a marker of bladder tumors by proteomic analysis. World J Urol 28(1):117–122. https://doi.org/10.1007/s00345-009-0432-y

78. Yang N et al (2011) Urinary glycoprotein biomarker discovery for bladder cancer detection using LC/MS-MS and label-free quantification. Clin Cancer Res 17(10):3349–3359. https://doi.org/10.1158/1078-0432.CCR-10-3121
79. Chen C-L et al (2013) Identification of potential bladder cancer markers in urine by abundant-protein depletion coupled with quantitative proteomics. J Proteome 85:28–43. https://doi.org/10.1016/j.jprot.2013.04.024
80. Li X et al (2021) Discovery of bladder cancer biomarkers in paired pre- and postoperative urine samples. Transl Androl Urol 10(8):3402–3414. https://doi.org/10.21037/tau-21-562
81. Suh J, Han D, Ku HJ, Kim HH, Kwak C, Jeong WC (2022) Next-generation proteomics-based discovery, verification, and validation of urine biomarkers for bladder cancer diagnosis. Cancer Res Treat 54(3):882–893. https://doi.org/10.4143/crt.2021.642
82. Chang Q et al (2024) Comprehensive urinary proteome profiling analysis identifies diagnosis and relapse surveillance biomarkers for bladder cancer. J Proteome Res 23(6):2241–2252. https://doi.org/10.1021/acs.jproteome.4c00199
83. Mohan C, Vanarsa K, Castillo J, Wang L, Lotan Y (2024) Abstract 3762: 11,000-plex aptamer based proteomics of bladder cancer urine for the discovery of novel disease biomarkers. Cancer Res 84(6_Supplement):3762–3762. https://doi.org/10.1158/1538-7445.AM2024-3762
84. Li H et al (2011) Identification of Apo-A1 as a biomarker for early diagnosis of bladder transitional cell carcinoma. Proteome Sci 9(1):21. https://doi.org/10.1186/1477-5956-9-21
85. Shi X et al (2025) UPLC-Q-TOF/MS-based urine metabolomics for the diagnosis and staging of bladder cancer. Clin Chim Acta 565:120022. https://doi.org/10.1016/j.cca.2024.120022
86. Tang X et al (2020) Urine metabolomics and proteomics for bladder cancer prediction by LC/MS based strategy. FASEB J 34(S1):1–1. https://doi.org/10.1096/fasebj.2020.34.s1.01764
87. Woodland B et al (2025) Sample preparation for multi-omics analysis: considerations and guidance for identifying the ideal workflow. Proteomics. https://doi.org/10.1002/pmic.13983

Metaproteomics for Water Biotechnology: Considerations and Study Cases

Ana C. Afonso, Juan M. Lema, and Alba Trueba-Santiso

Abstract This chapter summarizes the current knowledge on the practical, methodological, and interpretative aspects of applying metaproteomics in water biotechnology. We outline the full metaproteomic workflow—from sampling and protein extraction to LC-MS/MS acquisition, database construction, quantitative analysis, and bioinformatic interpretation—and emphasize critical considerations specific to complex matrices such as EPS-rich biofilms, granular sludge, and low-biomass drinking water. Case studies illustrate how metaproteomics can clarify mechanisms of micropollutant degradation, nitrogen-transforming pathways, biofilm functional architecture, and microbial resilience under operational stress. Recent advances in data-independent acquisition, metagenome-informed databases, and integrative multi-omics are shown to substantially improve depth, reproducibility, and functional resolution. Finally, we discuss emerging applications in wastewater-based epidemiology, where metaproteomics complement nucleic-acid-based surveillance by enabling the detection of large biomolecule biomarkers of population health and industrial activity. Although metaproteomics is already being applied across a wide range of water cycle contexts and is producing promising, robust results, several challenges, including limitations in analytical chemistry, database completeness, and bioinformatics workflows, continue to hinder its broader implementation. Continued technical research and innovation are therefore essential to fully unlock its potential in water biotechnology.

Keywords Metaproteomics · Environmental Proteomics · Engineered Water Treatment Facilities · Integral Water Cycle · Biotechnology

---

A. C. Afonso
LEPABE, ALICE, Faculty of Engineering, University of Porto,
Rua Dr. Roberto Frias, Porto, Portugal

J. M. Lema · A. Trueba-Santiso (✉)
Centre for Cross-Disciplinary Research in Environmental Technologies (CRETUS), Research Group on Environmental Technologies (BIOGROUP), Santiago de Compostela, Spain
e-mail: alba.truebas@udc.es

J.-L. Capelo-Martínez (ed.), *Emerging Sample Treatments in Proteomics*,
Advances in Experimental Medicine and Biology 1510,
https://doi.org/10.1007/978-3-032-21638-0_2

1 Introduction

Metaproteomics is built upon proteomics (the analysis of the complete set of proteins in a sample) yet applied to mixed and often complex microbial communities. In this chapter, we focus specifically on engineered microbial populations within the integral water cycle. That is, the holistic management of all water stages—from withdrawal to treatment, distribution, wastewater collection, reuse, and final return to the environment—as one interconnected system. Accordingly, our attention is limited to engineered systems relevant to water biotechnology and does not extend to natural aquatic environments such as oceans, rivers, and lakes.

From wastewater treatment plants (WWTPs) to drinking water distribution systems (DWDSs), to engineered bioreactors for bioresources recovery, the link between microbial community structure and function plays a decisive role in the integral water cycle process performance and stability. While classical microbiological methods and more recent metagenomics approaches have revealed what microorganisms are present in these systems and what genetic potential do they possess, they do not necessarily reveal what the microbial community is actively doing in situ.

Microbial communities play essential and diverse roles across the entire integral water cycle, driving many of the biological processes that make modern water treatment and resource recovery possible. In DW treatment, microorganisms contribute to the removal of contaminants such as ammonium, iron, manganese, methane, and natural organic matter, especially in biologically active filtration systems. Within DW distribution networks, they help maintain ecological stability but can also influence pipe integrity, disinfectant decay, and biofilm formation.

In WW treatment, microbial consortia are responsible for the biological degradation of organic pollutants and the removal of nitrogen, phosphorus, and other nutrients. Specialized microbial guilds, including nitrifiers, denitrifiers, anammox bacteria, phosphate-accumulating organisms, and methanogens, enable efficient purification and resource recovery. Microbes also support advanced processes such as anaerobic digestion and energy generation from biogas.

In water reuse and environmental discharge stages, microbial communities continue to transform residual contaminants, attenuate pathogens, and stabilize ecosystems. As water moves through natural environments (soil, aquifers, rivers), microbes drive carbon, nitrogen, sulfur, and metal cycling, ultimately closing the loop of the integral cycle. Overall, microbial communities are the biological engine of the integral water cycle, underpinning water quality, environmental protection, and the sustainability of water infrastructure.

In water biotechnology, applying metaproteomic methods offers several attractive opportunities: (i) linking microbial community activities directly to process outcomes (e.g., nitrification, denitrification, micropollutant degradation); (ii) detecting stress responses, biofilm formation, or functional redundancy; (iii) discovering new biotechnologically relevant enzymes or functional biomarkers; (iv) supporting the optimization and monitoring of engineered biological systems in real time or near real time.

However, metaproteomics also poses significant methodological and interpretative challenges: the complexity of microbial communities, the dynamic nature of aqueous and biofilm-associated systems, the need for tailor-made databases, quantitative issues, and integration with other omics data. While metaproteomics has been extensively demonstrated in aquatic ecology (e.g., ocean water column samples) [1, 2], its application in engineered water biotechnology systems is still incipient.

This chapter is organized to move progressively from principles to applications. We begin by outlining the metaproteomic workflow and the key considerations that arise when analyzing complex microbial communities in engineered water systems. We then address methodological constraints and system-specific challenges, such as low-biomass environments, EPS-rich biofilms, database construction, quantitative normalization, and functional interpretation, which are particularly relevant for water biotechnology. Finally, we illustrate how these concepts are implemented in practice through case studies encompassing wastewater treatment, anammox reactors, drinking water treatment and distribution systems, and engineered biofilm processes. This structure reflects the dual aim of the chapter: to provide a clear methodological foundation and to demonstrate, through real applications, how metaproteomics can advance both scientific understanding and operational performance in water biotechnology.

2 Metaproteomics Workflow and Key Considerations for Water Biotechnology

Metaproteomics is defined as the large-scale identification and quantification of proteins from microbial communities in environmental samples, and it is an umbrella term that covers a wide range of methodologies adapted to different matrixes, available equipment, and software.

No universal metaproteomic protocol exists. Instead, workflows must be adapted to the physicochemical properties, biomass density, and structural complexity of each system, as methodological choices made at early stages can strongly influence downstream proteome coverage and interpretation.

The metaproteomics workflow involves the following key common steps (Fig. 1): sampling and biomass collection, protein extraction, protein/peptide digestion, liquid chromatography–tandem mass spectrometry (LC-MS/MS), database search or de novo peptide identification, functional annotation, and quantitative analysis [3, 4]. In water biotechnology, each of these steps is influenced by the specific characteristics of the studied systems (e.g., biofilms, granular sludge, low-biomass drinking water, highly dynamic reactor operation), requiring context-aware experimental design.

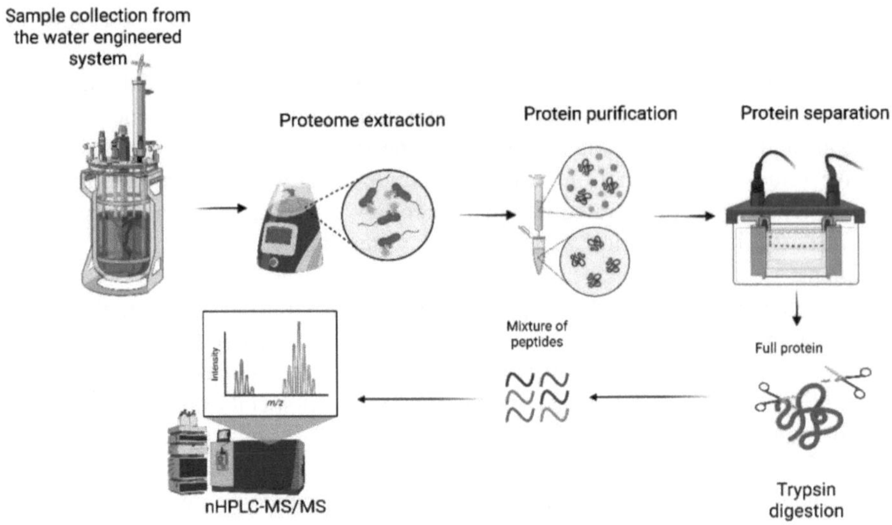

Fig. 1 Schematic representation of the metaproteomic workflow. Biomass collected from engineered water systems undergoes protein extraction, purification, and gel-based separation. Proteins are digested with trypsin into peptides, which are then analyzed by nano-HPLC-MS/MS to determine protein identities and abundances

In the following subsections, we discuss each workflow step in detail, highlighting matrix-dependent constraints and decision points. The following sections detail the individual workflow steps, while Table 1 summarizes key methodological bottlenecks and decision points across representative water biotechnology systems.

2.1 Sampling, Biomass Collection, and Experimental Design in Engineered Water Systems

Sampling engineered water systems (e.g. pipes, bioreactors, biofilms, sludge flocs, granules, microbial mats, or suspended biomass) require careful design to preserve in-situ protein expression patterns. Because proteins are labile biomolecules, delays in fixation or improper preservation can lead to degradation or shifts in community composition. For instance, biomass removal or exposure to atmospheric oxygen may rapidly alter the proteome profile of DW biofilms. Good practice therefore involves immediate flash-freezing of samples or the on-site addition of protease inhibitors, followed by storage at −80 °C [5, 6]. Storage and preservation protocols are especially important in water systems where biomass is diluted and sensitive to manipulation.

Certain engineered water systems are inherently dynamic due to influent fluctuations, biofilm formation, and hydraulic changes. An effective experimental design must control spatial and temporal heterogeneity [7]. In systems such as WWTP

Table 1 Methodological bottlenecks and decision points in metaproteomics of engineered water systems

Matrix type	Dominant structural constraint	Sampling priority	Extraction bottleneck	Database strategy	Quantitative risk	Typical mitigation strategy
EPS-rich biofilms	Dense EPS, strong spatial gradients (O_2, substrates)	Spatially resolved or stratified sampling	Incomplete cell lysis; protein entrapment in EPS	Metagenome-informed, reduced databases	Functional bias toward loosely bound proteins	Aggressive mechanical disruption + tailored detergents; replicate spatial zones
Granular sludge	Compact aggregates; micro-niches	Representative granule selection	Limited penetration of lysis buffers	Metagenome-informed databases essential	Underrepresentation of slow-growing taxa	Pre-fragmentation of granules; pooling multiple granules
Activated sludge	High taxonomic complexity	Temporal replication	Co-extraction of humics and inhibitors	Large, curated databases	High variance in LFQ	Strict normalization; high replicate number
Low-biomass DW (groundwater, filters)	Ultralow protein content	Large-volume sampling	Protein loss during handling	Broad but shallow databases	Limited depth, missing low-abundance functions	Concentration/enrichment prior to digestion; DIA acquisition

biofilm carriers or granular reactors, sampling at multiple time points and locations may be necessary to capture functional variability. Replication is critical: technical replicates (LC-MS/MS runs), biological replicates (independent samples), and process replicates (parallel reactors) enhance statistical power. Storage and preservation protocols are especially important in water systems where biomass is diluted and sensitive to manipulation.

A well-designed example of such a strategy was presented by Quiton-Tapia et al. [8], who optimized sampling in a bioreactor containing a submerged hollow-fiber ultrafiltration membrane module enriched in nitrite-driven anaerobic methane-oxidizing (N-Damo) bacteria. Because these microorganisms are strictly anaerobic and metabolically sensitive to oxygen, all manipulations were performed under oxygen-limited conditions. Two sampling campaigns were carried out during steady-state operation of the membrane bioreactor. Prior to sampling, nitrogen gas was bubbled through the system for 10 minutes to detach biofilms and microorganisms adhering to the membrane surface. Subsequently, liquid aliquots of suspended biomass were collected from the upper and lower reactor sections and combined in a 1:1 (v/v) ratio to yield duplicate 10 mL samples. All samples were immediately frozen at −20 °C and later processed for proteomic analysis. This procedure was repeated at three independent time points, separated by at least 3 days, across both sampling campaigns. Proteins were extracted individually from each sample, and the resulting six extracts per campaign were pooled prior to LC-MS/MS analysis. This meticulous approach minimized oxidative artefacts, ensured reproducible biomass representation across reactor zones, and exemplified how rigorous sampling design directly supports quantitative and comparative metaproteomic workflows.

2.2 *Protein Extraction and Sample Preparation in Complex Water Matrices*

Environmental microbial communities from water technologies often consist of mixed bacterial and archaeal populations embedded in extracellular polymeric substances (EPS), biofilm matrices, or granular aggregates. Efficient protein extraction must therefore overcome challenges of cell disruption, EPS interference, and co-extraction of humic substances or detergents. Common approaches include bead beating in urea/thiourea buffers, sonication, and precipitation (e.g., acetone/TCA) of proteins before digestion [6, 7]. The presence of salts, detergents, and EPS must be minimized to avoid interference in LC-MS/MS.

To illustrate how extraction protocols critically influence metaproteomic outcomes in complex engineered matrices, Abram et al. [9] systematically compared two protein extraction strategies for anaerobic granular sludge: French-press lysis versus a sonication-based method traditionally used for pure cultures. Their analysis, performed by 2-D gel electrophoresis, demonstrated that sonication substantially increased both the number and quality of protein spots obtained from tightly compacted granules. This study provided one of the earliest quantitative evaluations of extraction efficiency in anaerobic biofilm systems and underscored the need for tailored mechanical disruption strategies capable of breaching highly structured granules typically found in wastewater treatment.

In groundwater and DW matrices, biomass and protein concentrations are often extremely low, requiring specialized enrichment procedures. For such low-biomass systems, solid-phase extraction, filtration concentration, or cartridge-based enrichment approaches can be essential prior to digestion and LC-MS/MS, as further discussed in Sect. 4.3.

2.3 LC–MS/MS Analysis

After protein extraction and digestion, peptides are separated via liquid chromatography and introduced into a mass spectrometer where tandem MS (MS/MS) identifies peptide sequences based on their fragmentation spectra. Advances in instrumentation (e.g., high-resolution Orbitraps, time-of-flight, trapped ion mobility spectrometry) have improved sensitivity, depth of coverage, and speed [10]. However, compared with pure culture proteomics, metaproteomic samples are more complex (hundreds to thousands of species), require longer chromatographic gradients, and may yield insufficient protein for comprehensive analysis. Method optimization (e.g., gradient length, column chemistry, loading capacity) is therefore particularly important when analyzing sludge, biofilms, and other heterogeneous matrices.

2.4 Database Search, Data Acquisition Strategies, and Bioinformatics

Peptides identification relies on matching MS/MS fragmentation spectra to sequences in a protein (or translated gene) database. A major challenge in metaproteomics is that environmental and engineered microbial communities are only partially characterized: many organisms lack reference genomes, and homologous proteins often cause ambiguous taxonomic assignment [10]. Consequently, the "dark metaproteome," spectra that cannot be confidently mapped to known sequences, remains a substantial fraction of environmental datasets [10].

To improve identification rates and taxonomic resolution, metagenome-informed databases are commonly used: shotgun metagenomics or metatranscriptomics from the same sample are translated into custom protein search databases [11]. This approach substantially reduces search space complexity and decreases false discovery rates. As discussed earlier, the choice of protein database is critical. In water biotechnology systems, microbial communities may include engineered inocula (e.g., nitrifiers, anammox bacteria), indigenous sludge bacteria, archaea, viruses, and eukaryotes. A metagenome-informed approach (sequencing the same biomass) enables the construction of a sample-specific protein database, thereby improving identification rates and taxonomic specificity [11]. Moreover, given the potential for horizontal gene transfer events in engineered systems (e.g., antibiotic resistance), a comprehensive search space, including plasmid proteins, may be warranted [12]. De novo and hybrid sequencing strategies are also increasingly applied to recover peptides not present in reference databases, particularly in highly diverse or under-sequenced ecosystems [11].

In addition to database selection, the choice of MS/MS data acquisition strategy strongly impacts metaproteomic depth and reproducibility. Two primary workflows are currently used:

(i) *Data-Dependent Acquisition (DDA):* DDA selects the most intense precursor ions for fragmentation, providing high-quality spectra but introducing stochasticity, especially in complex samples where low-abundance peptides may be repeatedly missed [13]. Despite being the traditional workhorse of metaproteomics, DDA often under-samples the proteome of highly diverse microbial communities [13].

(ii) *Data-Independent Acquisition (DIA):* DIA fragments all ions within predefined mass-to-charge windows, enabling far more consistent sampling of the proteome. This approach improves reproducibility, quantitative accuracy, and detection of low-abundance peptides, attributes that are especially valuable for metaproteomics of complex microbiomes [14]. Recent DIA innovations such as DIA-PASEF (Parallel Accumulation–Serial Fragmentation) and advanced Orbitrap platforms, including the Orbitrap Astral mass spectrometer, now combine high sensitivity with unprecedented duty cycle and resolution, pushing metaproteomic coverage deeper into complex microbiomes [15, 16].

After peptide identification, functional annotation (e.g., KEGG, COG, GO, UniProt) and taxonomic assignment using Lowest Common Ancestor algorithms enable interpretation of microbial activity and linkage to system-level processes [17]. Annotation should include functional modules relevant to water biotechnology: nitrogen cycle enzymes, phosphorus binding/accumulation proteins, micropollutant-degrading enzymes (e.g., cytochrome P450s, dioxygenases), stress response factors, and biofilm matrix proteins [17].

2.5 Quantitative Analysis, Normalization, and Quality Control

Quantitative metaproteomics typically uses label-free quantification (LFQ) based on spectral counts or peptide ion intensities, though labeling methods (TMT, iTRAQ, SILAC) may be used in simpler systems [18]. In environmental samples, label-free is most common but requires careful normalization and replicates [7]. Because water biotechnology systems often deal with changing biomass concentrations (e.g., floc sludge vs. biofilm carriers), quantitative normalization is nontrivial.

Common approaches include normalizing total peptide intensity, total protein amount, or specific housekeeping proteins [19].

Data quality assurance (e.g., coefficient of variation across replicates, missing-value treatment) is essential in environmental proteomics. As with all high-throughput "omics'" rigorous quality control is crucial: this includes checking LC-MS/MS instrument performance, monitoring retention time reproducibility, verifying peptide identification false discovery rates, and assessing quantitation reproducibility across runs. Data should be deposited in public proteomics repositories (e.g., PRIDE, ProteomeXchange) to support reproducibility and meta-analysis [20, 21].

Statistical comparison of conditions (e.g., influent versus effluent, high versus low load) allows detection of differentially abundant proteins and thereby inference of shifts in microbial function or stress responses [4]. Functional redundancy and dynamic microbial turnover complicate interpretation, stable proteome patterns may mask taxonomic shifts [4]. Thus, combining metaproteomics with metagenomics/metatranscriptomics can strengthen conclusions.

3 Functional Interpretation and Integration with Process Performance

The ultimate goal of metaproteomics is to link microbial protein expression to system-level function (for example, nitrification kinetics, micropollutant removal, biofilm resilience). Interpretation requires mapping proteins to metabolic pathways, modules of interest (e.g., ammonia monooxygenase, nitrite oxidoreductase, dehalogenases, stress–response chaperones), and linking to operational data (e.g., sludge retention time, dissolved oxygen, micropollutant concentrations). A key insight is that protein expression reflects actual activity rather than genetic potential; thus, metaproteomics can bridge the gap between "what could happen" and "what is happening" [4, 22].

Beyond identifying active metabolic pathways, advanced approaches now allow the direct tracking of substrate utilization within microbial communities. A notable example is the use of protein-based stable isotope probing (protein-SIP), as demonstrated by Taubert et al. [23] in a sulfate-reducing, benzene-degrading consortium. By supplying (13)C-labeled benzene or carbonate and quantifying isotope incorporation into peptides over time, the authors resolved distinct functional guilds responsible for benzene degradation, syntrophic substrate transformation, and biomass scavenging. This approach overcomes a classical limitation of metaproteomics, its inability to infer carbon fluxes, by providing time-resolved functional allocation within mixed communities.

While metaproteomics offers unique functional insight, it is most powerful when integrated with metagenomics, metatranscriptomics, metabolomics, and engineering data. For instance, upregulation of proteins involved in denitrification can be corroborated with increased expression of *nirS* transcripts and corresponding nitrite/nitrate removal kinetics. Integration allows deeper mechanistic insight and helps overcome limitations of each individual omics layer [4].

Biofilms represent a dominant microbial lifestyle in many engineered water systems, including fixed-bed bioreactors, distribution pipelines, granular sludge systems, and membrane bioreactors [24]. From a metaproteomic perspective, biofilms differ fundamentally from planktonic communities in both structure and functional expression. Cells embedded within EPS typically exhibit lower growth rates, stronger stress–response signatures and higher production of adhesion proteins, matrix-associated enzymes, and transport systems adapted to diffusion-limited environments [25, 26]. These structural and physiological differences introduce several methodological challenges for metaproteomics. In particular, sampling must consider spatial heterogeneity, since biofilms can display pronounced vertical and horizontal gradients in oxygen, substrate concentration, redox state, and microbial composition [26]. Furthermore, protein extraction is considerably more complex in EPS-rich matrices; the EPS can entrap proteins, inhibit cell lysis, and interfere with peptide recovery and downstream LC-MS/MS analysis [9, 25].

These challenges underscore the importance of effective extraction strategies, with extraction chemistry being particularly influential. Huang et al. [27] systematically evaluated the impact of different detergents on protein extraction from terephthalate-degrading methanogenic biofilms and demonstrated that detergent choice dramatically alters both the quantity and diversity of extractable proteins. Using four extraction buffers coupled to nano-2D-LC-MS/MS, the authors showed that zwitterionic and anionic detergents achieved substantially higher recovery and deeper proteome coverage, particularly for archaeal and bacterial proteins embedded within dense EPS matrices. They also observed that each detergent enriched proteins with distinct physicochemical properties, resulting in different functional and taxonomic profiles. This study highlighted that biofilm metaproteomics is highly sensitive to extraction chemistry and reinforces the need to tailor sample preparation to the structural and chemical complexity of EPS-rich communities.

4 Study Cases

4.1 Wastewater Treatment Plants

Activated sludge and granular systems dominate modern WWTPs. Microbial communities drive nitrification, denitrification, anaerobic digestion, and removal of organic micropollutants. Applying metaproteomics in these systems enables the acquisition of diverse, topic-specific information.

A recent investigation addressing the co-metabolic degradation of organic micropollutants in full-scale wastewater treatment systems implemented an integrated multi-omics workflow, combining metaproteomics, shotgun metagenomics, and targeted enzyme assays [28]. To manage the complexity of mixed microbial biomass in an urban wastewater context, the authors optimized protein extraction and cleanup protocols suited to suspended biomass and matrix interference. They constructed a metagenome-informed protein database from the same biomass, significantly improving peptide identification and taxonomic resolution. Their quantitative metaproteomic analysis revealed substantial upregulation of cytochrome P450 monooxygenases, glutathione-S-transferases, and aromatic ring-cleaving dioxygenases under increased pharmaceutical loads. Importantly, these protein abundance changes significantly correlated with removal rates of micropollutants and operational parameters such as hydraulic retention time and electron donor supply, evidencing a direct mechanistic link between microbial function and remediation performance.

In a landmark study, Roothans et al. [29] applied a long-term multi-meta-omics strategy to a full-scale biological wastewater treatment plant, combining metagenome-resolved metaproteomics, ex situ kinetic assays, and detailed operational monitoring over ~18 months. They leveraged a metagenome-informed protein database and maintained rigorous replication and normalization, enabling them to unambiguously identify nitrifier denitrification as the dominant N_2O-producing pathway and to establish dissolved O_2 as the primary operational driver of seasonal emission peaks. Their approach resolved the ecophysiological cascade linking temperature-driven AOB/NOB growth shifts, nitrite accumulation, and N_2O emissions, thereby bridging the gap between microbial protein expression and full-scale process performance.

In their comparative meta-omics study of aerobic granular sludge from three different wastewater treatment plants, Kleikamp et al. [4] applied a synchronized workflow combining metaproteomics, whole-metagenome sequencing, and 16S rRNA amplicon profiling on uniform size granules. To enhance taxonomic comparability across methods, they aligned all taxonomic assignments to the Genome Taxonomy Database (GTDB), thereby minimizing inconsistencies in classification. Despite consistent detection of major taxa across all platforms, the authors observed substantial divergences at lower taxonomic levels, which in turn impacted interpretation of nutrient-removal functions. Importantly, metaproteomic analysis revealed that the genera detected across all omics techniques accounted for the majority of the expressed protein biomass, highlighting how functional proteomic data provide a biomass-weighted perspective that complements but differs from genomic and amplicon views. This case study thus exemplifies how rigorous adherence to multi-omics design and database alignment enables more nuanced understanding of microbial ecology in engineered granular systems and underscores the methodological value of integrating metaproteomics into wastewater treatment monitoring frameworks.

Beyond process engineering, WW metaproteomics is emerging as a tool for public-health monitoring [30]. Unlike conventional wastewater surveillance approaches that focus on small molecules, RNA or DNA, raw sewage also contains a rich mixture of microbial, viral, and human-derived proteins. A recent study developed a dedicated metaproteomic workflow for raw sewage, incorporating robust protein extraction from highly heterogeneous matrices and de novo sequencing–guided refinement of large public protein databases. Applied across two urban wastewater sites over 3 months, the approach revealed a diverse core microbiome including gut-associated bacteria and opportunistic pathogens, as well as nearly 200 human proteins, such as immunoglobulins, uromodulin, and cancer-associated biomarkers [30]. These findings demonstrate the potential of metaproteomics as a complementary tool for population-level health surveillance and chemical exposure assessment through wastewater analysis.

4.2 The Anammox Systems

Anaerobic ammonium oxidation (anammox) reactors play a critical role in energy-efficient nitrogen removal, particularly in mainstream and sidestream wastewater treatment. Because anammox bacteria exhibit slow growth rates, niche specialization, and strong sensitivity to environmental fluctuations, especially temperature, metaproteomics have been demonstrated to be a useful tool to assess their functional activity, stress responses, and pathway-level regulation within complex microbial consortia [31].

A comprehensive metaproteomic investigation of temperature effects on anammox communities was conducted by Lin et al. [32], who analyzed proteome changes following a stepwise decrease from 35 to 20 °C and subsequently to 15 °C. Their results showed that cooling from 35 to 20 °C triggered substantial restructuring of the community proteome, with marked modulation of proteins involved in energy conversion, transcription and translation machinery, and inorganic ion transport. At 15 °C, anammox activity was distinctly inhibited, and the proteomes revealed clear evidence of energy limitation and stress, particularly in *Candidatus* Kuenenia, while *Candidatus* Brocadia was moderately affected, and *Candidatus* Jettenia exhibited the most extensive proteome reorganization. Elevated levels of HAO-like proteins and catalase indicated that reactive oxygen species and toxic nitrogen intermediates formed major stressors at low temperature. This study provided some of the clearest evidence to date that anammox taxa differ markedly in their capacity to cope with temperature decline, highlighting the importance of species-specific physiological strategies in sustaining anammox performance in mainstream applications.

Complementary mechanistic insight was provided by Jiang et al. [33], who combined metaproteomics with molecular dynamics and density functional theory (DFT) to map catalytic steps within the anammox pathway. Their proteomic analysis revealed that *Kuenenia stuttgartiensis* contributed more than 88% of the detected proteins in the reactor, confirming its dominance as the functional keystone taxon. Hydrazine synthase (HZS) and hydrazine dehydrogenase (HDH) accounted for 15.94% and 3.45% of the total expressed proteins, respectively, and were identified as critical enzymes in nitrogen removal. DFT calculations showed that binding of NO to the HZSγ subunit had the lowest free-energy requirement

(-4.91 ± 1.33 kJ mol^{-1}), while the proton-transfer step catalyzed by HZSα exhibited the highest activation energy (up to 190.29 kJ mol^{-1}), designating it as the rate-limiting step of the anammox pathway. Through this combined proteomic–computational approach, the authors provided unprecedented mechanistic resolution of enzymatic bottlenecks, offering molecular targets for improving anammox reactor efficiency.

Further insight into anammox resilience under operational stress was obtained from the work of Wang et al. [34], who investigated the response of a granular anammox reactor to rapid temperature decreases from 35 to 15 °C. Although the temperature drops caused transient spikes in effluent nitrite, the reactor maintained a stable nitrogen removal rate of 2.71 ± 0.23 g N L^{-1} d^{-1} at 15 °C over 100 days. Metaproteomic and metagenomic analyses revealed that the overall abundance of anammox bacteria remained stable (~21% of the community), and that anammox proteins accounted for roughly 76% of all detected proteins. Nevertheless, noticeable proteomic adjustments occurred: Ca. *Brocadia* downregulated key enzymes involved in nitrogen metabolism, carbohydrate processing, and biosynthesis at 15 °C, while heterotrophic *Chloroflexi* spp. increased the expression of proteins linked to carbohydrate and metabolite utilization, suggesting a shift in community-level metabolic cooperation. Importantly, AnAOB showed stronger proteomic responses to temperature stress than heterotrophs, underscoring their physiological sensitivity and the importance of monitoring their functional state under fluctuating conditions.

These studies collectively highlight several important methodological considerations for metaproteomic applications in anammox systems. Temperature-sensitive sampling must minimize oxygen exposure and handling artefacts, as inadequate preservation can mimic oxidative-stress proteome signatures. The strong reliance on metagenome-informed databases is essential to resolve paralogous enzymes such as HAO-like and HZS isoforms, though database incompleteness still limits full taxonomic resolution. Furthermore, functional responses may shift independently of community composition, emphasizing the need for temporal sampling, multi-omics integration, and careful interpretation of proteomic plasticity. These insights underscore both the power and the current limitations of metaproteomics for elucidating anammox ecology under operational stress.

4.3 Drinking Water Distribution Systems

DW systems, including groundwater catchments, rapid sand filters, and distribution system biofilms, operate under oligotrophic conditions and are frequently exposed to disinfectant stress. Despite the low nutrient availability, microbial communities in these systems play key roles in ammonium oxidation, metal transformation, and the control of disinfection by-product precursors.

In addition, groundwater and DW matrices present extremely low protein concentrations, often requiring specialized enrichment procedures. Capriotti et al. [35] addressed this limitation by developing an innovative extraction and concentration workflow based on C18 extra-wide-pore cartridges, optimized specifically for dissolved organic matter in groundwater. By combining two sequential elution steps, one with a denaturing agent and one with an acidified organic solvent, the method maximized protein recovery from ultra-dilute samples and enabled the identification of 239 proteins from bacteria, protists, and ciliates. This proof-of-concept study demonstrated that targeted pretreatment is indispensable for metaproteomic characterization of drinking water systems and broadened the feasibility of applying functional proteomics to low-biomass aquatic environments.

Beyond the technical constraints of protein extraction, a second major challenge in oligotrophic groundwater systems is distinguishing which microorganisms are metabolically active, as low-nutrient settings often contain many dormant or slow-growing taxa. Taubert et al. [36] addressed this functional limitation by combining heavy-water (D_2O) metabolic labeling, single-cell Raman microspectroscopy, and metaproteomics. This integrative strategy enabled the authors to track deuterium incorporation at the single-cell level and identify the active microbial groups and metabolic pathways stimulated by different substrates. Their work illustrates how coupling activity-based labeling with metaproteomics provides a powerful framework to reveal functional heterogeneity in groundwater microbiomes that are invisible to conventional proteomics alone.

Building on these advances, Corbera-Rubio et al. [37] tackled a third challenge specific to drinking water treatment systems: functional stratification within treatment units, particularly rapid sand filters. By combining depth-resolved sampling, metagenome-guided database construction, and quantitative metaproteomics, they demonstrated that expression of key nitrification proteins varied by up to two orders of magnitude along the filter bed, despite relatively homogeneous microbial community composition. These findings reveal that strong spatial gradients in microbial activity can persist even when taxonomic profiles appear uniform, underscoring the value of metaproteomics in linking functional protein expression to treatment performance.

4.4 Engineered Biofilm Systems and Bioreactors

Engineered biofilm reactors (i.e., moving-bed biofilm reactors, membrane bioreactors, and granular sludge systems) offer high biomass retention and resilience, but rely on intricate microbial interactions and spatial metabolic organization. Previous studies have shown that metaproteomics is well suited to resolving these functional architectures.

Herschend et al. [38] investigated a defined multispecies model biofilm that mimics functional interactions in engineered wastewater communities. To achieve high coverage and accurate taxonomic resolution, they built reduced reference proteomes specific to each member of the consortium, ensuring that database size was minimized without losing relevant protein diversity, a direct implementation of the metagenome-informed database design. Furthermore, they optimized protein extraction for biofilm structures, rich in EPS, through mechanical disruption and stringent cleanup, improving peptide recovery and reducing matrix interference. By maintaining controlled biofilm development conditions and performing biological replication, the authors achieved robust quantitative comparison of protein abundances across single species versus co-culture biofilms, reflecting good experimental design practices. This methodological rigor enabled the detection of cross-feeding, altered amino-acid metabolism, and niche partitioning during multispecies biofilm formation.

In a long-term novel study of an autotrophic nitrifying biofilm reactor operated over 840 days, Mastroleo et al. [39] applied a comprehensive workflow combining metaproteomics, 16S rRNA amplicon sequencing, and qPCR to assess the spatial distribution and functional dynamics of the model nitrifiers *Nitrosomonas europaea* and *Nitrobacter winogradskyi* within a biofilm carrier system. They employed strategic sampling at five vertical and three horizontal positions within the packed-bed reactor, thereby addressing the spatial heterogeneity of biofilm processes. Their extraction and LC-MS/MS workflow were adapted for EPS-rich biofilm matrices, enabling reliable peptide recovery and quantification across reactor zones. A metagenome-informed approach supported a customized reference proteome that enhanced taxonomic resolution and functional interpretation of nitrifier proteins. Importantly, they found that although DNA-based abundance estimates suggested similar proportions of the two nitrifier species, protein-based estimates revealed substantially higher representation of *N. europaea*, highlighting the critical value of metaproteomics in capturing actual microbial activity rather than mere presence.

4.5 Wastewater-Based Epidemiology: Predicting the Future

The COVID-19 pandemic fundamentally changed the way we perceive WW systems: what was once considered merely a waste product became a powerful mirror of population-level health. During SARS-CoV-2 outbreaks, wastewater-based epidemiology proved invaluable for anticipating infection waves, often days to weeks before clinical case surges. These early surveillance programs relied primarily on DNA/RNA-based detection of viral fragments. While being effective, they provided limited insight into the functional biological footprint contained in complex WW matrices [30].

Today, the field is entering a new era in which metaproteomics can be applied to WW streams with promising results. The rationale is clear: proteins offer complementary and often more functionally informative biomarkers, reflecting not only microbial community composition but also human activity, industrial processes, health status, and physiological responses.

Recent studies illustrate how metaproteomics can elevate WW surveillance beyond nucleic-acid profiling. In their large-scale shotgun proteomics study, Sánchez-Jiménez et al. [40] demonstrated that WW proteomes differ strikingly across catchments, reflecting differences in population structure, lifestyle, and industrial activity. They established a streamlined method combining fractionation of soluble and particulate phases with high-resolution mass spectrometry, enabling the identification of a wide array of proteins transported through WW. The resulting protein signatures, ranging from human-derived proteins to livestock residues and plant material, open avenues for environmental health monitoring, industrial discharge assessment, and population-level behavioral studies.

Complementing this approach, Carrascal and coworkers [41] employed polymeric probes to concentrate proteins from different stages of a full-scale WWTP. Their method enabled the identification of 690 proteins originating from bacteria, plants, animals, and humans. Human proteins such as keratins, uromodulin, defensins, and S100A8, some of them clinically relevant disease biomarkers, were abundant in influent samples and efficiently removed in the effluent. This work offered the first comprehensive insight into the diversity, origin, and fate of WW proteins, highlighting the potential of proteomics for tracking pathogen indicators, inflammatory biomarkers, rodent pest activity, and livestock-processing residues.

A third study, analyzing WW from ten municipalities in Catalonia, further expanded the conceptual framework of "sewage protein information mining." Across multiple sampling campaigns, Carrascal et al. [42] profiled the soluble proteome using high-resolution LC-MS/MS. Their results confirmed that human excreta and livestock-related residues dominate the protein pool in wastewater and showed that the protein signatures correlate strongly with population density, industrial activities, and seasonal variation. The study proposed specific proteins, such as immunoglobulins, albumins, and amylases, as potential bioindicators for population monitoring, pest surveillance, and industrial activity assessment. This represents an important evolution in wastewater-based epidemiology, moving beyond chemical markers to large biomolecule biomarkers that reflect real-time population biology.

Together, these advances signal a paradigm shift: WW metaproteomics is poised to become a key component of environmental and public-health monitoring. By integrating the lessons learned from the pandemic, rapid detection, decentralized surveillance, and the power of wastewater as a population-representative sample, future monitoring systems could evolve toward multilayered surveillance combining nucleic acids, metabolites, and proteins. The inclusion of protein-based markers provides access to immune responses, inflammation signals, pathogen-secreted proteins, and industrial signatures that DNA-based methods cannot capture.

Nonetheless, several challenges must be addressed to unlock the full potential of metaproteomic surveillance: robust sampling and preservation protocols, improved protein extraction from chemically and biologically complex matrices, standardized databases for protein annotation, and scalable quantitative workflows. Despite these obstacles, the trajectory is clear: by learning from past experiences and embracing functional proteomics, WW-based surveillance can evolve from a reactive detection tool into a predictive, multiscale, and integrative platform for monitoring human and environmental health.

4.6 Metaproteomics as a Tool to Infer Taxonomic Composition in Mixed Bacterial Communities

Metaproteomics has emerged as a powerful approach for characterizing microbial communities because it directly measures the proteins expressed by organisms in complex environments, thereby linking taxonomy with function. Unlike DNA-based methods, which detect both active and inactive cells and are influenced by gene copy-number biases, metaproteomics focus on the metabolically active fraction of a community and provide taxonomic assignments based on peptide sequences that can be highly specific to particular microbial groups. This offers a more ecologically meaningful snapshot of which organisms are contributing to community function at a given moment.

A methodological framework for taxonomic inference through metaproteomics was articulated in Géron et al. [43], who emphasize that peptide identification relies on robust protein databases, often constructed from metagenomes associated with the same samples. With sufficiently complete databases, peptide-to-protein matching can achieve species-level or even strain-level resolution. Géron et al. [43] highlight that this approach not only resolves which taxa are present and active but also simultaneously provides functional profiles, yielding a unified picture of community structure and metabolism.

Experimental validation of metaproteomics-based community profiling was presented in Kleiner et al. [44]. Working with both defined synthetic consortia and natural microbial communities, they showed that taxon-specific peptide abundances accurately reflected the biomass contributions of each species. Their results revealed a tight quantitative correlation between metaproteomic estimates and known species inputs, outperforming metagenomics for relative abundance estimation. This work established that the proteome could serve as a reliable proxy for microbial biomass and taxonomic composition in mixed communities.

The relevance of metaproteomics to calculate the protein contribution of each member of the community was also recently highlighted by Kleikamp et al. [45], who developed a database-independent de novo metaproteomics approach and systematically evaluated its quantitative performance using synthetic and natural microbial communities comprising dozens of taxonomic families. Among its highlights, it enables a rapid quantitative profiling and capturing unsequenced community members that otherwise remain inaccessible for further interpretation.

Mass spectrometry–based proteomics, specifically matrix-assisted laser desorption/ionization time-of-flight mass spectrometry (MALDI-TOF MS), was employed as an additional bacterial identification approach. Single colonies on the TSA dishes were removed with a disposable tip micropipette and resuspended. The samples were analyzed with a Biotyper MALDI-TOF/TOF instrument (Bruker Daltonik). Peak lists were generated from the MS spectra using Bruker FlexAnalysis software (Version 3.4). The mass spectrum of the test isolate was compared to a database of reference spectra (RUO MALDI Biotyper Reference Library v3.3.1.2, SR library v1.0) software (Bruker Daltonik) to determine relatedness to spectra in the database. The closest related organisms are identified with a value provided as to the level of confidence in identification. According to the manufacturer's criteria for their RUO application, a score of ≥2.0 represents a well identification at species level, a score of 1.70–1.99 at genus level, and a score of <1.70 no reliable identification [44].

Recent applications of metaproteomics to highly complex aerobic activated sludge communities were published in Azizan et al. [46]. The study investigated lab-scale WWTP during start-up and found that pH fluctuations impaired ammonium removal by reducing AOB activity. Shotgun proteomics revealed a sharp decline in the key AOB enzyme hydroxylamine oxidase from *Nitrosomonas* during low performance. The findings show that quantitative proteomics can monitor process health, though database limitations still challenge protein identification in activated sludge.

In another aerobic engineered system, by Kennes-Veiga et al. [47], an enhanced activity of the genus *Corynebacterium* and specifically of five of its enzymes involved in its central carbon metabolism was detected when the antibiotic sulfamethoxazole was increased suggesting a role of *Corynebacterium* genus on SMX risks mitigation during aerobic heterotrophic conditions.

Together, these studies demonstrate that metaproteomics is not only feasible but also uniquely suited to inferring the taxonomic structure of microbial assemblages, particularly when the aim is to identify active community members and their functional roles.

4.7 Cross-System Synthesis: Methodological Lessons, Practical Constraints, and Transferability

The case studies presented collectively demonstrate that metaproteomics has matured into a powerful and versatile tool for interrogating microbial function across the integral water cycle. Despite the diversity of systems, scales, and objectives, several consistent patterns and transferable methodological lessons emerge.

First, across all engineered systems, metaproteomics repeatedly proved its strongest added value when used to link microbial protein expression directly to process performance. In WWTPs and anammox reactors, changes in protein abundance of key functional enzymes correlated more tightly with nitrogen removal efficiency, micropollutant degradation, and stress responses than did DNA- or RNA-based metrics. Similarly, in DW treatment and biofilm reactors, metaproteomics revealed strong functional stratification and activity gradients that were not apparent from taxonomic composition alone. These observations reinforce a central insight: functional activity can shift independently of community structure, and metaproteomics is uniquely positioned to capture this decoupling.

Second, the case studies highlight that no universal metaproteomic protocol exists, nor should one be expected. Instead, successful applications consistently relied on matrix-specific adaptation of sampling, extraction chemistry, database construction, and quantification strategies. EPS-rich biofilms required aggressive yet selective extraction approaches; granular sludge demanded effective mechanical disruption; drinking water and groundwater systems necessitated protein enrichment and activity-based labeling; anammox systems required strict oxygen control and temperature-sensitive handling. Across systems, metagenome-informed databases emerged as a prerequisite for robust identification and interpretation, particularly in communities with high diversity or functional redundancy.

Third, while metaproteomics offers unrivalled functional insight, the studies also illustrate clear practical constraints that currently limit routine deployment in operational plants. High instrumentation demands, long LC gradients, complex data analysis pipelines, and the need for expert interpretation still restrict metaproteomics primarily to research and advanced diagnostic contexts. In situations requiring rapid operational decisions, metatranscriptomics or targeted molecular assays may still offer faster turnaround. However, when the goal is mechanistic understanding, root-cause analysis, or system optimization, metaproteomics provides a depth of information that other omics approaches cannot match.

Importantly, recent technological advances are progressively lowering these barriers. DIA, ion mobility separation, improved spectral libraries, and increasingly standardized bioinformatics workflows are enhancing reproducibility and depth while reducing stochasticity. In parallel, innovative enrichment strategies and activity-based labeling approaches are extending metaproteomics into previously inaccessible low-biomass and oligotrophic environments. These developments suggest that the current limitations are technical rather than conceptual and, therefore, likely to diminish over time.

However, there is still room for improvement, as currently, there are some unresolved challenges. Table 2 schematizes both the opportunities and the limitations related to metaproteomic application to the water integral cycle. This synthesis underscores that the balance between opportunity and limitation is system-dependent and should guide informed decisions on when and how metaproteomics can be effectively deployed along the integral water cycle.

5 Conclusions

Metaproteomics has the potential to progress from a niche analytical technique into a core methodology for understanding and engineering microbial processes in water biotechnology. Across DW and WW treatment systems, the analyses of the proteomes reveal what cannot be captured through genomics or transcriptomics alone:

Table 2 Opportunities and limitations related to the application of metaproteomics in engineered water systems

Operational benefits	
Direct measurement of microbial activity	Reveals actual microbial functions; links protein expression to nitrification, denitrification, anammox, micropollutant degradation, biofilms, stress responses.
Strong relevance across the integral water cycle	Applicable to WW, DW, distribution systems, groundwater, reuse and biofilm reactors; explains functional bases of nitrogen removal, metal oxidation, biological filtration, biogas production.
Supports engineering optimization	Detects functional bottlenecks, stress signatures, redox gradients, enzyme limitations; aids troubleshooting (e.g., inhibited nitrifiers, stressed anammox); correlates protein trends with operational parameters (HRT, DO, pollutant load).
Enables discovery	Identifies new enzymes, biomarkers, uncharacterized proteins; reveals metabolic cooperation, niche partitioning, and functional redundancy.
Multi-omics integration	Validates genomic and transcriptomic predictions; enhanced when combined with SIP, metagenomics, metatranscriptomics, metabolomics and engineering data.
Wastewater-based epidemiology potential	Detects human proteins, pathogens, immune markers, industrial residues; adds functional biomarkers absent from DNA/RNA methods.
Advances in instrumentation and workflows	DIA, DIA-PASEF, Orbitrap Astral, ion mobility, customized databases increase depth and reproducibility; new enrichment methods allow ultra-low-biomass drinking water proteomics.
Limitations and current challenges	
Complexity of engineered microbial communities	High diversity expands search space and reduces IDs; many organisms lack genomes → large "dark metaproteome".
Database dependence	Requires metagenome-informed, sample-specific databases; incomplete databases cause misassignments, weak taxonomic resolution, hindered quantification.
Sample-specific constraints in water systems	Biofilms: EPS-rich, heterogeneous. Granules: tightly structured. Drinking water: ultralow biomass. Anammox: extremely oxygen/temperature-sensitive.
Extraction and sample-prep challenges	EPS, humics, detergents, salts interfere with protein recovery and MS; extraction chemistry biases proteome coverage; contaminants hinder digestion and LC-MS/MS.
Quantification and reproducibility issues	Label-free quantification complicated by biomass variability and missing values; requires many replicates and careful normalization; high variability in environmental samples.
Interpretation difficulties	Functional redundancy and turnover decouple taxonomy from function; spatial/temporal gradients complicate comparisons; preservation artefacts (e.g., oxygen exposure) distort proteomes.
Instrumentation demands	Requires long gradients, high-sensitivity MS, advanced platforms; high cost and low throughput.
Lack of standardization	No universal protocols for sampling, extraction, database building, normalization, or reporting; limits comparability across studies.

real-time functional activity, stress responses, enzymatic bottlenecks, and metabolic cooperation within complex microbial consortia. The case studies reviewed throughout this chapter demonstrate that, when supported by rigorous sampling, metagenome-informed databases, and robust quantification strategies, metaproteomics can directly link protein expression patterns to contaminant removal, nitrogen cycling performance, biofilm organization, and system resilience.

From a practical perspective, however, the routine implementation of metaproteomics in operational water systems remains constrained by several factors. Sample preparation is labor-intensive and often matrix-specific, mass spectrometry infrastructure requires substantial investment, and data analysis demands advanced bioinformatics expertise. These constraints currently limit its deployment as a rapid diagnostic tool in full-scale plants, particularly when immediate operational decisions are required. In this context, metatranscriptomics may offer faster turnaround times and lower analytical complexity for short-term monitoring or early warning applications. Nevertheless, transcript abundance does not always correlate with enzymatic activity, and metaproteomics provides a uniquely robust measure of functional expression that is indispensable for mechanistic interpretation and long-term process optimization.

Several technological advances are expected to lower these barriers in the coming years. Improvements in data-independent acquisition, ion mobility separation, high-speed mass spectrometers, and standardized spectral libraries are already enhancing depth, reproducibility, and throughput. Parallel progress in automated sample preparation, streamlined extraction protocols, and curated, system-specific protein databases will further improve scalability and comparability across studies. Together, these developments are likely to shift metaproteomics from a predominantly research-driven technique toward a more accessible tool for applied water biotechnology. Looking ahead, the integration of metaproteomics into advanced monitoring frameworks holds particular promise. Coupled with process sensors, operational data, and complementary omics, proteome-based functional biomarkers could support predictive control strategies, early detection of performance deterioration, and improved resilience of engineered systems. Beyond treatment optimization, the emerging application of metaproteomics in wastewater-based epidemiology illustrates its broader societal relevance, enabling the detection of population-level health signals, industrial activity markers, and functional indicators that extend beyond nucleic-acid-based surveillance.

In this light, metaproteomics should be viewed not as a standalone solution, but as a central component of integrated, multilayered monitoring strategies. By capturing what microbial communities are actively doing, metaproteomics provides a mechanistic foundation for next-generation treatment technologies, more resilient engineered systems, and innovative public-health surveillance strategies.

Acknowledgments Alba Trueba-Santiso and Juan M. Lema belong to the Galician Competitive Research Group (GRC)_ED431C-2021/37 and to the Cross-disciplinary Research in Environmental Technologies (CRETUS Research Center, ED431G 2023/12). Ana Afonso acknowledge projects InnovAntiBiofilm (ref. 101157363) financed by European Commission (Horizon-Widera

2023-Acess-02/Horizon-CSA), and HCAI_Disinfect (ref. COMPETE2030-FEDER-00752300; N° 16360) co-funded by European Union through COMPETE 2030 and by national funds through FCT/MECI; national funds through FCT/MECI: LEPABE, UID/00511/2025 (https://doi.org/10.54499/UID/00511/2025) and UID/PRR/00511/2025 (https://doi.org/10.54499/UID/PRR/00511/2025) and ALiCE, LA/P/0045/2020 (https://doi.org/10.54499/LA/P/0045/2020).

References

1. Saito MA, Bertrand EM, Duffy ME et al (2019) Progress and challenges in ocean metaproteomics and proposed best practices for data sharing. J Proteome Res 18:1461–1476. https://doi.org/10.1021/ACS.JPROTEOME.8B00761
2. Kong LF, Yan KQ, Xie ZX et al (2021) Metaproteomics reveals similar vertical distribution of microbial transport proteins in particulate organic matter throughout the water column in the Northwest Pacific Ocean. Front Microbiol 12:629802. https://doi.org/10.3389/FMICB.2021.629802/BIBTEX
3. Armengaud J (2023) Metaproteomics to understand how microbiota function: the crystal ball predicts a promising future. Environ Microbiol 25:115–125. https://doi.org/10.1111/1462-2920.16238
4. Kleikamp HBC, Grouzdev D, Schaasberg P et al (2023) Metaproteomics, metagenomics and 16S rRNA sequencing provide different perspectives on the aerobic granular sludge microbiome. Water Res 246:120700. https://doi.org/10.1016/J.WATRES.2023.120700
5. Jagtap PD, Blakely A, Murray K et al (2015) Metaproteomic analysis using the Galaxy framework. Proteomics 15:3553–3565. https://doi.org/10.1002/PMIC.201500074
6. Van Den Bossche T, Armengaud J, Benndorf D et al (2025) The microbiologist's guide to metaproteomics. iMeta 4:e70031. https://doi.org/10.1002/IMT2.70031;WGROUP:STRING:PUBLICATION
7. Nebauer DJ, Pearson LA, Neilan BA (2024) Critical steps in an environmental metaproteomics workflow. Environ Microbiol 26. https://doi.org/10.1111/1462-2920.16637
8. Quiton-Tapia S, Trueba-Santiso A, Garrido JM et al (2023) Metalloenzymes play major roles to achieve high-rate nitrogen removal in N-damo communities: lessons from metaproteomics. Bioresour Technol 385:129476. https://doi.org/10.1016/J.BIORTECH.2023.129476
9. Abram F, Gunnigle E, O'Flaherty V (2009) Optimisation of protein extraction and 2-DE for metaproteomics of microbial communities from anaerobic wastewater treatment biofilms. Electrophoresis 30:4149–4151. https://doi.org/10.1002/ELPS.200900474
10. Xian F, Brenek M, Krisp C et al (2025) Ultra-sensitive metaproteomics redefines the dark metaproteome, uncovering host-microbiome interactions and drug targets in intestinal diseases. Nat Commun 16:1–16. https://doi.org/10.1038/s41467-025-61977-7
11. Rodríguez-Ramos JA, Borton MA, McGivern BB et al (2022) Genome-resolved metaproteomics decodes the microbial and viral contributions to coupled carbon and nitrogen cycling in river sediments. mSystems 7. https://doi.org/10.1128/MSYSTEMS.00516-22;PAGE:STRING:ARTICLE/CHAPTER
12. Martinez JL (2009) The role of natural environments in the evolution of resistance traits in pathogenic bacteria. Proc R Soc B Biol Sci 276:2521. https://doi.org/10.1098/RSPB.2009.0320
13. Ahmed S, Altman J, Jones G et al (2025) Tear fluid proteomics: a comparative study of DIA and DDA mass spectrometry. J Mass Spectrom Adv Clin Lab 38:26–36. https://doi.org/10.1016/J.JMSACL.2025.10.001
14. Rajczewski AT, Blakeley-Ruiz JA, Meyer A et al (2025) Data-independent acquisition mass spectrometry as a tool for metaproteomics: interlaboratory comparison using a model microbiome. Proteomics 25:e202400187. https://doi.org/10.1002/PMIC.202400187;ISSUE:ISSUE:DOI

15. Guzman UH, Martinez-Val A, Ye Z et al (2024) Ultra-fast label-free quantification and comprehensive proteome coverage with narrow-window data-independent acquisition. Nat Biotechnol 42(12):1855–1866. https://doi.org/10.1038/s41587-023-02099-7
16. Meier F, Brunner AD, Frank M et al (2020) diaPASEF: parallel accumulation–serial fragmentation combined with data-independent acquisition. Nat Methods 17(12):1229–1236. https://doi.org/10.1038/s41592-020-00998-0
17. Van Den Bossche T, Verschaffelt P, Vande Moortele T et al (2024) Biodiversity analysis of metaproteomics samples with Unipept: a comprehensive tutorial. Methods Mol Biol 2836:183–215. https://doi.org/10.1007/978-1-0716-4007-4_11
18. Sivanich MK, Gu TJ, Tabang DN, Li L (2022) Recent advances in isobaric labeling and applications in quantitative proteomics. Proteomics 22:2100256. https://doi.org/10.1002/PMIC.202100256;WEBSITE:WEBSITE:ANALYTICALSCIENCEJOURNALS;REQUESTEDJOURNAL:JOURNAL:16159861;ISSUE:ISSUE:DOI
19. Ting L, Cowley MJ, Hoon SL et al (2009) Normalization and statistical analysis of quantitative proteomics data generated by metabolic labeling. Mol Cell Proteomics 8:2227. https://doi.org/10.1074/MCP.M800462-MCP200
20. Orchard S, Hermjakob H (2008) The HUPO proteomics standards initiative—easing communication and minimizing data loss in a changing world. Brief Bioinform 9:166–173. https://doi.org/10.1093/BIB/BBM061
21. Vizcaíno JA, Deutsch EW, Wang R et al (2014) ProteomeXchange provides globally coordinated proteomics data submission and dissemination. Nat Biotechnol 32(3):223–226. https://doi.org/10.1038/nbt.2839
22. Heyer R, Wolf M, Benndorf D et al (2025) Metaproteomics in the One Health framework for unraveling microbial effectors in microbiomes. Microbiome 13. https://doi.org/10.1186/S40168-025-02119-5
23. Taubert M, Vogt C, Wubet T et al (2012) Protein-SIP enables time-resolved analysis of the carbon flux in a sulfate-reducing, benzene-degrading microbial consortium. ISME J 6:2291–2301. https://doi.org/10.1038/ISMEJ.2012.68
24. Flemming HC, Wingender J, Szewzyk U et al (2016) Biofilms: an emergent form of bacterial life. Nat Rev Microbiol 14:563–575. https://doi.org/10.1038/NRMICRO.2016.94
25. Flemming HC, Wingender J (2010) The biofilm matrix. Nat Rev Microbiol 8:623–633. https://doi.org/10.1038/NRMICRO2415
26. Stewart PS, Franklin MJ (2008) Physiological heterogeneity in biofilms. Nat Rev Microbiol 6:199–210. https://doi.org/10.1038/NRMICRO1838
27. Huang HJ, Chen WY, Wu JH (2014) Total protein extraction for metaproteomics analysis of methane producing biofilm: the effects of detergents. Int J Mol Sci 15:10169–10184. https://doi.org/10.3390/IJMS150610169
28. Lyu B, Manna B, Zhou X et al (2025) Microbial metabolic enzymes, pathways and microbial hosts for co-metabolic degradation of organic micropollutants in wastewater. Water Res 276:123229. https://doi.org/10.1016/J.WATRES.2025.123229
29. Roothans N, Pabst M, van Diemen M et al (2025) Long-term multi-meta-omics resolves the ecophysiological controls of seasonal N2O emissions during wastewater treatment. Nat Water 3(5):590–604. https://doi.org/10.1038/s44221-025-00430-x
30. Tugui CG, Cordesius F, van Holthe W et al (2025) Wastewater metaproteomics: tracking microbial and human protein biomarkers. bioRxiv 2025.02.08.637285. https://doi.org/10.1101/2025.02.08.637285
31. Kartal B, van Niftrik L, Keltjens JT et al (2012) Anammox—growth physiology, cell biology, and metabolism. Adv Microb Physiol 60:211–262. https://doi.org/10.1016/B978-0-12-398264-3.00003-6
32. Lin X, Wang Y, Ma X et al (2018) Evidence of differential adaptation to decreased temperature by anammox bacteria. Environ Microbiol 20:3514–3528. https://doi.org/10.1111/1462-2920.14306

33. Jiang Z, He Y, Zeng M et al (2024) Revealing critical functional enzymes in anammox nitrogen removal and rate-limiting step in catalytic pathways: insight into metaproteomics and density functional theory. Bioresour Technol 406:131090. https://doi.org/10.1016/J.BIORTECH.2024.131090
34. Wang S, Ishii K, Yu H et al (2021) Stable nitrogen removal by anammox process after rapid temperature drops: insights from metagenomics and metaproteomics. Bioresour Technol 320:124231. https://doi.org/10.1016/J.BIORTECH.2020.124231
35. Capriotti AL, Aita SE, Cavaliere C et al (2021) A rapid and innovative extraction and enrichment method for the metaproteomic characterization of dissolved organic matter in groundwater samples. J Sep Sci 44:1612–1620. https://doi.org/10.1002/JSSC.202001025
36. Taubert M, Stöckel S, Geesink P et al (2018) Tracking active groundwater microbes with D2 O labelling to understand their ecosystem function. Environ Microbiol 20:369–384. https://doi.org/10.1111/1462-2920.14010
37. Corbera-Rubio F, Laureni M, Koudijs N et al (2023) Meta-omics profiling of full-scale groundwater rapid sand filters explains stratification of iron, ammonium and manganese removals. Water Res 233:119805. https://doi.org/10.1016/J.WATRES.2023.119805
38. Herschend J, Damholt ZBV, Marquard AM et al (2017) A meta-proteomics approach to study the interspecies interactions affecting microbial biofilm development in a model community. Sci Rep 7:1–13. https://doi.org/10.1038/s41598-017-16633-6
39. Mastroleo F, Arnau C, Verbeelen T et al (2022) Metaproteomics, heterotrophic growth, and distribution of Nitrosomonas europaea and Nitrobacter winogradskyi after long-term operation of an autotrophic nitrifying biofilm reactor. Appl Microbiol 2:272–287. https://doi.org/10.3390/APPLMICROBIOL2010020/S1
40. Sánchez-Jiménez E, Abian J, Ginebreda A et al (2023) Shotgun proteomics to characterize wastewater proteins. MethodsX 11:102403. https://doi.org/10.1016/J.MEX.2023.102403
41. Carrascal M, Abian J, Ginebreda A, Barceló D (2020) Discovery of large molecules as new biomarkers in wastewater using environmental proteomics and suitable polymer probes. Sci Total Environ 747:141145. https://doi.org/10.1016/J.SCITOTENV.2020.141145
42. Carrascal M, Sánchez-Jiménez E, Fang J et al (2023) Sewage protein information mining: discovery of large biomolecules as biomarkers of population and industrial activities. Environ Sci Technol 57:10929–10939. https://doi.org/10.1021/ACS.EST.3C00535
43. Géron A, Werner J, Wattiez R et al (2019) Deciphering the functioning of microbial communities: Shedding light on the critical steps in metaproteomics. Front Microbiol 10:485079. https://doi.org/10.3389/FMICB.2019.02395/BIBTEX
44. Kleiner M, Thorson E, Sharp CE et al (2017) Assessing species biomass contributions in microbial communities via metaproteomics. Nat Commun 8:1558. https://doi.org/10.1038/S41467-017-01544-X
45. Kleikamp HBC, Pronk M, Tugui C et al (2021) Database-independent de novo metaproteomics of complex microbial communities. Cell Syst 12:375–383.e5. https://doi.org/10.1016/J.CELS.2021.04.003
46. Azizan A, Kaschani F, Barinas H et al (2020) Using proteomics for an insight into the performance of activated sludge in a lab-scale WWTP. Int Biodeterior Biodegradation 149:104934. https://doi.org/10.1016/J.IBIOD.2020.104934
47. Kennes-Veiga DM, Trueba-Santiso A, Gallardo-Garay V et al (2022) Sulfamethoxazole enhances specific enzymatic activities under aerobic heterotrophic conditions: a metaproteomic approach. Environ Sci Technol 56:13152–13159. https://doi.org/10.1021/ACS.EST.2C05001

Advanced (Nano)material-Based Enrichment Strategies for Quantitative Phosphoproteomics Analysis by Mass Spectrometry

Inmaculada Ortiz-Gómez and Francisco Calderón Celis

Abstract Phosphorylation is one of the most prevalent and dynamic post-translational modifications. It regulates aspects of cellular signaling, metabolism, and disease progression. Comprehensive characterization of phosphoproteins and their phosphorylation remain analytically challenging due to their low abundance, the dynamic nature of the phosphorylation, substoichiometric modification levels, and the complexity of biological matrices. However, recent advancements in enrichment strategies have substantially increased the depth and precision of phosphoproteomics analyses using mass spectrometry. Strategies such as immobilized metal ion affinity chromatography and metal oxide affinity chromatography refine the selective isolation of phosphorylated peptides from complex mixtures. Emerging materials, such as advanced metal nanoparticles, MXenes, and carbon-based nanostructures, are increasingly being used in phosphoproteomics enrichment due to their inherent features such as high surface areas, easily tunable surface chemistry, and strong structural stability, which provide enhanced enrichment efficiency and selectivity. Here, we outline strategies and innovations in phosphoprotein enrichment materials in quantitative proteomics MS platforms.

Keywords Phosphoproteomics · Phosphorylation · Enrichment strategies · Nanomaterials · Mass spectrometry · Chromatography

---

I. Ortiz-Gómez
Centro Tecnológico de Componentes-CTC, Scientific and Technological Park of Cantabria (PCTCAN), Santander, Spain
e-mail: iortiz@centrotecnologicoctc.com

F. C. Celis (✉)
Institute of Biomedicine and Biotechnology of Cantabria (IBBTEC), University of Cantabria—Spanish National Research Council (CSIC), Santander, Spain
e-mail: calderonf@unican.es

J.-L. Capelo-Martínez (ed.), *Emerging Sample Treatments in Proteomics*,
Advances in Experimental Medicine and Biology 1510,
https://doi.org/10.1007/978-3-032-21638-0_3

1 Introduction

Protein phosphorylation is one of the most prevalent and relevant post-translational modifications (PTMs) in biological systems. It plays a core role in regulating almost every cellular process, such as signal transduction, cell growth and differentiation, metabolism, cell cycle control, and apoptosis [1–3]. It acts as a reversible molecular switch, with the addition or removal of the phosphate group, catalyzed by kinases and reversed by phosphatases [4]. Phosphorylation alters the protein's conformation, physicochemical properties, and binding affinities, thereby modulating its activity, stability, subcellular localization, and interactions with other biomolecules [5, 6]. Through these effects, phosphorylation serves as a finely tuned regulatory mechanism that coordinates cellular responses to internal and external stimuli and controls cellular function in a highly dynamic and coordinated manner [6].

In eukaryotic systems, phosphorylation is catalyzed by kinases and reversed by phosphatases, forming complex and highly interconnected signaling networks that regulate nearly all aspects of cell physiology. About one-third of all eukaryotic proteins are estimated to be phosphorylated under physiological conditions, with canonical phosphorylation occurring mainly on serine, threonine, and tyrosine residues [7–12]. Although less common, noncanonical phosphorylation on residues such as histidine, lysine, arginine, aspartic acid, glutamic acid, and cysteine has also been reported [13]. Multi-site phosphorylation and multi-type phosphorylation generate diverse proteoforms, adding further layers of complexity to the regulation of cellular signaling pathways [14].

Despite its importance, the study of phosphorylation remains technically challenging. Many phosphoproteins exist at low abundance, and phosphorylation frequently occurs substoichiometrically, meaning that only a small fraction of the protein pool is modified at any given time [15, 16]. These characteristics make it difficult to detect and quantify phosphorylated species, especially within the complex proteomic background of unmodified peptides. Moreover, phosphorylation and dephosphorylation events are highly transient, often occurring within milliseconds, hence enabling cells to quickly modulate signaling cascades and adjust their physiological state [17, 18].

In addition to biological challenges, analytical obstacles further complicate phosphoproteomic analysis. Despite currently being the workhorse in (phospho) proteomics analysis, mass spectrometry analysis of phosphoproteins remains challenging due to precisely the low abundance of phosphoproteins and the substoichiometric nature of the modification [19]. Furthermore, phosphopeptides often exhibit poor ionization efficiency in mass spectrometry (MS) and are subject to signal suppression by more abundant non-phosphorylated peptides [20, 21].

Consequently, enrichment strategies that selectively isolate phosphopeptides prior to MS analysis are essential to improve detection sensitivity and coverage. Techniques such as immobilized metal affinity chromatography (IMAC), metal oxide affinity chromatography (MOAC), immunoprecipitation using anti-phosphotyrosine antibodies, and molecularly imprinted polymers (MIPs) have become indispensable tools in modern phosphoproteomics [22, 23]. Recent developments in nanomaterial-based affinity materials have further enhanced enrichment selectivity and recovery, addressing some of the persistent bottlenecks in phosphoproteomic workflows [24, 25].

Quantitative phosphoproteomics aims not only to identify phosphorylation sites but also to determine their relative or absolute abundance under different conditions. The quantification of phosphopeptides provides the foundation for biological interpretation, enabling the mapping of dynamic phosphorylation changes that occur in response to signaling cues or perturbations. In this sense, dysregulated phosphorylation can lead to aberrant signaling and is implicated in numerous diseases, including cancer, metabolic disorders, and neurodegenerative conditions [1]. Therefore, mapping the phosphoproteome and quantifying dynamic phosphorylation changes under different physiological and pathological conditions are critical for elucidating cellular regulation and disease mechanisms.

2 Enzymatic Digestion

The most common approach in (phospho)proteomics analysis (*bottom-up* proteomics) (Fig. 1) entails the enzymatic digestion of proteins into peptides prior to LC-MS analysis [26]. This proteolytic digestion produces peptides more suitable for mass spectrometric analysis than the intact protein [27]. In contrast, any

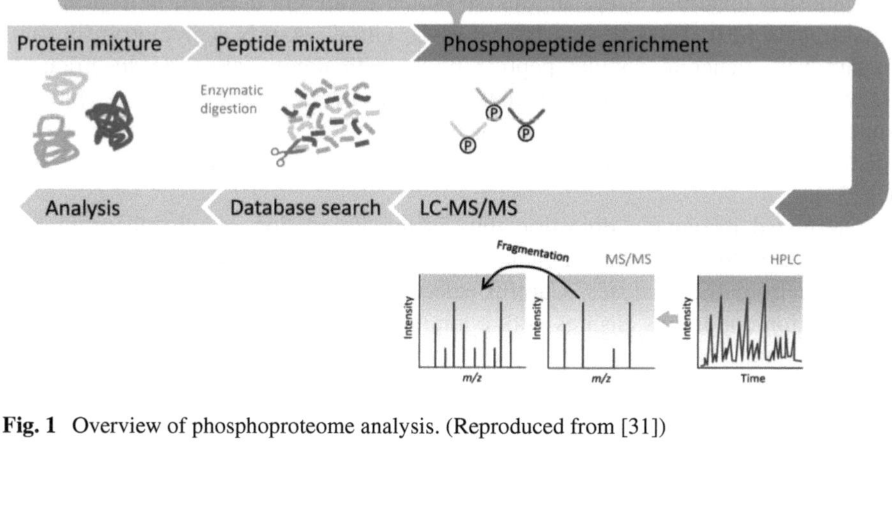

Fig. 1 Overview of phosphoproteome analysis. (Reproduced from [31])

inefficiency or bias at this stage can lead to incomplete sequence coverage and the underrepresentation of relevant phosphorylation events [28]. Furthermore, due to the low abundance of phosphoproteins in complex matrices, efficient digestion is critical for phosphopeptide recovery and accurate quantification. Among the most critical parameters in proteolytic digestion, besides the enzyme, are protein denaturation, reduction, and alkylation steps. These steps result in the unfolding of the protein tertiary structures, cleave existing disulfide bonds and cysteine alkylation to prevent oxidation, ensuring optimal protease accessibility. In the case of phosphopeptides, it is crucial to preserve the phosphate integrity and stability. Thus, extreme acidity or high temperatures can cause hydrolysis of phosphoester bonds, compromising both identification and quantification accuracy. Therefore, digestion protocols in phosphoproteomics must balance efficiency, reproducibility, and phosphorylation preservation [29, 30].

Among proteases, trypsin is the workhorse in proteomics due to its high specificity for cleavage at the C-termini of lysine (K) and arginine (R) residues. This results normally in highly compatible peptides with MS analysis in terms of ionization and fragmentation, because of their consistent small size (around 1 kDa) and the presence of at least two positive charges (the N-term and the K/R at C-term). However, the trypsin proteolytic efficiency is hindered in the case of phosphorylated proteins. The negatively charged phosphate group can form hydrogen bonds and salt bridges with adjacent basic amino acid residues, reducing substrate accessibility and leading to increased numbers of missed cleavages [30]. This incomplete digestion may affect phosphopeptide enrichment (as phosphopeptides with miscleavages may bind strongly to affinity matrices) and result in a decreased sequence coverage and lower detectability of phosphopeptides. To overcome this limitation, several strategies have been developed, like the use of organic solvents, elevated protease concentrations, or add phosphate-binding additives like gallium complexes to unwanted salt bridge formation [28, 30, 32].

While trypsin remains the protease of choice for most proteomic workflows, complementary enzymes can substantially expand sequence coverage and improve phosphopeptide analysis. Comparative analyses have shown that up to two-thirds of phosphosites identified using alternative proteases are not detected in trypsin-only experiments [29]. Enzymes such as Lys-C, Glu-C, Asp-N, Chymotrypsin, and LysargiNase generate distinct peptide pools with unique cleavage specificities, thereby enhancing phosphosite localization. In particular, the combined use of trypsin and LysargiNase to produce mirror peptide spectra significantly improved the identification accuracy and localization confidence in phosphoproteome studies [33].

Optimized enzymatic digestion thus is the foundation of phosphoproteomic workflows, influencing the efficiency of subsequent enrichment and quantitative mass-spectrometric analyses. Careful selection of proteases, buffer composition, and reaction conditions ensures both comprehensive sequence coverage and accurate representation of the dynamic phosphorylation landscape.

3 Enrichment Strategies in Phosphoproteomics

As previously commented, phosphorylation is often present at substoichiometric levels and occurs only on a small portion of protein molecules. Combined with the fact that proteomics analysis of biological samples results in pools of thousands of peptides, phosphosite MS analysis entails an enormous level of complexity. Even more as phosphorylated residues occur at different frequencies, with pSer being the most abundant, followed by pThr, and pTyr occurring at lower rates [34]. In addition, phosphopeptides ionize less efficiently and are often masked by more abundant, non-phosphorylated peptides. Consequently, enrichment of phosphorylated proteins or peptides is an essential step to reduce sample complexity and improve detection sensitivity for comprehensive quantitative mass spectrometry phosphoproteomics analysis.

Large scale phosphoproteomic studies often include sample fractionation before or after enrichment to provide additional separation and reduce sample complexity [35]. Traditional strategies carry out sample fractionation prior to phosphopeptide enrichment. This strategy, however, requires larger amounts of sample material, hence not suitable for limited samples. In contrast, recent adaptations have reversed the workflow order to work with smaller sample quantities, carrying out firstly phosphopeptide enrichment followed by an enrichment step [36].

Phosphopeptide enrichment takes advantage of the specific chemical properties of the phosphate group, particularly its negative charge and strong affinity for metal ions or oxides. Based on these interactions, phosphosite-selective enrichment or separation methods can be grouped into four main types:

(i) Affinity-based chromatography
(ii) Liquid chromatography fractionation
(iii) Proteic and synthetic antibody–based capture (especially for phosphotyrosine)
(iv) Nanomaterials-based approaches.

Although enrichment strategies differ in detail, they generally follow the same basic steps: (1) binding of phosphopeptides to a selective matrix, (2) washing to remove non-phosphorylated peptides, and (3) elution of the bound phosphopeptides by changing pH or adding competing ligands that release them from the matrix. Manual enrichment can be time-consuming and less reproducible, but recent advances in automation using robotic liquid handlers have improved precision, throughput, and consistency [37]. Combined with new materials and improved enrichment chemistries, these developments continue to expand the depth, reproducibility, and scalability of phosphoproteomic analyses.

3.1 Immobilized Metal Affinity Chromatography (IMAC)

Immobilized metal affinity chromatography is one of the earliest and most widely used techniques for phosphopeptide enrichment and remains a core component of phosphoproteomic workflows [38]. The method is based on the interaction between negatively charged phosphate groups on peptides and positively charged metal ions immobilized on a solid support. Commonly used metals include Fe^{3+}, Ga^{3+}, Ti^{4+}, Zr^{4+}, Nb^{5+}, and occasionally, rare earth elements such as Gd^{3+} or Ce^{4+}, which act as Lewis acids and form coordination complexes with the phosphate groups of phosphorylated peptides [39–45].

A typical IMAC matrix contains three main components: a chelating ligand (usually nitrilotriacetic acid, NTA, or iminodiacetic acid, IDA), a solid support such as agarose, silica, or magnetic nanoparticles, and the immobilized metal ions. The ligand anchors the metal ions through coordination bonds, creating an active surface that selectively binds phosphopeptides. During enrichment, phosphorylated peptides are captured by coordination and electrostatic interactions with the metal centers, while non-phosphorylated peptides are removed through washing steps. Bound phosphopeptides are then eluted under acidic or chelating conditions.

IMAC is a robust and efficient technique; however, it suffers from limited specificity and enrichment efficiency due to inadequate binding capacity [22, 46, 47]. Different metal ions also influence enrichment selectivity. Fe^{3+}- and Ga^{3+}-based IMAC typically favor mono-phosphorylated peptides, while Ti^{4+} and Zr^{4+} systems show higher affinity for multi-phosphorylated species [48]. To overcome these limitations, novel IMAC materials that use more hydrophilic ligands and stronger chelating groups like phosphate, arsenate, ATP, or dopamine are being developed [45, 49–51]. Furthermore, sequential use of different metal ions, such as Ga^{3+}–Fe^{3+}, has been shown to increase phosphoproteome coverage by capturing complementary peptide subsets [52].

3.2 Metal Oxide Affinity Chromatography (MOAC)

Metal oxide affinity chromatography is conceptually similar to IMAC but uses solid metal oxides instead of chelated metal ions as the binding material [53, 54]. Phosphopeptides interact with metal oxide surfaces through a combination of electrostatic and Lewis acid–base interactions, forming bidentate or bridging coordination bonds between phosphate groups and metal centers [55]. Among the available materials, titanium dioxide (TiO_2) is the most commonly used sorbent, followed by zirconium dioxide (ZrO_2), iron oxides, and various lanthanide oxides. Under acidic conditions, TiO_2 surfaces become positively charged, promoting selective binding of negatively charged phosphate groups while allowing unmodified peptides to be removed through washing steps.

The typical MOAC workflow includes peptide binding under acidic conditions, washing to minimize nonspecific adsorption, and elution under basic or chelating conditions. TiO_2-based MOAC exhibits excellent chemical stability, reproducibility, and specificity, making it one of the most widely adopted techniques for phosphopeptide enrichment and, in many cases, a preferred alternative to IMAC [56, 57]. Several studies indicate that TiO_2 generally favors multi-phosphorylated peptides, whereas ZrO_2 tends to enrich mono-phosphorylated species [58–60]. Beyond TiO_2 and ZrO_2, a variety of other metal oxides have been explored for their phosphate-binding properties, as reviewed by Leitner [55]. The performance of these materials depends on their isoelectric point, with those with isoelectric point around 6 (e.g., TiO_2, ZrO_2, and In_2O_3) being the ones with higher enrichment efficiency [60].

The different performance and affinity of metal oxides entail that advances and enhancements of MOAC efficiency in phosphoproteomics enrichment are related to the development of novel materials (see next section) or the combination of different oxides such as TiO_2, ZrO_2, SnO_2, or SiO_2 [61–63]. These materials show enhanced specificity and higher peptide recovery.

MOAC and IMAC are often used together because of their complementary selectivity profiles [64, 65]. IMAC tends to favor multi-phosphorylated peptides, while MOAC more efficiently captures mono-phosphorylated species. This principle is the basis of sequential or tandem enrichment strategies such as SIMAC (sequential IMAC-MOAC), where Fe^{3+}-IMAC first isolates multi-phosphorylated peptides, and TiO_2 subsequently enriches the remaining monophosphorylated fraction [66]. Such combined workflows have become standard in large-scale phosphoproteomic analyses, providing broader site coverage and more complete representation of phosphorylation dynamics across diverse biological samples.

3.3 Liquid Chromatography

Chromatographic fractionation plays an important role in phosphoproteomics by reducing sample complexity and improving the detection of low abundant phosphopeptides before mass spectrometry analysis. These methods separate peptides based on physicochemical properties such as charge, hydrophobicity, hydrophilicity, or isoelectric point, providing orthogonal separation dimensions that complement affinity-based enrichment strategies. Fractionation can be applied either before phosphopeptide enrichment to simplify complex peptide mixtures or after enrichment to further increase phosphoproteome coverage.

Strong cation exchange (SCX) separates peptides according to their charge states under acidic conditions [67]. Tryptic peptides are generally doubly or triply protonated, while phosphorylated peptides carry additional negative charges that reduce their net positive charge. As a result, phosphopeptides (especially multi-phosphorylated ones) interact more weakly with the negatively charged stationary phase and elute early in the gradient. Conversely, in strong anion exchange (SAX) chromatography, phosphopeptides bind more strongly to the positively charged resin and elute later with increasing ionic strength [68]. The two methods are complementary, as SCX often favors mono-phosphorylated peptides, whereas SAX more efficiently recovers multi-phosphorylated species. Moreover, these methods can be combined with common reversed-phase separation for orthogonal multi-dimensional separations [69].

In addition to ion-exchange systems, other chromatographic techniques such as hydrophilic interaction liquid chromatography (HILIC) and high-pH reversed-phase (HpH-RP) chromatography are increasingly used to improve depth and orthogonality [70]. HILIC employs a polar stationary phase and a high-organic mobile phase, favoring the retention of hydrophilic phosphopeptides that elute later in the gradient. HpH-RP chromatography, on the other hand, separates peptides based on hydrophobicity under basic conditions and is highly orthogonal to the low-pH reversed-phase chromatography typically used in LC-MS/MS analysis.

Electrostatic repulsion–hydrophilic interaction chromatography (ERLIC) represents a hybrid technique that combines weak anion exchange with hydrophilic interaction mechanisms [71]. In ERLIC, phosphopeptides are electrostatically attracted to the positively charged stationary phase, while non-phosphorylated peptides are repelled. This configuration allows simultaneous enrichment and fractionation in a single step. ERLIC shows high selectivity for acidic and multi-phosphorylated peptides and provides coverage complementary to SCX-IMAC workflows [72].

In practice, these diverse chromatographic modes are often combined with metal-affinity enrichment (IMAC or MOAC) in multidimensional workflows [73–76]. Such two-dimensional (2D) strategies distribute complex peptide mixtures across complementary separation principles, substantially increasing phosphoproteome coverage and enabling large-scale, high-throughput characterization of phosphorylation events in complex biological samples.

3.4 Immunoprecipitation

Immunoprecipitation using phospho-specific antibodies is a highly specific affinity-based approach, particularly valuable for isolating phosphotyrosine (pTyr) peptides. Tyrosine phosphorylation accounts for a small fraction of the phosphoproteome (less than 1–2%), yet it plays a central role in cellular signaling, growth, and differentiation [77–79]. Anti-phosphotyrosine (anti-pTyr) antibodies enable the selective enrichment of thousands of pTyr peptides, facilitating deep profiling of tyrosine kinase signaling networks [80].

Despite their specificity, antibody-based enrichment methods face several limitations, including the requirement for high amounts of sample, and inconsistent specificity and selectivity inherent to the use of antibodies [81]. To improve sensitivity and coverage, alternatives include the integration of IP into multi-step affinity enrichment workflows [78, 82], the modification of binding motifs [83], or the use of synthetic antibody-mimic molecularly imprinted polymers [84].

Molecularly imprinted polymers are synthetic polymeric materials manufactured to exhibit affinity for a target molecule. MIPs are also defined as synthetic "plastic antibodies" engineered to mimic the molecular recognition properties of natural antibodies through a template-driven polymerization process. During synthesis, functional monomers form a pre-polymerization complex with a phosphate-containing template molecule. After polymerization and template removal, the resulting material contains binding cavities that are structurally and chemically complementary to phosphorylated peptides, allowing highly selective recognition [85].

Compared with biological antibodies, MIPs are cost-effective, chemically robust, and tolerant to variations in pH, temperature, and solvent conditions. They can also be regenerated and reused without significant loss of performance, making them suitable for high-throughput or field-deployable phosphoproteomic applications. Their tunable selectivity and structural versatility enable the design of materials targeting specific phosphorylation motifs or residue classes, positioning MIPs as promising alternatives or complements to antibody-based enrichment strategies [86].

Enrichment strategies for proteins based on the use of molecularly imprinted polymers allow for the custom synthesis of polymers tailored to enrich peptides (Fig. 2). Synthetic strategies of MIPs involve the use of fumed silica as sacrificial support acting as solid porogen with the template (phosphotyrosine) immobilized on its surface. These MIPs enable the enrichment of protein histidine

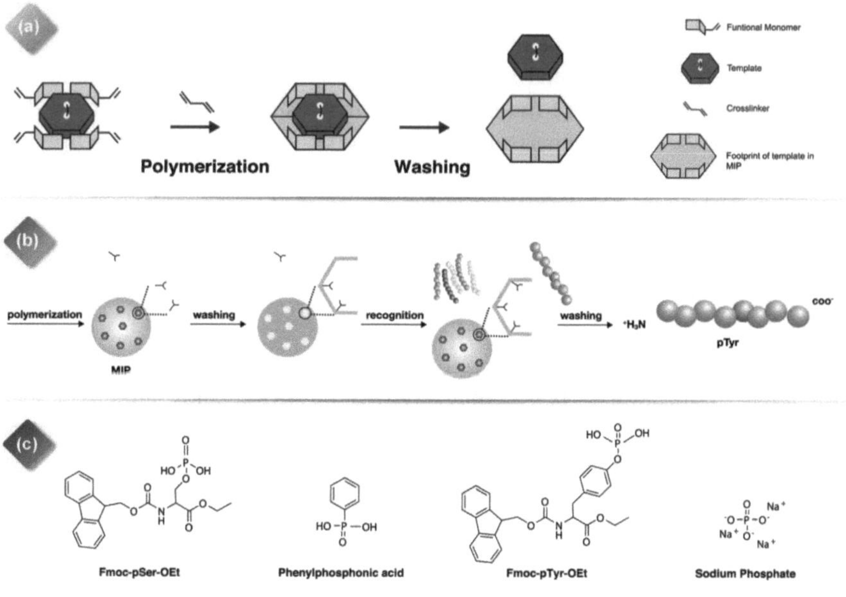

Fig. 2 (**a**) Synthesis of MIPs. (**b**) Preparation process for a pY-MIP and its application for isolating pTyr-peptides. (**c**) Template molecules that are typically used to generate pY-MIP. (Reproduced from [23])

phosphorylation (pHis) from tryptic digests of protein mixtures prior to analysis and sequencing by liquid chromatography–mass spectrometry [87]. During the synthesis of MIPs, the template and the functional monomers form a complex based on different covalent or non-covalent interactions such as hydrogen bonding, ion pairing, other electrostatic interactions, and hydrophobic bonds (Fig. 2). In addition, MIPs can be readily incorporated in a wide variety of analytical setups, including chromatography columns or micro-extraction devices. In this regard, MIPs can be combined with micro-solid-phase extraction, using the MIPs (μMISPE) packed in a pipette tip for phosphopeptide enrichment [86, 88, 89]. They can also be combined with chromatographic SCX fractionation, resulting in enhanced performance in phosphopeptide enrichment compared to TiO_2 [86].

4 Advanced Nanoparticles in Phosphoproteomics Enrichment

Recent advances in nanotechnology have greatly expanded the range of materials available for phosphoproteomics enrichment. A wide range of nanomaterials, including magnetic nanoparticles, metal oxide nanoparticles, MXenes, carbon-based nanoparticles, metal–organic frameworks (MOFs), and covalent organic frameworks (COFs), provide exceptionally high surface areas, easily tunable surface chemistry, and strong structural stability. These properties collectively enhance the selective capture of phosphopeptides and support efficient separation. Nanoscale affinity materials have higher interaction efficiency due to the unique physical and chemical properties associated with their size and shape, which contribute to the specificity and sensitivity of the enrichment of phosphorylated proteins and peptides. A comprehensive comparison of nanomaterial-based strategies and previously described enrichment methodologies is provided in Table 1.

4.1 Magnetic Nanoparticles

Magnetic nanoparticles have emerged as highly promising candidates for protein enrichment applications. Their conjugates with affinity ligands and target species can be readily isolated from sample solutions by applying an external magnetic field, enabling convenient and selective enrichment of phosphopeptides from tryptic digests of protein mixtures. In addition, the magnetic separation process is fully compatible with a wide range of analytical techniques, thereby facilitating the characterization of complex samples. Given these advantages, it is anticipated that magnetic nanoparticles further functionalized with complementary affinity materials could provide enhanced selectivity and thus be effectively applied to enrich phosphopeptides from highly complex biological matrices [90].

For example, functionalized magnetic Fe_3O_4 nanoparticles can be modified with phosphate-binding ligands such as Zn(II)-DPA [91] or silane and glutaric acid [92]

Table 1 Comparison of the advantages and disadvantages of the described phosphoprotein enrichment methods

Method	Advantages	Disadvantages
IMAC	Robust and well established High affinity for phosphopeptides Can be tuned with improved ligands or sequential metal usage	Limited specificity Non-specific binding of acidic peptides Metal ion leaching Metal-dependent selectivity
MOAC	High chemical stability and reproducibility Oxide combination provides higher specificity and peptide recovery	Bias toward multi-phosphorylated peptides Requires careful control of pH conditions Performance and selectivity depend on metal oxide
Liquid Chromatography	High separation efficiency and reproducibility Reduces sample complexity before MS Orthogonal separation to IMAC/MOAC	Limited selectivity for phosphopeptides alone Lower sensitivity for low-abundance sites Requires downstream affinity enrichment
Immunoprecipitation		
pTyr Antibodies	Very high specificity for phosphotyrosine peptides Excellent sensitivity for low-abundant pTyr	Expensive antibodies Limited to targeted residues (mainly pTyr) Requires large sample amounts Antibody bias and batch variability
MIPs	High selectivity and tunability High chemical and thermal stability Reusable and cost-effective	Lower specificity than antibodies Complex design and optimization Template leakage risk
Nanoparticles		
Magnetic Nanoparticles	Easy and fast separation Good selectivity when functionalized Compatible with many analytical methods	Selectivity largely depends on surface functionalization Require additional modification steps to achieve high specificity
Metal Oxide Nanoparticles	High affinity and selectivity via chelation and electrostatic adsorption Tunable morphology and size Good hydrophilicity and controlled surface chemistry	Some have poor aqueous dispersibility Often need composite formation to improve stability and handling
MXenes	Rich surface chemistry and high hydrophilicity Formation of Ti–O sites upon oxidation enhances phosphate binding High enrichment efficiency and specificity Performance further improved via functionalized or magnetic composites	Susceptible to oxidation and degradation Synthesis requires careful handling Often require composite design for dimensional control and long-term stability

(continued)

Table 1 (continued)

Method	Advantages	Disadvantages
Carbon-Based Nanostructures	Large surface area and excellent adsorption capacity Reduce aggregation of metal oxide nanoparticles High compatibility with MALDI-MS Flexible functionalization strategies	Affinity may rely on specific functional groups rather than the carbon framework itself Some ligands are expensive Variable selectivity
MOFs	Extremely high surface area and ordered porosity Tunable pore size and chemical functionality Abundant Lewis acid metal sites enable high selectivity Can be combined with magnetic nanoparticles for efficient magnetic separation	Poor stability in aqueous and extreme pH conditions Structure relies on coordination bonds that can degrade during enrichment
COFs	High chemical and water stability due to covalent bonding Large surface area and intrinsic porosity Multiple affinity interactions (e.g., hydrogen bonding, HILIC) Enable simultaneous enrichment of phosphopeptides and glycopeptides Magnetic COF composites allow fast and simplified workflows	Synthesis requires expensive catalysts or harsh conditions Preparation complexity limits broader application

for the selective isolation of phosphoproteins. Hybrid nanocomposites like Ti^{4+}-immobilized magneto-porous networks (TA–Ti–PA@Fe_3O_4) took advantage of magnetic properties for easier recovery, showing improved selectivity in phosphopeptide enrichment [93]. Similarly, Fe_3O_4@SiO_2@TiO_2 nanoparticles with ordered mesoporous architectures enhance phosphopeptide binding and can act as microreactors that accelerate proteolytic digestion [94].

4.2 Metal Oxide Nanoparticles

The use of metal oxide nanoparticles in sample preparation has grown rapidly, and the body of literature on the subject is constantly expanding. Amphoteric metal oxides exhibit Lewis acid properties and can immobilize phosphopeptides through an ion-exchange mechanism. Various studies focus on the development of graphene oxides, hafnium oxides, and titanium oxide nanoparticles functionalized with amines for highly selective phosphopeptide enrichment. The interaction mechanism with phosphopeptides can be attributed to a synergistic effect of chelation adsorption and electrostatic adsorption. The prepared metal oxides typically present lower Lewis acidity, higher hydrophilicity, and strong affinity for phosphopeptides. These oxides are synthesized via hydrothermal reactions, which allow good control over the morphology and size of the metal structures, enabling the formation of nanoparticles, nanorods, or nanowires.

Beyond titanium-based systems, cerium oxide (Ce/CeO_2) nanomaterials have been tailored to bind phosphorylated peptides selectively, differentiating different phosphorylation grades, while magnetic composites facilitate rapid, low-loss separation [95]. Another of the most used structures as a chelating metal oxide is MoO_3 [88]. However, due to the poor dispersibility of MoO_3 in water, it is combined with graphene oxide as a stable substrate for MoO_3 loading to facilitate enrichment and separation. The prepared graphene oxide provides a large surface area with abundant reaction sites for the homogeneous growth of MoO_3 and improves the dispersion of the composite in water.

4.3 MXenes

Recent works have demonstrated the potential of MXenes as highly promising nanomaterials for protein enrichment, owing to their rich surface chemistry, hydrophilicity, and metallic conductivity [96]. MXenes consist of transition metal carbides, carbonitrides, and nitrides with the universal formula $M_{n+1}X_nT_x$, where M represents transition metals (e.g., Sc, Ti, and Zr), X is either carbon, nitrogen, or both, and T is surface-terminating functional groups (i.e., –OH, –F, –O, etc.), of which the most representative is $Ti_3C_2T_x$ MXene. $Ti_3C_2T_x$ is susceptible to oxidation under aqueous or atmospheric conditions, leading to the formation of TiO_2 and thereby introducing numerous Ti–O binding sites. Combined with its inherent hydrophilicity and tunable structural features, MXene $Ti_3C_2T_x$ is regarded as an ideal candidate for efficient phosphopeptide enrichment (Fig. 3).

Although MXenes have been extensively studied for phosphopeptide enrichment, combining them with metal–organic frameworks (e.g., magnetic Ti_3C_2 MXene

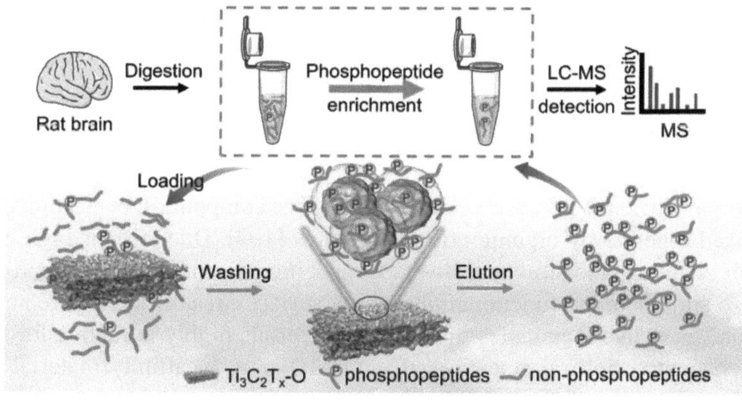

Fig. 3 Schematic diagram of the workflow of phosphopeptides enrichment by the MXene Ti_3C_2Tx-O. (Reproduced from [96])

nanosheets) [97] or constructing functionalized composites (e.g., PAMAM dendrimer-modified MXenes) [98] enables precise dimensional control and markedly enhances enrichment efficiency and specificity.

4.4 Carbon-Based Nanostructures

Carbon-based nanostructures, such as oxidized carbon nanotubes (CNTs), graphene oxide (GO), oxidized carbon nanohorns, and oxidized graphitized carbon black, have attracted interest as scaffolds for functional nanocomposites, which are also highly promising for phosphopeptide enrichment. In this regard, carbon-based (nano)composites coupled to a MOAC strategy avoid common pitfalls, such as aggregation of MO nanoparticles (NPs), while adding the advantageous increased contact between the target biomolecule and the affinity material. Carbon nanotubes and graphene or its derivatives have been used in isolated reports [99]. Polyarginine (PA)-coated diamond nanoparticles have recently been developed for the selective enrichment of multiphosphopeptides, on the basis of the non-covalent complexes formed by PA and phosphopeptides. However, the affinity of PA toward phosphopeptides originates mainly from the interaction of phosphate groups with positively charged guanidyl groups, rather than from its peptide framework. On the other hand, the introduction of consecutive arginine residues in addition to a peptide framework is not cost-efficient, and this may limit its application to phosphopeptide enrichment; the guanidyl group may therefore be regarded as a more suitable functional group for phosphopeptide enrichment [100].

Graphene-based materials have attracted extensive research interest in recent years due to their exceptional electrical and mechanical properties. In addition, graphene oxide is characterized by easy preparation and has a large surface area available for adsorption. Its surface can be modified with a variety of functional groups and is compatible with direct matrix-assisted laser desorption ionization (MALDI) MS analysis. Graphene oxide has thus become an ideal functionalization support in the development of novel graphene-based materials for sample preparation [101].

4.5 Metal–Organic Frameworks (MOFs)

Metal–organic frameworks are a class of crystalline compounds, consisting of metal ions linked together by organic bridging ligands [102]. Due to their high porosity and their chemical and structural diversity, specific adsorption properties (tuneable pore size and chemical functionalities), high surface area, ordered pore structure, and abundant active metal sites make these materials highly attractive for bioanalytical chemistry. Several studies deal with MOFs as an affinity material for the selective enrichment of phosphopeptides [103].

Recent works explore the use of MOFs as a highly selective resin for phosphopeptide enrichment. For this purpose, the selection of high-valent metal ions (e.g.,

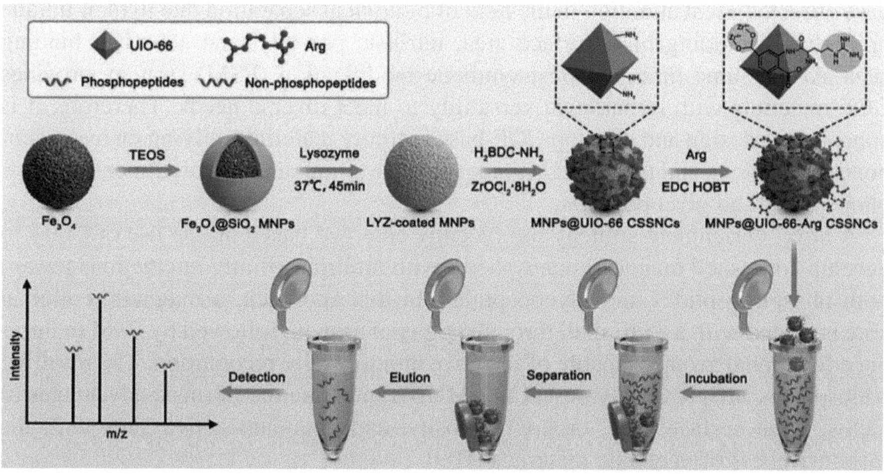

Fig. 4 Schematic illustration of the synthesis of bifunctional MNPs@UIO-66-Arg CSSNCs and the enrichment procedure of phosphopeptides. (Reproduced from [107])

Zr^{4+}, Ti^{4+}, Sn^{4+}, Fe^{3+}, Ga^{3+}, Al^{3+}, and Co^{2+}) and chelating ligands (e.g., iminodiacetic acid, nitrilotriacetic acid, iminodiacetic acid, and adenosine triphosphate) is the priority for the design of these materials [104].

Furthermore, compounds such as $[Er_2(PDA)_3(H_2O)]\cdot 2H_2O$ were selected, owing to its remarkable chemical stability and the presence of coordinatively unsaturated metal sites. Trivalent lanthanide ions, and Er (III) in particular, have been previously demonstrated to exhibit strong and selective binding toward phosphopeptides in precipitation studies. These characteristics make MOFs a promising candidate for advanced phosphopeptide enrichment strategies [105].

Many studies reported in the literature on MOFs combine these nanostructures with magnetic nanoparticles as novel strategies for the preparation of functional materials aimed at phosphopeptide enrichment [93]. The introduced Fe_3O_4 particles exhibit good superparamagnetism, ensuring efficient magnetic separation [106].

Furthermore, owing to the outstanding properties of MOFs, such as their high specific surface area and the abundance of Lewis acid sites, MOF-based IMAC materials have demonstrated efficient and highly selective enrichment of phosphopeptides (Fig. 4). However, since their assembly mainly relies on coordination bonds between metal ions and organic ligands, the stability of most MOFs in water is often insufficient [108].

4.6 Covalent Organic Frameworks (COFs)

Covalent organic frameworks constructed by rigid building units via strong covalent bonds are a new type of crystalline organic porous material, which show better stability in water. COFs, a type of periodic and crystalline porous organic polymer, have attracted great attention in the field of biological separation due to their unique properties, including high surface area, intrinsic porosity, and abundant binding sites. At the same time, the post-synthetic modification (PSM) strategy provides COF materials with remarkable versatility to meet diverse needs. Therefore, it is appealing to design and develop COF-based affinity materials, relying on hydrogen-bond interactions and the HILIC strategy, for the simultaneous enrichment of phosphopeptides and glycopeptides.

A novel enrichment strategy has recently emerged based on the use of COFs to develop core–shell magnetic nanospheres with multiple affinity interactions toward both phosphopeptides and glycopeptides. In this approach, a magnetic colloidal core is coated with a COF shell through epitaxial growth, followed by PSM to introduce functional groups capable of selective biomolecular recognition. The resulting nanospheres MCNC@Polymer@COF-MUBA integrate the intrinsic advantages of COFs, large surface area, ordered porosity, and abundant active sites with the responsiveness of magnetic materials [109].

The functional molecule MUBA plays a key role in the selective capture of phosphopeptides through hydrogen-bond interactions between its carboxyl and carbamido groups and the phosphate moieties of phosphorylated peptides. In parallel, the hydrophilic ligands incorporated during COF formation, together with the modified functional groups, endow the nanospheres with excellent hydrophilicity, enhancing their affinity for glycopeptides.

Overall, this multifunctional nanoplatform provides an optimal material for the simultaneous enrichment of phosphopeptides and glycopeptides due to: (i) the presence of multiple affinity sites, (ii) the structural benefits of the COF shell that improve adsorption efficiency, (iii) the size-exclusion effect associated with its mesoporous channels, and (iv) the fast magnetic response that enables a simplified and rapid enrichment workflow while preserving peptide integrity.

Despite organic porous framework materials, including MOFs and COFs, having shown promising results in phosphopeptide enrichment [110], these porous materials still have some drawbacks. For example, the structure of MOFs can be easily destroyed under acidic or alkaline conditions. The synthesis of COFs requires expensive catalysts or harsh reaction conditions. The above drawbacks limit their further application in the enrichment field.

5 MS-Based Quantitative Phosphoproteomics

The reliable identification and precise quantification of phosphorylation sites are crucial for elucidating molecular mechanisms and functional outcomes. Methods for phosphorylation analysis have evolved from early phosphorus content assays to isotope labeling, phospho-specific antibodies, and modern MS-based techniques. Each approach offers distinct advantages and limitations, differing in throughput, specificity, and ability to determine site stoichiometry. Today, MS-based methods dominate large-scale phosphoproteomic studies, enabling the simultaneous identification and quantification of thousands of phosphosites [111].

Nevertheless, several analytical and instrumental challenges remain, including low stoichiometry, low peptide abundance, sample complexity, and the reduced ionization efficiency of phosphopeptides compared with their unmodified counterparts [19]. Although the substoichiometry and low abundance of phosphopeptides as well as complexity of the samples can be addressed through the enrichment and fractionation strategies described above, it is necessary to use appropriate conditions in mass spectrometry analysis. This is required to achieve more efficient ionization of the peptides and to enable fragmentation that results in the least possible loss of information. It is also essential to apply an adequate strategy for quantification, standardization, and data analysis. In this context, purifying the sample through selective enrichment of phosphopeptides entails the removal of most non-phosphorylated peptides, reducing the so-called ionization suppression effect that hinders MS detectability of phosphopeptides. Furthermore, the ionization of phosphopeptides can be optimized by adjusting the sample pH; either increasing it to promote deprotonation and perform detection in negative mode or using very low pH values to prevent phosphate loss and carry out detection in positive mode [112].

5.1 Fragmentation Strategy

Accurate identification of phosphorylation sites in shotgun proteomics depends strongly on how peptide ions fragment inside the mass spectrometer. Because the phosphoester bond in phosphoserine and phosphothreonine is very unstable, phosphorylated peptides often break differently from unmodified ones, which makes choosing the right fragmentation method important [113].

In collision-induced dissociation (CID), the repeated low-energy collisions inside an ion trap tend to break the phosphoester bond before the peptide backbone fragments, producing a dominant neutral-loss peak of phosphoric acid and often only limited b- and y-ion information. Because this can complicate localization when multiple possible sites are present, methods such as pseudo-MS^3 or neutral-loss-triggered MS^3 were introduced to generate additional sequence ions from the neutral-loss species and improve the overall interpretability of the spectrum [136]. Higher-energy collisional dissociation (HCD), sometimes considered a beam-type CID method, mitigates this problem by accelerating ions more strongly prior to collision, creating fewer but higher-energy impacts that favor backbone cleavage over neutral loss. As a result, HCD typically yields more complex and informative spectra for phosphoserine- and phosphothreonine-containing peptides and is widely adopted in modern phosphoproteomics workflows [137].

In contrast to collision-based techniques, electron-based methods such as electron capture dissociation (ECD) and electron transfer dissociation (ETD) induce peptide cleavage by electron addition or transfer, generating c- and z-type ions while generally preserving labile modifications like phosphorylation. This characteristic makes them especially valuable for unambiguous phosphosite assignment, although their efficiency depends on precursor charge state and can be lower for typical doubly charged tryptic peptides. To address these limitations, alternative proteases or decision-tree acquisition strategies are sometimes used to increase ETD performance. More recently, hybrid fragmentation approaches such as EThcD, which apply supplemental HCD following electron transfer, combine the strengths of both mechanisms by producing complementary b/y and c/z ions within the same spectrum [138]. This hybrid mode enhances sequence coverage and often provides clearer evidence for correct phosphosite localization, making it an increasingly attractive option as instrumentation and acquisition speed continue to improve.

5.2 *Quantification Approaches*

The quantification of protein phosphorylation, that is, determining how the phosphorylation state of a protein changes under specific conditions, can be achieved by comparing the phosphopeptides present in a sample with those in a reference condition. This comparison can be carried out using isotopic labeling strategies or label-free approaches.

Isotopic labeling can be introduced chemically, enzymatically, or metabolically, but the principle is the same in all cases: an isotopic tag is incorporated into the peptide, producing a predictable mass shift. In the mass spectrum, both the labeled and unlabeled phosphopeptides appear simultaneously but with a defined mass difference. This allows the intensities of the two forms to be compared directly, providing a quantitative measurement of the phosphorylation change.

Enzymatic labeling typically involves introducing the isotopic tag during enzymatic reactions. One strategy to quantify how the phosphorylation state of a peptide varies involves the use of labeled water ($H_2{}^{18}O$) during proteolytic digestion. During this process, two ^{18}O atoms are introduced into the carboxyl terminus of the peptide after hydrolysis of the peptide bond, resulting in a peptide mass that is 4 Da higher than that of its non-labeled counterpart. This isotopic labeling is then followed by the common phosphopeptide-analysis workflow previously described. Finally, phosphatase treatment produces the dephosphorylation prior to MS analysis, so that eventually the intensity of both non-labeled and labeled peptide signals is compared to quantify the variation in the phosphorylation [90].

Metabolic labeling is based on the use of cell culture media that contains isotopically labeled substrates (with ^{13}C and/or ^{15}N, enabling multiplexing), so that newly synthesized proteins incorporate this label. This labeling is exemplified by SILAC (Stable Isotope Labeling by Amino acids in Cell culture) [114]. This strategy is mainly applied to cell-based systems and enables quantitative comparison of phosphorylation between experimental conditions, reflecting the true in vivo dynamics of protein modification. After SILAC labeling, samples are mixed, digested, enriched/fractionated, and analyzed by LC–MS/MS. Phosphopeptides are quantified with standard approaches, through comparison of MS signals between the differently SILAC-labeled phosphopeptides [115]. Then, phosphorylation changes can be determined after normalization against total protein level ratios [116].

Chemical labeling strategies, such as iTRAQ (Isobaric tag for relative and absolute quantitation) or TMT (tandem mass tags), involve the addition of isotopic or isobaric chemical groups to peptides after digestion. These groups allow for both peptide identification and relative quantification by comparing reporter ion intensities during mass spectrometric analysis. These methods can suffer from ratio compression due to co-isolation of peptides, although MS^3-based TMT workflows largely mitigate this artifact [117, 118]. Isobaric labeling approaches have expanded multiplexing capabilities, allowing simultaneous quantification across multiple samples [119–121].

While isotopic labeling provides relative quantification between samples or conditions, absolute quantification strategies also exist. These approaches allow not only the determination of phosphorus content in peptides but also the calculation of phosphorylation stoichiometry. One common strategy uses synthetic peptides, often referred to as AQUA (absolute quantification) peptides, which are produced in both phosphorylated and non-phosphorylated forms and introduced into the sample at known concentrations [122]. By measuring these standards alongside endogenous peptides in the mass spectrometer, it is possible to determine the absolute level of phosphorylation for a given protein or peptide [123–125].

Recent advances in mass spectrometry instrumentation, including higher sensitivity and resolution, along with improved computational tools for data analysis, have increased the applicability of label-free quantification (LFQ) strategies. In these approaches, quantification is based on the comparison of mass signal intensities (e.g., area under the curve or MS1 intensities) or spectral counts (e.g., number of peptides or spectra events observed for each protein). Quantitative values are then determined, based on the assumption of a correlation between the intensity, height or peak area, or number of peptides observed, respectively, with protein quantities, hence inferring protein abundances with computational tools. Abundance of phosphopeptides is quantified by measuring ion intensities or spectral counts directly from the sample without prior labeling. LFQ of protein phosphorylation can be carried out through the comparison of phosphopeptide signals across samples, whereas correlation of the signals of phosphorylated peptides and their unphosphorylated counterpart provides phosphorylation stoichiometry [126]. Furthermore, thanks to the high-throughput nature of LFQ, it is possible to quantify protein phosphorylation variations with time, providing valuable insights into phosphorylation dynamics [127]. However, since phosphorylation alters peptide hydrophobicity and protonation behavior, this approach assumes that both forms ionize equally well, which may lead to quantitative bias. These differences in ionization efficiency often necessitate empirical corrections using adequate peptide standards and careful normalization to account for total protein content and the use of statistical methods to ensure reliable comparisons across samples [128].

6 Elemental Mass Spectrometry

Inductively Coupled Plasma Mass Spectrometry (ICP-MS) is a fundamental MS technique that involves the formation of atomic ions in an argon plasma at approximately 8000 K from the chemical species introduced into it. A key feature of ICP-MS is that ionization is virtually independent of the chemical species and matrix. Consequently, unlike molecular mass spectrometry with electrospray ionization, the signal can be considered quantitative, i.e., directly proportional to the amount of the element present in the sample. This property offers significant advantages for quantitative analysis of elements of interest, particularly metals.

Recent advances in instrumentation and methodology over the past decade have expanded the application of ICP-MS to the study of organic molecules [129, 130]. Improvements such as tandem analyzer integration and more efficient radio frequency ionization systems have enabled sensitive and effective analysis of non-metallic elements, including selenium, arsenic, chlorine, sulfur, and phosphorus. In this context, the quantitative determination of sulfur and phosphorus has represented a significant breakthrough for protein phosphorylation analysis [131].

Protein phosphorylation involves the addition of a phosphate group, i.e., phosphorous. Determining phosphorylated proteins or peptides using ICP-MS produces a P signal proportional to its content in the protein. Therefore, comparing P levels between two samples allows the determination of the phosphorylation degree or its variation. This quantification can also be performed using non-specific standards to establish a calibration factor for phosphorus, enabling absolute quantification [132]. This approach is useful not only for assessing phosphorylation but also for monitoring phosphorus recovery during protein or peptide purification and enrichment, facilitating optimization of enrichment strategies from a quantitative perspective [133].

One limitation of ICP-MS for phosphorus analysis of peptides is the loss of one dimension of resolution compared to molecular mass spectrometry. The elemental phosphorus signal is independent of its molecular context, so it cannot distinguish the phosphorus source unless the compounds are separated prior to ionization, typically using reverse-phase chromatography. This limitation is particularly relevant for complex samples, as enzymatic digestion of phosphorylated proteins can generate dozens or thousands of peptides.

Analyzing intact proteins (*top-down* proteomics) can help overcome this complexity by integrating all phosphorus from a protein into a single measurable entity. However, this approach is suitable only for global phosphorylation analysis, not for site-specific determination, which requires digestion.

To determine the phosphorylation degree, it is necessary to correlate phosphorus content with protein quantity [134]. This can be achieved by measuring the sulfur content in methionine and cysteine residues. Knowing the protein sequence (or determining it in parallel via electrospray MS) allows calculation of the number of sulfur atoms. Using a non-specific sulfur calibration, the protein quantity can be determined, and the phosphorus-to-protein ratio provides the phosphorylation degree, along with the absolute protein amount [131, 135]. This strategy can also be applied to peptides, but it is more limited because not all peptides contain sulfur, and not all sulfur-containing peptides are phosphorylated. Therefore, some peptides may be lost during enrichment strategies.

7 Concluding Remarks

Phosphoproteomics has evolved from a niche analytical challenge into a cornerstone of systems biology. Continuous innovation in enrichment strategies, chromatographic workflows, and mass spectrometry instrumentation has enabled the identification and quantification of tens of thousands of phosphorylation sites in a single experiment. Despite this progress, several challenges persist. The inherently low stoichiometry of many phosphorylation events, losses during sample preparation, and incomplete site coverage still limit analytical depth and reproducibility.

Affinity-based enrichment strategies remain the foundation of modern phosphoproteomics, each offering distinct advantages and trade-offs in selectivity, scalability, and cost. Immunoprecipitation provides unparalleled specificity for phosphorylation types like pTyr but is less suited for global analyses. Molecularly imprinted polymers represent a new generation of synthetic affinity materials that combine the selectivity of antibodies with the durability and reusability of engineered polymers. Meanwhile, advanced nanomaterial-based platforms introduce unprecedented versatility, coupling magnetic manipulation, mesoporous architectures, and multi-ligand chemistries to enhance capture efficiency and automation. Furthermore, the numerous potential possibilities in the metal centers and metal oxides provide excellent performance in enrichment selectivity and efficiency.

The diversity of available enrichment strategies reflects the complexity of the phosphoproteome and the analytical challenges inherent to its characterization. No single method achieves complete coverage; each displays unique selectivity biases toward certain phosphorylation types or peptide characteristics. Consequently, combining complementary enrichment techniques has become standard practice to improve depth, reproducibility, and quantitative accuracy. Emerging hybrid materials and nanocomposites further increase binding capacity and specificity, while automation and robotics are accelerating workflow standardization. Collectively, these innovations have transformed phosphopeptide enrichment into a robust, scalable, and integrative component of quantitative phosphoproteomics, supporting more comprehensive and reproducible mapping of phosphorylation dynamics in complex biological systems.

In terms of quantification, MS remains the method of choice for site-specific phosphorylation profiling. Quantitative approaches such as SILAC, TMT, and AQUA provide precise relative and absolute measurements, although label-free approaches now rival them in scalability and cost efficiency. Nonetheless, determining true phosphorylation stoichiometry, particularly in low-abundance proteins or complex tissues, remains a significant analytical hurdle.

Acknowledgments FCC acknowledges the "Ramón y Cajal" grant (RYC2024-048391-I), funded by MICIU/AEI/https://doi.org/10.13039/501100011033 and the FSE+.

References

1. Kolch W, Halasz M, Granovskaya M, Kholodenko BN (2015) The dynamic control of signal transduction networks in cancer cells. Nat Rev Cancer 15(9):515–527. https://doi.org/10.1038/nrc3983
2. Zhao P, Malik S (2022) The phosphorylation to acetylation/methylation cascade in transcriptional regulation: how kinases regulate transcriptional activities of DNA/histone-modifying enzymes. Cell Biosci 12(1):83. https://doi.org/10.1186/s13578-022-00821-7
3. O'Connor DS, Grossman D, Plescia J et al (2000) Regulation of apoptosis at cell division by $p34^{cdc2}$ phosphorylation of survivin. Proc Natl Acad Sci USA 97(24):13103–13107. https://doi.org/10.1073/pnas.240390697
4. Olsen JV, Blagoev B, Gnad F et al (2006) Global, in vivo, and site-specific phosphorylation dynamics in signaling networks. Cell 127(3):635–648. https://doi.org/10.1016/j.cell.2006.09.026
5. Martinez-Val A, Bekker-Jensen DB, Steigerwald S et al (2021) Spatial-proteomics reveals phospho-signaling dynamics at subcellular resolution. Nat Commun 12(1):7113. https://doi.org/10.1038/s41467-021-27398-y
6. Humphrey SJ, James DE, Mann M (2015) Protein phosphorylation: a major switch mechanism for metabolic regulation. Trends Endocrinol Metabol 26(12):676–687. https://doi.org/10.1016/j.tem.2015.09.013
7. Cohen P (2000) The regulation of protein function by multisite phosphorylation – a 25 year update. Trends Biochem Sci 25(12):596–601. https://doi.org/10.1016/S0968-0004(00)01712-6
8. Cohen P (2002) The origins of protein phosphorylation. Nat Cell Biol 4(5):E127–E130. https://doi.org/10.1038/ncb0502-e127
9. Vlastaridis P, Kyriakidou P, Chaliotis A, Van De Peer Y, Oliver SG, Amoutzias GD (2017) Estimating the total number of phosphoproteins and phosphorylation sites in eukaryotic proteomes. GigaScience 6(2):giw015. https://doi.org/10.1093/gigascience/giw015
10. Chen M, Gou Y, Lei M et al (2025) EPSD 2.0: an updated database of protein phosphorylation sites across eukaryotic species. Song J, ed. Genomics Proteomics Bioinformatics 23(3):qzaf057. https://doi.org/10.1093/gpbjnl/qzaf057
11. Hornbeck PV, Zhang B, Murray B, Kornhauser JM, Latham V, Skrzypek E (2015) PhosphoSitePlus, 2014: mutations, PTMs and recalibrations. Nucleic Acids Res 43(D1):D512–D520. https://doi.org/10.1093/nar/gku1267
12. Hornbeck PV, Kornhauser JM, Latham V et al (2019) 15 years of PhosphoSitePlus®: integrating post-translationally modified sites, disease variants and isoforms. Nucleic Acids Res 47(D1):D433–D441. https://doi.org/10.1093/nar/gky1159
13. Adam K, Hunter T (2018) Histidine kinases and the missing phosphoproteome from prokaryotes to eukaryotes. Lab Investig 98(2):233–247. https://doi.org/10.1038/labinvest.2017.118
14. The Consortium for Top Down Proteomics, Smith LM, Kelleher NL (2013) Proteoform: a single term describing protein complexity. Nat Methods 10(3):186–187. https://doi.org/10.1038/nmeth.2369
15. Smith JC, Figeys D (2008) Recent developments in mass spectrometry-based quantitative phosphoproteomics. This paper is one of a selection of papers published in this Special Issue, entitled CSBMCB — Systems and Chemical Biology, and has undergone the Journal's usual peer review process. Biochem Cell Biol 86(2):137–148. https://doi.org/10.1139/O08-007

16. Chandramouli K, Qian PY (2009) Proteomics: challenges, techniques and possibilities to overcome biological sample complexity. Godwin AK, ed. Hum Genom Proteom 1(1). https://doi.org/10.4061/2009/239204
17. Humphrey SJ, Azimifar SB, Mann M (2015) High-throughput phosphoproteomics reveals in vivo insulin signaling dynamics. Nat Biotechnol 33(9):990–995. https://doi.org/10.1038/nbt.3327
18. Blazek M, Santisteban TS, Zengerle R, Meier M (2015) Analysis of fast protein phosphorylation kinetics in single cells on a microfluidic chip. Lab Chip 15(3):726–734. https://doi.org/10.1039/C4LC00797B
19. Solari FA, Dell'Aica M, Sickmann A, Zahedi RP (2015) Why phosphoproteomics is still a challenge. Mol BioSyst 11(6):1487–1493. https://doi.org/10.1039/C5MB00024F
20. Kratzer R, Eckerskorn C, Karas M, Lottspeich F (1998) Suppression effects in enzymatic peptide ladder sequencing using ultraviolet - matrix assisted laser desorption/ionization - mass spectrometry. Electrophoresis 19(11):1910–1919. https://doi.org/10.1002/elps.1150191109
21. Janek K, Wenschuh H, Bienert M, Krause E (2001) Phosphopeptide analysis by positive and negative ion matrix-assisted laser desorption/ionization mass spectrometry. Rapid Commun Mass Spectr 15(17):1593–1599. https://doi.org/10.1002/rcm.417
22. Leitner A (2016) Enrichment strategies in phosphoproteomics. In: Von Stechow L (ed) Phospho-proteomics. Vol 1355. Methods in molecular biology. Springer New York, pp 105–121. https://doi.org/10.1007/978-1-4939-3049-4_7
23. Low TY, Mohtar MA, Lee PY, Omar N, Zhou H, Ye M (2021) Widening the bottleneck of phosphoproteomics: evolving strategies for phosphopeptide enrichment. Mass Spectrom Rev 40(4):309–333. https://doi.org/10.1002/mas.21636
24. Sun N, Yu H, Wu H, Shen X, Deng C (2021) Advanced nanomaterials as sample technique for bio-analysis. TrAC Trends Anal Chem 135:116168. https://doi.org/10.1016/j.trac.2020.116168
25. Rejeeth C, Sharma A (2022) Label-free designed nanomaterials enrichment and separation techniques for phosphoproteomics based on mass spectrometry. Front Nanotechnol 4:1047055. https://doi.org/10.3389/fnano.2022.1047055
26. Jiang Y, Rex DAB, Schuster D et al (2024) Comprehensive overview of bottom-up proteomics using mass spectrometry. ACS Meas Sci Au 4(4):338–417. https://doi.org/10.1021/acsmeasuresciau.3c00068
27. Dupree EJ, Jayathirtha M, Yorkey H, Mihasan M, Petre BA, Darie CC (2020) A critical review of bottom-up proteomics: the good, the bad, and the future of this field. Proteomes 8(3):14. https://doi.org/10.3390/proteomes8030014
28. Dickhut C, Feldmann I, Lambert J, Zahedi RP (2014) Impact of digestion conditions on phosphoproteomics. J Proteome Res 13(6):2761–2770. https://doi.org/10.1021/pr401181y
29. Giansanti P, Aye TT, van den Toorn H, Peng M, van Breukelen B, Heck AJR (2015) An augmented multiple-protease-based human phosphopeptide atlas. Cell Rep 11(11):1834–1843. https://doi.org/10.1016/j.celrep.2015.05.029
30. Bubis JA, Gorshkov V, Gorshkov MV, Kjeldsen F (2020) PhosphoShield: improving trypsin digestion of phosphoproteins by shielding the negatively charged phosphate moiety. J Am Soc Mass Spectrom 31(10):2053–2060. https://doi.org/10.1021/jasms.0c00171
31. Imamura H, Wakabayashi M, Ishihama Y (2012) Analytical strategies for shotgun phosphoproteomics: status and prospects. Semin Cell Dev Biol 23(8):836–842. https://doi.org/10.1016/j.semcdb.2012.05.007
32. Gonczarowska-Jorge H, Loroch S, Dell'Aica M, Sickmann A, Roos A, Zahedi RP (2017) Quantifying missing (phospho)proteome regions with the broad-specificity protease subtilisin. Anal Chem 89(24):13137–13145. https://doi.org/10.1021/acs.analchem.7b02395
33. Xu F, Yu L, Peng X et al (2020) Unambiguous phosphosite localization through the combination of Trypsin and LysargiNase mirror spectra in a large-scale phosphoproteome study. J Proteome Res 19(6):2185–2194. https://doi.org/10.1021/acs.jproteome.9b00562

34. Sharma K, D'Souza RCJ, Tyanova S et al (2014) Ultradeep human phosphoproteome reveals a distinct regulatory nature of Tyr and Ser/Thr-based signaling. Cell Rep 8(5):1583–1594. https://doi.org/10.1016/j.celrep.2014.07.036
35. Urban J (2022) A review on recent trends in the phosphoproteomics workflow. From sample preparation to data analysis. Anal Chim Acta 1199:338857. https://doi.org/10.1016/j.aca.2021.338857
36. Li J, Wang J, Yan Y et al (2022) Comprehensive evaluation of different TiO2-based phosphopeptide enrichment and fractionation methods for phosphoproteomics. Cells 11(13):2047. https://doi.org/10.3390/cells11132047
37. Birk MS, Charpentier E, Frese CK (2021) Automated phosphopeptide enrichment for gram-positive bacteria. *J Proteome Res* 20(10):4886–4892. https://doi.org/10.1021/acs.jproteome.1c00364
38. Andersson L, Porath J (1986) Isolation of phosphoproteins by immobilized metal (Fe3+) affinity chromatography. *Anal Biochem* 154(1):250–254. https://doi.org/10.1016/0003-2697(86)90523-3
39. Posewitz MC, Tempst P (1999) Immobilized gallium(III) affinity chromatography of phosphopeptides. Anal Chem 71(14):2883–2892. https://doi.org/10.1021/ac981409y
40. Seeley E, Riggs L, Regnier F (2005) Reduction of non-specific binding in Ga(III) immobilized metal affinity chromatography for phosphopeptides by using endoproteinase glu-C as the digestive enzyme. J Chromatogr B 817(1):81–88. https://doi.org/10.1016/j.jchromb.2004.03.024
41. Tape CJ, Worboys JD, Sinclair J et al (2014) Reproducible automated phosphopeptide enrichment using magnetic TiO_2 and Ti-IMAC. Anal Chem 86(20):10296–10302. https://doi.org/10.1021/ac5025842
42. Thingholm TE, Larsen MR (2016) Phosphopeptide enrichment by immobilized metal affinity chromatography. In: Von Stechow L (ed) Phospho-proteomics. Vol 1355. Methods in molecular biology. Springer New York, pp 123–133. https://doi.org/10.1007/978-1-4939-3049-4_8
43. Huang J, Liu X, Wang D et al (2021) Dual-functional Ti(IV)-IMAC material enables simultaneous enrichment and separation of diverse glycopeptides and phosphopeptides. *Anal Chem* 93(24):8568–8576. https://doi.org/10.1021/acs.analchem.1c01324
44. Arribas Diez I, Govender I, Naicker P, Stoychev S, Jordaan J, Jensen ON (2021) Zirconium(IV)-IMAC revisited: improved performance and phosphoproteome coverage by magnetic microparticles for phosphopeptide affinity enrichment. J Proteome Res 20(1):453–462. https://doi.org/10.1021/acs.jproteome.0c00508
45. Tang R, Bai Q, Ma S, Ou J (2023) Materials, workflows and applications of IMAC for phosphoproteome profiling in the recent decade: a review. TrAC Trends Anal Chem 158:116862. https://doi.org/10.1016/j.trac.2022.116862
46. Kånge R, Selditz U, Granberg M et al (2005) Comparison of different IMAC techniques used for enrichment of phosphorylated peptides. J Biomol Tech 16(2):91–103
47. Thingholm TE, Jensen ON (2009) Enrichment and characterization of phosphopeptides by immobilized metal affinity chromatography (IMAC) and mass spectrometry. In: Graauw MD (ed) Phospho-proteomics. Vol 527. Methods in molecular biology™. Humana Press, pp 47–56. https://doi.org/10.1007/978-1-60327-834-8_4
48. Lai AC, Tsai C, Hsu C, Sun Y, Chen Y (2012) Complementary Fe^{3+} – and Ti^{4+} -immobilized metal ion affinity chromatography for purification of acidic and basic phosphopeptides. Rapid Commun Mass Spectr 26(18):2186–2194. https://doi.org/10.1002/rcm.6327
49. Shi C, Lin Q, Deng C (2015) Preparation of on-plate immobilized metal ion affinity chromatography platform via dopamine chemistry for highly selective isolation of phosphopeptides with matrix assisted laser desorption/ionization mass spectrometry analysis. Talanta 135:81–86. https://doi.org/10.1016/j.talanta.2014.12.041
50. Wang D, Huang J, Zhang H et al (2023) ATP-coated dual-functionalized titanium(IV) IMAC material for simultaneous enrichment and separation of glycopeptides and phosphopeptides. J Proteome Res 22(6):2044–2054. https://doi.org/10.1021/acs.jproteome.3c00118

51. Yan Y, Lu Y, Chen M, Liang H (2018) A novel IMAC platform – adenosine coupled functional magnetic microspheres for phosphoproteome research. Anal Methods 10(10):1190–1195. https://doi.org/10.1039/C7AY02931D
52. Tsai CF, Hsu CC, Hung JN et al (2014) Sequential phosphoproteomic enrichment through complementary metal-directed immobilized metal ion affinity chromatography. Anal Chem 86(1):685–693. https://doi.org/10.1021/ac4031175
53. Pinkse MWH, Uitto PM, Hilhorst MJ, Ooms B, Heck AJR (2004) Selective isolation at the femtomole level of phosphopeptides from proteolytic digests using 2D-NanoLC-ESI-MS/MS and titanium oxide precolumns. Anal Chem 76(14):3935–3943. https://doi.org/10.1021/ac0498617
54. Sano A, Nakamura H (2004) Chemo-affinity of Titania for the column-switching HPLC analysis of phosphopeptides. Anal Sci 20(3):565–566. https://doi.org/10.2116/analsci.20.565
55. Leitner A (2010) Phosphopeptide enrichment using metal oxide affinity chromatography. TrAC Trends Anal Chem 29(2):177–185. https://doi.org/10.1016/j.trac.2009.08.007
56. Larsen MR, Thingholm TE, Jensen ON, Roepstorff P, Jørgensen TJD (2005) Highly selective enrichment of phosphorylated peptides from peptide mixtures using titanium dioxide micro-columns. Mol Cell Proteomics 4(7):873–886. https://doi.org/10.1074/mcp.T500007-MCP200
57. Negroni L, Claverol S, Rosenbaum J, Chevet E, Bonneu M, Schmitter JM (2012) Comparison of IMAC and MOAC for phosphopeptide enrichment by column chromatography. J Chromatogr B 891-892:109–112. https://doi.org/10.1016/j.jchromb.2012.02.028
58. Kweon HK, Håkansson K (2006) Selective zirconium dioxide-based enrichment of phosphorylated peptides for mass spectrometric analysis. Anal Chem 78(6):1743–1749. https://doi.org/10.1021/ac0522355
59. Aryal UK, Ross ARS (2010) Enrichment and analysis of phosphopeptides under different experimental conditions using titanium dioxide affinity chromatography and mass spectrometry. Rapid Comm Mass Spectr 24(2):219–231. https://doi.org/10.1002/rcm.4377
60. Leitner A, Sakeye M, Zimmerli CE, Smått JH (2017) Insights into chemoselectivity principles in metal oxide affinity chromatography using tailored nanocast metal oxide microspheres and mass spectrometry-based phosphoproteomics. Analyst 142(11):1993–2003. https://doi.org/10.1039/C7AN00570A
61. Yan J, Li X, Cheng S, Ke Y, Liang X (2009) Facile synthesis of titania–zirconia monodisperse microspheres and application for phosphopeptides enrichment. Chem Commun 20:2929. https://doi.org/10.1039/b901424a
62. Wang M, Deng C, Li Y, Zhang X (2014) Magnetic binary metal oxides affinity probe for highly selective enrichment of phosphopeptides. ACS Appl Mater Interfaces 6(14):11775–11782. https://doi.org/10.1021/am502530c
63. Wang M, Sun X, Li Y, Deng C (2016) Design and synthesis of magnetic binary metal oxides nanocomposites through dopamine chemistry for highly selective enrichment of phosphopeptides. Proteomics 16(6):915–919. https://doi.org/10.1002/pmic.201500277
64. Carrascal M, Ovelleiro D, Casas V, Gay M, Abian J (2008) Phosphorylation analysis of primary human T lymphocytes using sequential IMAC and titanium oxide enrichment. J Proteome Res 7(12):5167–5176. https://doi.org/10.1021/pr800500r
65. Yue X, Schunter A, Hummon AB (2015) Comparing multistep immobilized metal affinity chromatography and multistep TiO_2 methods for phosphopeptide enrichment. *Anal Chem* 87(17):8837–8844. https://doi.org/10.1021/acs.analchem.5b01833
66. Thingholm TE, Jensen ON, Robinson PJ, Larsen MR (2008) SIMAC (sequential elution from IMAC), a phosphoproteomics strategy for the rapid separation of monophosphorylated from multiply phosphorylated peptides. Mol Cell Proteomics 7(4):661–671. https://doi.org/10.1074/mcp.M700362-MCP200
67. Edelmann MJ (2011) Strong cation exchange chromatography in analysis of posttranslational modifications: innovations and perspectives. Makhatadze G, ed. BioMed Res Int 2011(1):936508. https://doi.org/10.1155/2011/936508

68. Han G, Ye M, Zhou H et al (2008) Large-scale phosphoproteome analysis of human liver tissue by enrichment and fractionation of phosphopeptides with strong anion exchange chromatography. Proteomics 8(7):1346–1361. https://doi.org/10.1002/pmic.200700884
69. Lim KB, Kassel DB (2006) Phosphopeptides enrichment using on-line two-dimensional strong cation exchange followed by reversed-phase liquid chromatography/mass spectrometry. Anal Biochem 354(2):213–219. https://doi.org/10.1016/j.ab.2006.04.027
70. Albuquerque CP, Smolka MB, Payne SH, Bafna V, Eng J, Zhou H (2008) A multidimensional chromatography technology for in-depth phosphoproteome analysis. Mol Cell Proteomics 7(7):1389–1396. https://doi.org/10.1074/mcp.M700468-MCP200
71. Alpert AJ (2008) Electrostatic repulsion hydrophilic interaction chromatography for isocratic separation of charged solutes and selective isolation of phosphopeptides. Anal Chem 80(1):62–76. https://doi.org/10.1021/ac070997p
72. Zarei M, Sprenger A, Rackiewicz M, Dengjel J (2016) Fast and easy phosphopeptide fractionation by combinatorial ERLIC-SCX solid-phase extraction for in-depth phosphoproteome analysis. Nat Protoc 11(1):37–45. https://doi.org/10.1038/nprot.2015.134
73. Villén J, Villén J, Gygi SP, Gygi SP (2008) The SCX/IMAC enrichment approach for global phosphorylation analysis by mass spectrometry. Nat Protoc 3(10):1630–1638. https://doi.org/10.1038/nprot.2008.150
74. Kettenbach AN, Kettenbach AN, Gerber SA, Gerber SA (2011) Rapid and reproducible single-stage phosphopeptide enrichment of complex peptide mixtures: application to general and phosphotyrosine-specific phosphoproteomics experiments. Anal Chem 83(20):7635–7644. https://doi.org/10.1021/ac201894j
75. Carrera M, Cañas B, Lopez-Ferrer D (2017) Fast global phosphoproteome profiling of Jurkat T cells by HIFU-TiO_2 -SCX-LC-MS/MS. Anal Chem 89(17):8853–8862. https://doi.org/10.1021/acs.analchem.7b01321
76. Chen CW, Lin PY, Lai YM, Lin MH, Lin SY, Hsu CC (2024) TIMAHAC: streamlined tandem IMAC-HILIC workflow for simultaneous and high-throughput plant phosphoproteomics and N-glycoproteomics. Mol Cell Proteomics 23(5):100762. https://doi.org/10.1016/j.mcpro.2024.100762
77. Cooper JA, Sefton BM, Hunter T (1983) Detection and quantification of phosphotyrosine in proteins. In: Methods in enzymology, vol 99. Elsevier, pp 387–402. https://doi.org/10.1016/0076-6879(83)99075-4
78. Salomon AR, Ficarro SB, Brill LM et al (2003) Profiling of tyrosine phosphorylation pathways in human cells using mass spectrometry. Proc Natl Acad Sci USA 100(2):443–448. https://doi.org/10.1073/pnas.2436191100
79. Alonso A, Sasin J, Bottini N et al (2004) Protein tyrosine phosphatases in the human genome. Cell 117(6):699–711. https://doi.org/10.1016/j.cell.2004.05.018
80. Rush J, Moritz A, Lee KA et al (2005) Immunoaffinity profiling of tyrosine phosphorylation in cancer cells. Nat Biotechnol 23(1):94–101. https://doi.org/10.1038/nbt1046
81. Van Der Mijn JC, Labots M, Piersma SR et al (2015) Evaluation of different phosphotyrosine antibodies for label-free phosphoproteomics. J Proteome 127:259–263. https://doi.org/10.1016/j.jprot.2015.04.006
82. Abe Y, Nagano M, Tada A, Adachi J, Tomonaga T (2017) Deep phosphotyrosine proteomics by optimization of phosphotyrosine enrichment and MS/MS parameters. J Proteome Res 16(2):1077–1086. https://doi.org/10.1021/acs.jproteome.6b00576
83. Bian Y, Li L, Dong M et al (2016) Ultra-deep tyrosine phosphoproteomics enabled by a phosphotyrosine superbinder. Nat Chem Biol 12(11):959–966. https://doi.org/10.1038/nchembio.2178
84. Helling S, Shinde S, Brosseron F et al (2011) Ultratrace enrichment of tyrosine phosphorylated peptides on an imprinted polymer. Anal Chem 83(5):1862–1865. https://doi.org/10.1021/ac103086v
85. BelBruno JJ (2019) Molecularly imprinted polymers. Chem Rev 119(1):94–119. https://doi.org/10.1021/acs.chemrev.8b00171

86. Chen J, Shinde S, Koch MH et al (2015) Low-bias phosphopeptide enrichment from scarce samples using plastic antibodies. Sci Rep 5(1):11438. https://doi.org/10.1038/srep11438
87. Incel A, Arribas Díez I, Wierzbicka C, Gajoch K, Jensen ON, Sellergren B (2021) Selective enrichment of histidine phosphorylated peptides using molecularly imprinted polymers. Anal Chem 93(8):3857–3866. https://doi.org/10.1021/acs.analchem.0c04474
88. Jagadeesan KK, Wierzbicka C, Laurell T, Sellergren B, Shinde S, Ekström S (2016) Multiplexed MALDI-MS arrays for screening of MIP solid phase extraction materials. J Chromatogr B 1021:213–220. https://doi.org/10.1016/j.jchromb.2015.10.033
89. Duarte M, Subedi P, Yilmaz E, Marcus K, Laurell T, Ekström S (2018) Molecularly imprinted polymers synthesized *via* template immobilization on fumed silica nanoparticles for the enrichment of phosphopeptides. J Mol Recogn 31(3):e2677. https://doi.org/10.1002/jmr.2677
90. Wei J, Zhang Y, Wang J et al (2008) Highly efficient enrichment of phosphopeptides by magnetic nanoparticles coated with zirconium phosphonate for phosphoproteome analysis. Rapid Comm Mass Spectr 22(7):1069–1080. https://doi.org/10.1002/rcm.3485
91. Hwang L, Ayaz-Guner S, Gregorich ZR et al (2015) Specific enrichment of phosphoproteins using functionalized multivalent nanoparticles. J Am Chem Soc 137(7):2432–2435. https://doi.org/10.1021/ja511833y
92. Roberts DS, Chen B, Tiambeng TN, Wu Z, Ge Y, Jin S (2019) Reproducible large-scale synthesis of surface silanized nanoparticles as an enabling nanoproteomics platform: enrichment of the human heart phosphoproteome. Nano Res 12(6):1473–1481. https://doi.org/10.1007/s12274-019-2418-4
93. Zhang K, Hao Y, Hu D et al (2021) Development of dual-ligand titanium (IV) hydrophilic network sorbent for highly selective enrichment of phosphopeptides. J Chromatogr A 1659:462648. https://doi.org/10.1016/j.chroma.2021.462648
94. Gao L, Tao J, Qi L et al (2022) Synthesis of a metal oxide affinity chromatography magnetic mesoporous nanomaterial and development of a one-step selective phosphopeptide enrichment strategy for analysis of phosphorylated proteins. Anal Chim Acta 1195:339430. https://doi.org/10.1016/j.aca.2022.339430
95. Long X, yu, Zhang Z jin, Li J yuan, Sheng D, Lian H zhen. (2017) A combination strategy using two novel cerium-based nanocomposite affinity probes for the selective enrichment of mono- and multi-phosphopeptides in mass spectrometric analysis. Chem Commun 53(33):4620–4623. https://doi.org/10.1039/C7CC00540G
96. Xu F, Shang D, Zhu C et al (2025) In situ MXene-controlled synthesis of polycrystalline TiO_2 for highly efficient enrichment of phosphopeptides. ACS Appl Mater Interfaces 17(1):260–268. https://doi.org/10.1021/acsami.4c14113
97. Zhang S, Yang X, Zhang F, Qiao J, qin, Zheng W juan, Lian H zhen. (2026) Potato-chip-like two-dimensional magnetic trimetallic oxides derived from MXene-MOF nanocomposites for fast enrichment of phosphopeptides. Sep Purif Technol 382:135918. https://doi.org/10.1016/j.seppur.2025.135918
98. Yu L, He J, Yan S, Liu Y, Lan F, Wu Y (2022) Magnetic MXene/PAMAM composites with flexible dimensional regulation for highly effective enrichment of phosphopeptides. ACS Sustain Chem Eng 10(7):2494–2508. https://doi.org/10.1021/acssuschemeng.1c07715
99. Piovesana S, Iglesias D, Melle-Franco M et al (2020) Carbon nanostructure morphology templates nanocomposites for phosphoproteomics. Nano Res 13(2):380–388. https://doi.org/10.1007/s12274-020-2620-4
100. Xu LN, Li LP, Jin L, Bai Y, Liu HW (2014) Guanidyl-functionalized graphene as a bifunctional adsorbent for selective enrichment of phosphopeptides. Chem Commun 50(75):10963–10966. https://doi.org/10.1039/C4CC04327H
101. Hua S, Wang B, Ding CF, Yan Y (2024) A novel carbon-based material with titanium and zirconium ions etched on hollow mesoporous carbon tubes for specific capture of phosphopeptides and exosomes. Talanta 266:125139. https://doi.org/10.1016/j.talanta.2023.125139
102. Ding F, Zhao Y, Liu H, Zhang W (2020) Core–shell magnetic microporous covalent organic framework with functionalized Ti(IV) for selective enrichment of phosphopeptides. Analyst 145(12):4341–4351. https://doi.org/10.1039/D0AN00038H

103. Meng L, Wang B, Wang B et al (2023) Post-synthesis of a titanium-rich magnetic COF nanocomposite with flexible branched polymers for efficient enrichment of phosphopeptides from human saliva and serum. Analyst 148(19):4738–4745. https://doi.org/10.1039/D3AN00989K
104. He Y, Huang W, Zheng Q et al (2021) Two-dimensional guanidinium-based covalent organic nanosheets for controllable recognition and specific enrichment of global/multi-phosphopeptides. Talanta 233:122497. https://doi.org/10.1016/j.talanta.2021.122497
105. Messner CB, Mirza MR, Rainer M et al (2013) Selective enrichment of phosphopeptides by a metal–organic framework. Anal Methods 5(9):2379. https://doi.org/10.1039/c3ay40308d
106. Jiang D, Li Y, Wu S, Lan L, Liu J (2025) Design of Ce3+ ions functionalized magnetic black phosphorus nanosheets for highly efficient enrichment of phosphopeptides. Anal Chim Acta 1350:343878. https://doi.org/10.1016/j.aca.2025.343878
107. Zhang Y, Li N, Li J, Fan M, Zhang Q, Dang F (2024) Bifunctional MNPs@UIO-66-Arg core-shell-satellite nanocomposites for enrichment of phosphopeptides. Microchim Acta 191(4):211. https://doi.org/10.1007/s00604-024-06177-8
108. Gao W, Zhang F, Zhang S, Li J, yuan, Lian H zhen. (2023) Ti(IV) immobilized bisphosphate fructose-modified magnetic Zr metal organic framework (MOF) for specific enrichment of phosphopeptides. Sep Purif Technol 305:122426. https://doi.org/10.1016/j.seppur.2022.122426
109. Luo B, Yan S, Zhang Y, Zhou J, Lan F, Wu Y (2021) Bifunctional magnetic covalent organic framework for simultaneous enrichment of phosphopeptides and glycopeptides. Anal Chim Acta 1177:338761. https://doi.org/10.1016/j.aca.2021.338761
110. Jiang D, Qi R, Wu S, Li Y, Liu J (2024) Zirconium-rich magnetic polyoxometalate-based metal–organic framework: tailored for phosphopeptide analysis from lung cancer A549 cells. J Colloid Interface Sci 663:123–131. https://doi.org/10.1016/j.jcis.2024.02.138
111. Mann M, Ong SE, Grønborg M, Steen H, Jensen ON, Pandey A (2002) Analysis of protein phosphorylation using mass spectrometry: deciphering the phosphoproteome. Trends Biotechnol 20(6):261–268. https://doi.org/10.1016/S0167-7799(02)01944-3
112. Pan N, Liu P, Cui W, Tang B, Shi J, Chen H (2013) Highly efficient ionization of phosphopeptides at low pH by desorption electrospray ionization mass spectrometry. Analyst 138(5):1321. https://doi.org/10.1039/c3an36737a
113. Boersema PJ, Mohammed S, Heck AJR (2009) Phosphopeptide fragmentation and analysis by mass spectrometry. *J Mass Spectrom* 44(6):861–878. https://doi.org/10.1002/jms.1599
114. Chen X, Wei S, Ji Y, Guo X, Yang F (2015) Quantitative proteomics using SILAC: principles, applications, and developments. Proteomics 15(18):3175–3192. https://doi.org/10.1002/pmic.201500108
115. Griffith AA, Callahan KP, King NG, Xiao Q, Su X, Salomon AR (2022) SILAC Phosphoproteomics reveals unique signaling circuits in CAR-T cells and the inhibition of B cell-activating phosphorylation in target cells. J Proteome Res 21(2):395–409. https://doi.org/10.1021/acs.jproteome.1c00735
116. Zhang H, Xu Y, Filipovic A et al (2013) SILAC-based phosphoproteomics reveals an inhibitory role of KSR1 in p53 transcriptional activity via modulation of DBC1. Br J Cancer 109(10):2675–2684. https://doi.org/10.1038/bjc.2013.628
117. Ting L, Rad R, Gygi SP, Haas W (2011) MS3 eliminates ratio distortion in isobaric multiplexed quantitative proteomics. Nat Methods 8(11):937–940. https://doi.org/10.1038/nmeth.1714
118. McAlister GC, Nusinow DP, Jedrychowski MP et al (2014) MultiNotch MS3 enables accurate, sensitive, and multiplexed detection of differential expression across cancer cell line proteomes. Anal Chem 86(14):7150–7158. https://doi.org/10.1021/ac502040v
119. Jiang X, Bomgarden R, Brown J et al (2017) Sensitive and accurate quantitation of phosphopeptides using TMT isobaric labeling technique. *J Proteome Res* 16(11):4244–4252. https://doi.org/10.1021/acs.jproteome.7b00610
120. Navarrete-Perea J, Yu Q, Gygi SP, Paulo JA (2018) Streamlined tandem mass tag (SL-TMT) protocol: an efficient strategy for quantitative (phospho)proteome profiling using tandem

mass tag-synchronous precursor selection-MS3. J Proteome Res 17(6):2226–2236. https://doi.org/10.1021/acs.jproteome.8b00217
121. Zhang T, Keele GR, Gyuricza IG et al (2023) Multi-omics analysis identifies drivers of protein phosphorylation. Genome Biol 24(1):52. https://doi.org/10.1186/s13059-023-02892-2
122. Gerber SA, Rush J, Stemman O, Kirschner MW, Gygi SP (2003) Absolute quantification of proteins and phosphoproteins from cell lysates by tandem MS. Proc Natl Acad Sci USA 100(12):6940–6945. https://doi.org/10.1073/pnas.0832254100
123. Kettenbach AN, Rush J, Gerber SA (2011) Absolute quantification of protein and post-translational modification abundance with stable isotope–labeled synthetic peptides. Nat Protoc 6(2):175–186. https://doi.org/10.1038/nprot.2010.196
124. Zhang M, Flury S, Kim CK, Chung WCJ, Kirk JA, Pak TR (2021) Absolute quantification of phosphorylated ERβ amino acids in the hippocampus of women and in a rat model of menopause. *Endocrinology* 162(9):bqab122. https://doi.org/10.1210/endocr/bqab122
125. Nguyen TT, Kane MA, Swaan PW (2024) Determination of site-specific phosphorylation occupancy using targeted mass spectrometry reveals the regulation of human apical bile acid transporter. ASBT ACS Omega 9(37):38477–38489. https://doi.org/10.1021/acsomega.4c02999
126. Steen H, Jebanathirajah JA, Springer M, Kirschner MW (2005) Stable isotope-free relative and absolute quantitation of protein phosphorylation stoichiometry by MS. Proc Natl Acad Sci USA 102(11):3948–3953. https://doi.org/10.1073/pnas.0409536102
127. De Graaf EL, Giansanti P, Altelaar AFM, Heck AJR (2014) Single-step enrichment by Ti4+-IMAC and label-free quantitation enables in-depth monitoring of phosphorylation dynamics with high reproducibility and temporal resolution. Mol Cell Proteomics 13(9):2426–2434. https://doi.org/10.1074/mcp.O113.036608
128. Calderón-Celis F, Encinar JR, Sanz-Medel A (2018) Standardization approaches in absolute quantitative proteomics with mass spectrometry. Mass Spec Rev 37(6):715–737. https://doi.org/10.1002/mas.21542
129. Fernández SD, Sugishama N, Encinar JR, Sanz-Medel A (2012) Triple quad ICPMS (ICPQQQ) as a new tool for absolute quantitative proteomics and phosphoproteomics. Anal Chem 84(14):5851–5857. https://doi.org/10.1021/ac3009516
130. Calderón-Celis F, Sanz-Medel A, Encinar JR (2018) Universal absolute quantification of biomolecules using element mass spectrometry and generic standards. Chem Commun 54(8):904–907. https://doi.org/10.1039/C7CC09059E
131. Calderón-Celis F, Sugiyama N, Yamanaka M et al (2019) Enhanced universal quantification of biomolecules using element MS and generic standards: application to intact protein and phosphoprotein determination. Anal Chem 91(1):1105–1112. https://doi.org/10.1021/acs.analchem.8b04731
132. Pereira Navaza A, Ruiz Encinar J, Sanz-Medel A (2007) Absolute and accurate quantification of protein phosphorylation by using an elemental phosphorus standard and element mass spectrometry. Angew Chem Int Ed 46(4):569–571. https://doi.org/10.1002/anie.200602517
133. Navaza AP, Encinar JR, Carrascal M, Abián J, Sanz-Medel A (2008) Absolute and site-specific quantification of protein phosphorylation using integrated elemental and molecular mass spectrometry: its potential to assess phosphopeptide enrichment procedures. Anal Chem 80(5):1777–1787. https://doi.org/10.1021/ac7022316
134. Ciavardelli D, Sacchetta P, Federici G, Di Ilio C, Urbani A (2010) Protein phosphorylation stoichiometry by simultaneous ICP-QMS determination of phosphorus and sulfur oxide ions: a multivariate optimization of plasma operating conditions. Talanta 80(4):1513–1525. https://doi.org/10.1016/j.talanta.2009.06.082
135. Calderón-Celis F, Diez-Fernández S, Costa-Fernández JM, Encinar JR, Calvete JJ, Sanz-Medel A (2016) Elemental mass spectrometry for absolute intact protein quantification without protein-specific standards: application to snake venomics. Anal Chem 88(19):9699–9706. https://doi.org/10.1021/acs.analchem.6b02585

136. Villén J, Beausoleil SA, Gygi SP (2008) Evaluation of the utility of neutral-loss-dependent MS3 strategies in large-scale phosphorylation analysis. Proteomics 8:4444–4452. https://doi.org/10.1002/pmic.200800283
137. Frese CK, Altelaar AF, Hennrich ML, Nolting D, Zeller M, Griep-Raming J, Heck AJ, Mohammed S (2011) Improved peptide identification by targeted fragmentation using CID, HCD and ETD on an LTQ-Orbitrap Velos. J Proteome Res 10(5):2377–2388. https://doi.org/10.1021/pr1011729
138. Brunner AM, Lössl P, Liu F, Huguet R, Mullen C, Yamashita M, Zabrouskov V, Makarov A, Altelaar AF, Heck AJ (2015) Benchmarking multiple fragmentation methods on an orbitrap fusion for top-down phospho-proteoform characterization. Anal Chem 87(8):4152–4158. https://doi.org/10.1021/acs.analchem.5b0016

Proteomic Analysis of Histone Sequence Variants and Post-translationally Modified Forms

Palina Ryzhaya, Pavlína Pírek, Radomír Pech, Karolína Kryštofová, Zbyněk Zdráhal, and Gabriela Lochmanová

Abstract Post-translational modifications (PTMs) of histone variants are the key mechanisms driving epigenetic regulation. They play a crucial role in modulating chromatin structure and gene activity. As epigenetics has become a foundational aspect of molecular biology, the precise characterization of histones is essential for understanding gene regulation, cellular plasticity, and the molecular basis of diseases.

Mass spectrometry in combination with liquid chromatography has emerged as the leading technology for mapping and quantification of histone peptidoforms. It offers exceptional sensitivity and resolution to unravel the complexity of the epigenetic landscape. Despite its advantages, the inherent diversity of histone isoforms and PTMs presents analytical challenges, and no universal workflow currently enables comprehensive and unbiased profiling of the overall state of histone modifications.

This chapter presents key methodologies for histone analysis using bottom-up proteomics, including protocols for histone extraction, enzymatic digestion, and chemical derivatization prior to LC-MS/MS. It also provides details of specific aspects of LC-MS/MS analysis and outlines strategies for data evaluation and interpretation. The clinical potential of histone proteomics is illustrated through a case study focusing on the characterization of histones in human sperm in the context of male infertility.

P. Ryzhaya · K. Kryštofová · Z. Zdráhal · G. Lochmanová (✉)
Mendel Center for Plant Genomics and Proteomics, Central European Institute of Technology, Masaryk University, Brno, Czech Republic

National Centre for Biomolecular Research, Faculty of Science, Masaryk University, Brno, Czech Republic
e-mail: gabriela.lochmanova@ceitec.muni.cz

P. Pírek · R. Pech
Mendel Center for Plant Genomics and Proteomics, Central European Institute of Technology, Masaryk University, Brno, Czech Republic

J.-L. Capelo-Martínez (ed.), *Emerging Sample Treatments in Proteomics*, Advances in Experimental Medicine and Biology 1510,
https://doi.org/10.1007/978-3-032-21638-0_4

Keywords Mass spectrometry · Liquid chromatography · Histone characterization · Post-translational modifications · Chemical derivatization

1 Introduction

Epigenetics is a rapidly growing field of biology focused on changes in gene expression that occur without alterations to the underlying DNA sequence. Among the key mechanisms driving epigenetic regulation are post-translational modifications (PTMs) of histones, which play a crucial role in modulating chromatin structure and gene activity.

Throughout much of the twentieth century, histones were primarily recognized for their strong interaction with negatively charged DNA, facilitating its tight packaging into chromatin and repressing RNA synthesis. They were described as crucial for compaction but largely irrelevant to gene regulation. This view began to change in the 1960s with the discovery that histones can undergo chemical modifications such as acetylation and methylation—modifications that were found to be associated with enhanced RNA synthesis [1].

Over the following decades, research revealed a broad spectrum of histone PTMs: acetylation, methylation, phosphorylation, ubiquitination, etc., and the importance of their combinations. This resulted in a "histone code" concept: a regulatory language written in chemical marks that fine-tune gene accessibility and transcriptional outcomes [2].

Histone PTMs are increasingly recognized as valuable clinical biomarkers, for example, in cancer and neurodegenerative diseases. Altered patterns, including elevated histone H3 lysine 27 trimethylation (H3K27me3) and reduced histone H4 lysine 16 or 20 acetylation (H4K16ac, H4K20ac), are consistently associated with tumor progression and poor prognosis across multiple cancer types [3]. In neurodegenerative disorders, impaired histone acetylation has been associated with diminished synaptic plasticity and cognitive dysfunction [4]. These disease-specific PTM signatures also point toward new therapeutic strategies. Histone-modifying enzymes—especially histone deacetylases (HDACs)—have already become targets in clinical therapies [5]. Several HDAC inhibitors have received approval from the U.S. Food and Drug Administration (FDA), including Vorinostat and Romidepsin for cutaneous and peripheral T-cell lymphomas, Belinostat for peripheral T-cell lymphoma, and Panobinostat for multiple myeloma [6].

In recent years, several studies have pointed toward aberrant epigenetic reprogramming of the genome as a potential contributory factor to male infertility [7–10]. Recent focus has shifted to molecular factors associated with spermatozoa that could adversely affect the fertilization process, with particular attention given to the proteome as a whole, and histone epigenetic profiles [11–16].

Beyond clinical settings, histone modifications also play a crucial role in plant biology, influencing development, stress tolerance, and productivity [17–19]. This highlights the potential of histone modifiers in agricultural biotechnology, aiming to engineer crops better suited to environmental challenges.

Despite their comparatively short length, histone tails represent some of the most densely modified regions of eukaryotic proteins. Before the rise of mass spectrometry (MS), the study of histone modifications relied heavily on antibody-based techniques such as Western blotting, immunofluorescence, and chromatin immunoprecipitation (ChIP). These methods enabled detection, localization, and genome-wide mapping of specific PTMs. However, they are limited by issues of cross-reactivity, low resolution, and an inability to profile combinatorial PTM patterns. Despite these constraints, non-MS methods remain widely used for functional and spatial validation of histone marks [20].

The advent of MS has revolutionized histone analysis by enabling high-throughput identification and quantification of PTMs. It is highly accurate, providing information on the total levels of histone PTMs, but without localizing the modification to specific genome regions. However, due to the complexity and diversity of histone isoforms, no single standardized workflow currently allows comprehensive, unbiased analysis of all histone peptides and their modifications [21]. High variability of PTM patterns leads to the formation of isobaric peptides—positional isomers with identical or nearly identical masses differing in the PTM location. It is important to distinguish these positional isomers because certain modification patterns are linked to different functions of chromatin. Another significant analytical challenge arises from the high degree of sequence conservation observed between numerous histone sequence variants.

In the most widely used MS approach—bottom-up proteomics—the standard pipeline for preparing mammalian histones involves acidic extraction followed by protein precipitation to remove contaminants and maximize the yield of pure histones. As standard trypsin digestion is disrupted by the high amount of lysine and arginine residues in histone's primary structure, different alternative protocols of sample preparation based on arginine-specific cleavage were developed to prevent the excessive generation of short peptide chains. Cleavage at the carboxyl side of arginine residues is typically achieved through the chemical derivatization of lysines followed by trypsin digestion. The recent introduction of Arg-C Ultra, a recombinant proteinase with a substantially improved digestion specificity, facilitates Arg-C-type peptides preparation by eliminating protein-level derivatization [22]. Subsequent peptide-level derivatization, i.e., labeling newly formed amine groups at peptide N-termini after the cleavage, is required for most histone peptides to enhance chromatographic separation and improve quantification of histone peptidoforms [23].

This chapter outlines essential methodologies for histone characterization using bottom-up proteomics, highlighting their distinct features and specific requirements. It covers protocols for histone preparation, analysis via LC-MS/MS, and strategies for data evaluation and interpretation (Fig. 1). To highlight the potential of histone proteomics in clinical research, the chapter concludes with a case study demonstrating its application in characterizing histones in human spermatozoa related to male fertility disorders.

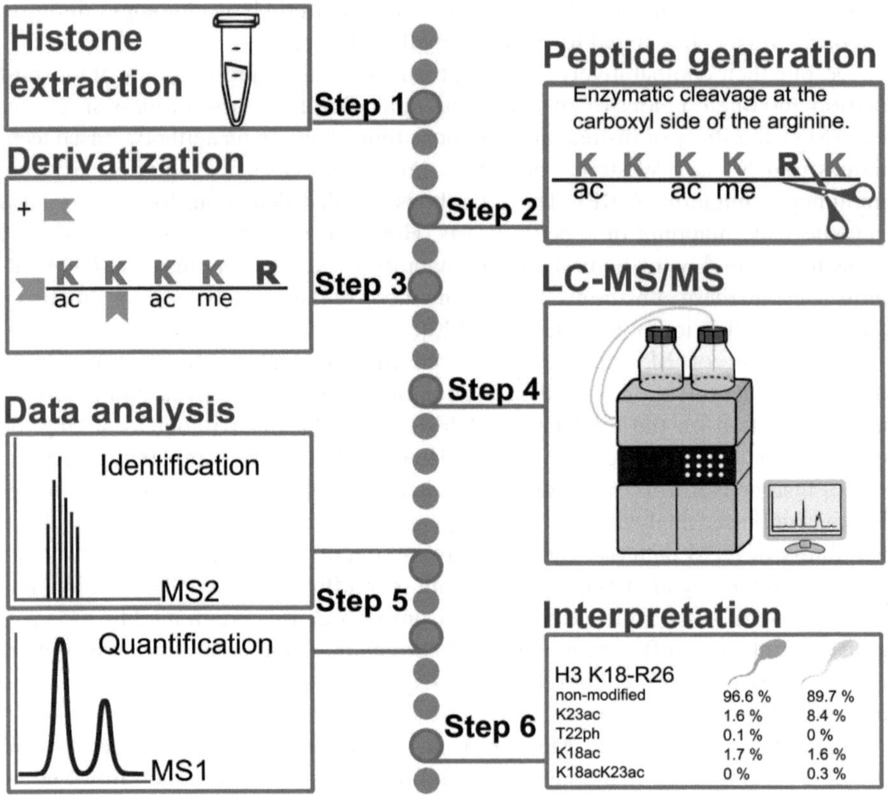

Fig. 1 Schematic overview of a representative bottom-up workflow for histone analysis

2 Histone Extraction and Preparation for LC-MS/MS

2.1 *Methods for Histone Extraction*

Histone enrichment prior to LC-MS/MS analysis is a prerequisite for detailed characterization of modified peptide forms, which occur in a wide range of concentrations within cells. Pure histones are required for maximum coverage of the amino acid sequence to identify histone marks in specific amino acid sites and histone sequence variants based on their unique peptides. Several protocols have been published, which are based on histone preparation from cells or nuclei using acidic extraction [24], high-salt or urea-salt extraction [25, 26], or acidic organic extraction [27]. During acidic extraction, histones are solubilized from precipitated chromatin, while most other nuclear proteins remain insoluble. High- and urea-salt extractions can serve as alternative methods for analysis of acid-labile PTMs at neutral pH. In urea-salt extraction, histones H2A/H2B can be separated from H3/H4 by gradually increasing the salt concentration. Although these three extraction methods can complement each other by isolating distinct histone proteoforms from

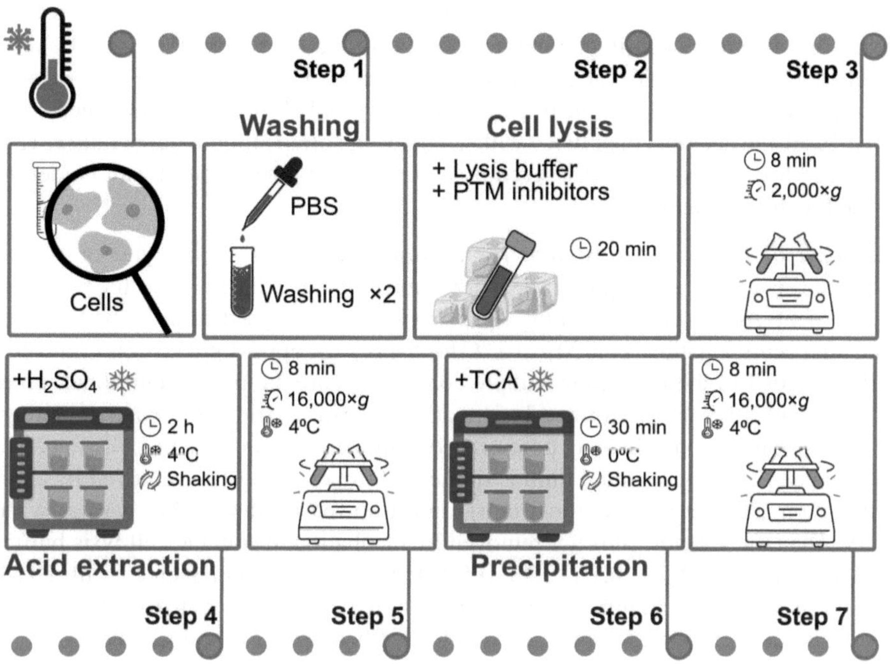

Fig. 2 Acidic extraction of histones from mammalian cell culture. Histones are solubilized from the chromatin in sulfuric acid and then precipitated with trichloroacetic acid. *PBS* phosphate-buffered saline, *PTM* inhibitors compounds that inhibit proteases and protein-modifying enzymes, *TCA* trichloroacetic acid

various chromatin environments, salt-based techniques are rarely used in experimental research.

Here, we describe acidic extraction as the most common method used for histone preparation from mammalian cells (Fig. 2) [24]. It is important to note that the following protocol may not be directly applicable to certain specialized cell types. For cells with unique sample matrices, such as adipocytes or sperm cells, modifications to the common histone extraction protocol may be necessary, as presented below. Previous studies have also shown that particular care must be taken when working with plant material, as species- and organ-specific cellular compositions can significantly affect histone extract purity. If not properly removed, natural contaminants may interfere with various steps of the histone sample preparation workflow and compromise subsequent LC-MS/MS analysis [28, 29].

Protocol: Acidic Extraction of Histone Proteins Cells (10^5–10^6) are washed twice in ice-cold phosphate-buffered saline (PBS) and incubated in lysis buffer (80 mM NaCl, 20 mM EDTA, 1% Triton X-100) supplemented with PTM inhibitors (e.g., 45 mM sodium butyrate, 0.1 mM PMSF, protease and phosphatase inhibitor cocktails) on ice for 20 min. After centrifugation (2000 *g* for 8 min), pelleted chromatin is resuspended in 250 μL of ice-cold H_2SO_4 and incubated at 4 °C for 2 h. The supernatant cleared by centrifugation (16,000 *g* and 4 °C for 8 min) contains histones that are subsequently precipitated during incubation with 250 μL of 50% ice-cold trichloroacetic acid (TCA) while shaking at 0 °C for 30 min. The resulting precipitate is harvested by centrifugation (16,000 *g* at 4 °C for 30 min), washed with 50 mM HCl in acetone, then twice with acetone, and dried at room temperature. The prepared histone extract is dissolved in water. Protein concentration can be determined with the Bradford assay.

Example of Protocol Adjustment for Histone Extraction Adipocytes or sperm cells can serve as an example of specialized mammalian cells that need the protocol adjustment. Adipocytes containing lipid droplets can be lysed as described above, and after centrifugation (20,000 *g* at 4 °C for 8 min), the upper lipid layer is removed, PBS is added to the remaining sample containing chromatin, and additional three washes with PBS are performed before adding sulfuric acid [30]. When isolating histones from sperm cells, any potential contamination from somatic cells must be eliminated [31]. The seminal fluid is diluted 2.5 times with PBS, removed by centrifugation (2000 *g* at 4 °C for 8 min), and followed by three other washes with PBS. Then the somatic cells are removed by incubation in somatic cell lysis buffer (10 mM Tris-Cl pH 7.6, 0.1% SDS, 0.5% Triton X-100) supplemented with PTM inhibitors at 4 °C for 30 min. After centrifugation, pelleted spermatozoa are lysed in the lysis buffer, followed by acidic extraction of histones as described above.

2.2 Histone Prefractionation

A bottom-up proteomic approach is typically used to analyze complex histone samples. However, if there is a need to reduce sample complexity and enrich specific histone sequence variants, prefractionation can be performed prior to LC-MS/MS analysis. This can be achieved through gel electrophoresis or offline reversed-phase separation.

SDS-PAGE is not efficient in separating histone sequence variants and their post-translationally modified forms. Separation of histone proteoforms can be enhanced using acid-urea (AU) or acid-urea-Triton (AUT) gel electrophoresis [25]. AU separates histones based on charge-to-mass ratio in acidic conditions. By the presence of Triton-X 100 in the gel, resolution is increased in AUT electrophoresis. AU and AUT electrophoresis might be performed in a two-dimensional system, either in combination with SDS-PAGE or as two-dimensional AUT/AU [24, 32], representing the gel technique with the highest resolution of histone proteoforms. Although effective, urea-based gel techniques are time- and labor-intensive and demand a high level of laboratory skills.

Histones are small proteins with a significant proportion of hydrophobic amino acids, making them amenable to reversed-phase separation. HPLC using C8 or C18 columns is employed to separate histone variants, with the resolution influenced by the sorbent properties, gradient duration, and column temperature [24, 25, 33].

2.3 Requirements for Histone Preparation in Bottom-up Proteomics

Trypsin is the standard proteinase used in shotgun proteomics due to its high specificity and efficiency. It cleaves at the C-terminus of lysine and arginine residues and, depending on the amino acid sequence, produces peptides with 5–20 amino acids [34, 35]. Prior to trypsin digestion, the reduction/alkylation step is usually performed to unfold proteins and break disulfide bonds between cysteine residues. For histone preparation, alkylation is unnecessary due to the very low cysteine content.

The high frequency of lysine and arginine residues in the amino acid sequences of many histone variants often results in the generation of short tryptic peptides. It is also important to note that trypsin skips lysine residues that have been modified. The length of the resulting histone peptides thus depends on the PTM state of the respective proteoform [36]. Tryptic digestion of histone proteoforms with low PTM levels generates a variety of short hydrophilic peptides that may be lost during reverse-phase chromatographic separation.

To minimize bias in the identification and quantification of post-translationally modified histone forms, generating peptides that terminate with arginine residues has proven advantageous. This approach is particularly beneficial for the most extensively studied histones H3 and H4, as it reproducibly generates longer peptides with a reasonable number of lysines. Coverage of defined sequence regions regardless of their modification status facilitates analysis of PTM dynamics. Given that the Arg-C enzyme cleavage has demonstrated suboptimal digestion specificity and reproducibility at the arginine C-termini, two primary strategies are currently employed to produce Arg-C-type peptides: (1) chemical derivatization of amine groups of lysines prior to trypsin digestion [21, 23, 37] and (2) histone digestion with an alternative proteinase Arg-C Ultra [22]. Chromatographic separation is advanced by adding higher hydrophobicity to the peptides. Derivatization at the protein level increases peptide hydrophobicity depending on the chemical derivatization reagent used. However, to further enhance chromatographic separation and improve the identification and quantification of histone peptidoforms, it is recommended to additionally label the amine groups at the newly generated peptide N-termini following enzymatic cleavage. In case of Arg-C Ultra digestion, all free amine groups present at N-termini and internal lysines are labeled during peptide-level derivatization [22].

Amine-specific reagents for the chemical derivatization of the lysine residues on histones under different conditions have been tested, including various acid anhydrides [21, 38], phenylisocyanate [37], and N-hydroxysuccinimide ester [39, 40]. To avoid distortion in the MS signals of variably modified peptide forms, the reagent should be selected with regard to (1) the labeling efficiency, (2) the ionization efficiency of resulting products, and (3) the chromatographic separation of products. In fact, the choice represents a compromise, as none of the commonly used reagents fulfill all the requirements.

Although propionic anhydride is the most established labeling agent for histones, offering high amine labeling efficiency, its use is accompanied by side reactions with the hydroxyl groups of serine, threonine, and tyrosine. Additionally, propionylation does not provide sufficient hydrophobicity to short, hydrophilic peptides or effectively separate positional isomers of multiply acetylated peptides on reversed-phase stationary phases.

Anhydrides containing aromatic rings, such as phenylisocyanate or benzoic anhydride, are inefficient for lysine derivatization at the protein level due to steric hindrance. However, phenylisocyanate can be applied to improve the detection of certain hydrophilic peptides by labeling their N-termini following prior propionylation of lysine residues at the protein level [29, 37, 41]. If propionylation at the protein level is incomplete during hybrid labeling, it can increase spectral complexity by producing a mixture of peptides with variably modified amine groups, including phenylcarbamoyl residues on lysine side chains.

Trimethylacetic anhydride (TMA) was introduced to enhance the separation of isobaric peptide forms, which facilitates quantification of commonly occurring positional isomers of histones [21]. However, to achieve efficient labeling, extended or microwave-assisted incubation and stepwise reagent addition are required.

Here, we present a facilitated method based on digestion with Arg-C Ultra and peptide labeling with TMA for the preparation of histone proteins for LC-MS/MS (Fig. 3), which eliminates protein-level derivatization, improves chromatographic

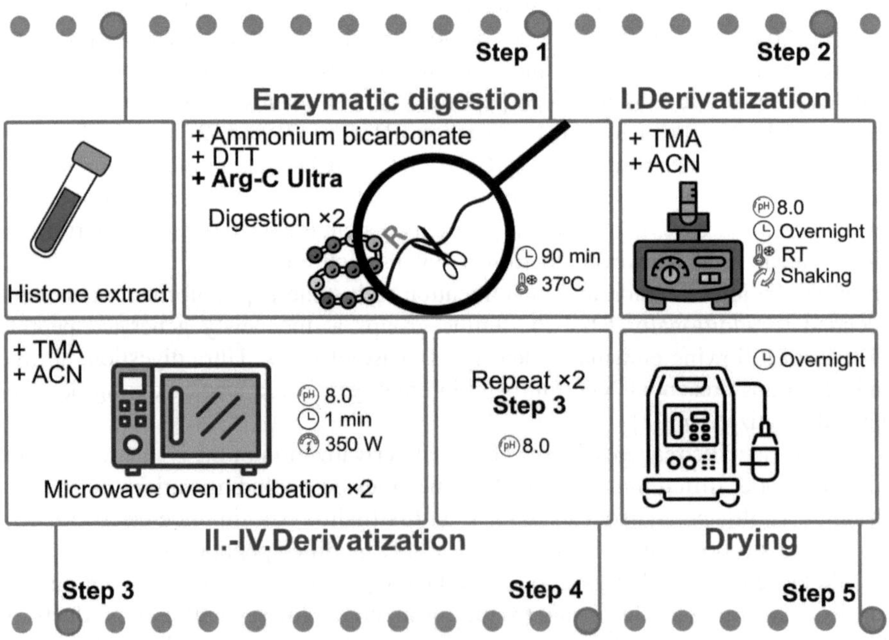

Fig. 3 Histone digestion with Arg-C Ultra followed by derivatization of peptides with TMA. Arg-C-type peptides after cleavage are derivatized with TMA at amine groups of N-termini and internal lysines to improve retention of short hydrophilic peptides and resolution of isobaric peptides. The efficiency of TMA labeling is supported by overnight incubation followed by microwave-assisted derivatization. *DTT* dithiothreitol, *TMA* trimethylacetic anhydride, *ACN* acetonitrile

retention of short hydrophilic peptides, enhances resolution of isobaric peptides, and enables quantification of post-translationally modified forms from MS1-level spectral information [22].

Preparation of Arg-C-Type Histone Peptides

Previously, the Arg-C proteinase showed limited cleavage site specificity and poor reproducibility [42]. Thus, derivatization of lysine residues in histone proteins with acid anhydrides, followed by trypsin digestion, has been used to mimic cleavage with Arg-C [23]. Recent improvements in proteolytic enzymes are reflected in the introduction of Arg-C Ultra, a novel recombinant proteinase with an increased arginine cleavage specificity, which simplifies the preparation of histone proteins for LC-MS/MS by omitting derivatization at the protein level [22].

Long histone peptides containing a high number of lysines (i.e., the N-terminal peptide of certain H2B variants) are recommended to be analyzed immediately after the cleavage. However, for most peptides derived from histones H2A, H3, and H4, derivatization at the peptide level after cleavage with Arg-C Ultra is recommended to improve chromatographic behavior through increased hydrophobicity.

The following protocol can be used to digest histone proteins at the carboxyl side of arginine residues.

Protocol: Digestion of Histone Proteins with Arg-C Ultra Histone extract containing 5 μg of total protein (prepared as described in Sect. 2.1) is diluted with H_2O to a final concentration of 1 $\mu g.\mu L^{-1}$. Ammonium bicarbonate buffer (pH 8.0) is added to a final concentration of 50 mM. Then, dithiothreitol (DTT) is added to a final concentration of 10 mM. Arg-C Ultra (VA1831; Promega) is added at a 1:100 (w/w) enzyme-to-substrate ratio. The mixture is incubated at 37 °C for 1.5 h. Following the first incubation, an additional Arg-C Ultra aliquot is added at the same ratio, and the sample is incubated for an additional 1.5 h at 37 °C. The digestion is either stopped by adding formic acid to a final concentration of 1%, dried in a vacuum concentrator, and desalted, or subjected to subsequent derivatization steps.

Chemical Derivatization of Histone Peptides

Chemical derivatization of Arg-C-type peptides enables labeling of all free amine groups, including those on lysine residues and at the peptide N-terminus. By eliminating derivatization at the protein level, Arg-C Ultra cleavage potentially expands the portfolio of compounds for chemical derivatization of amine groups [22].

However, the derivatization reagents must be carefully selected with regard to subsequent LC-MS/MS analysis. An increase in hydrophobicity at optimal retention time shift and chromatographic resolution of peptidoforms should be considered in conjunction with the reagent's effect on ionization efficiency [38]. The adduct mass of commonly used derivatization agents is as follows: +56.0262 Da for propionyl, +119.0371 Da for phenylisocyanate, and +84.0575 Da for TMA. It is important to note that the chromatographic behavior of a particular peptide sequence varies depending on the number of lysines and the degree of PTM.

For instance, up to five labels can be introduced to the N-terminal Arg-C-type peptide G4KGGKGLGKGGAKR17 (G4–R17) of histone H4, as it possesses amine residues in four lysines and at the peptide N-terminus. In a naturally fully acetylated form, only one label is added during chemical derivatization, whereas a non-modified form would incorporate five labels, resulting in a substantial shift in retention time. Importantly, peptides with multiple lysines can occur in forms of isobaric peptides, i.e., the same amino acid sequence with the same kind of PTM at different positions. The resolution of such positional isomers is a major challenge in the quantification of post-translationally modified histone peptides.

In this regard, we present here a protocol for histone peptide derivatization using TMA that provides efficient chromatographic retention and separation of histone peptidoforms without compromising MS signal quality [21].

Protocol: Chemical Derivatization of Histone Peptides with TMA The pH of the peptide mixture (prepared as described in section "Preparation of Arg-C-Type Histone Peptides") is adjusted to 8.0 using NH_4OH. Subsequently, 2.25 μL of freshly prepared derivatization reagent consisting of TMA (Sigma–Aldrich) and acetonitrile (ACN) in a 1:3 (v/v) ratio is added to the sample. The sample is incubated in the thermomixer overnight at room temperature with shaking at 1000 rpm and then subjected to two rounds of microwave-assisted derivatization, as follows.

The microtube is capped and placed in a glass beaker that is covered with another glass beaker during incubation in a microwave oven for the first round of microwave-assisted derivatization. This consists of three subcycles with the following steps: adjustment of the sample's pH to 8.0 with NH_4OH, addition of 2.25 μL of derivatization reagent, and two 1-min incubations in the microwave oven at 350 W with a short centrifugation between them. The sample is then concentrated, and the second round of microwave-assisted derivatization is carried out with the same protocol. The labeled peptides are dried overnight in a vacuum concentrator and desalted.

Protocol: Sample Desalting and Preparation for LC-MS/MS Histone peptides of the Arg-C type, either directly after cleavage or chemically derivatized, must be desalted before LC-MS/MS. The dried sample is reconstituted in 50 μl of 0.1% trifluoroacetic acid (TFA). Desalting can be performed using C18 sorbent, e.g., AttractSPE® Tips C18 (Affinisep, France). Peptides are sequentially eluted with 2 × 10 μL of 0.1% TFA in 50% ACN and 2 × 20 μL of 0.1% TFA in 75% ACN. The pooled eluates are transferred to an LC vial, concentrated to 10 μl, and acidified with formic acid to a final concentration of 1%.

3 Analysis of Arg-C-Type Histone Peptidoforms

The frequent occurrence of positional isomers, along with the wide dynamic range of histone peptidoform abundances originating from highly similar histone sequence variants, poses a significant challenge for histone characterization using bottom-up proteomics. The impact of these factors extends beyond LC-MS/MS analysis and affects the workflows of identification and quantification. Implications and ways to address the associated challenges for each step are described in the following sections.

3.1 LC-MS/MS Analysis of Histone Peptidoforms

Technological improvements in the development of high-resolution MS instrumentation enabled us to distinguish histone peptidoforms with small differences in mass-to-charge ratio, e.g., forms carrying acetylation (+42.0106 Da) and trimethylation (+42.0470 Da) [43]. Still, both LC and MS/MS methods must be carefully optimized to improve the detection of isobaric peptides and low-abundance post-translationally modified forms. In histone samples, co-elution of peptides with multiple lysines occurs quite frequently. Chemical derivatization enhances retention and separation selectivity on reversed-phase columns. The extent to which histone peptidoforms are shifted in retention time and chromatographically resolved depends on the specific chemical properties of the introduced adduct [21]. The LC gradient length and steepness must be carefully optimized to accommodate the increased hydrophobicity of derivatized peptides, which elute later than their non-derivatized counterparts (Fig. 4). Both the choice of analytical column and gradient profile play a critical role in achieving optimal peak resolution.

In data-dependent acquisition (DDA) mode, peptide identification is limited by intensity-based precursor selection. Because only a finite number of spectra can be acquired during elution, some precursors may not be selected or may produce sub-optimal fragmentation, preventing confident identification of all peptide forms present. Further, identification and quantification of co-eluting isobaric peptides at the MS1 level are often compromised due to resulting chimeric MS/MS spectra.

Data-independent acquisition (DIA) and targeted approaches such as Selective Reaction Monitoring (SRM) and Parallel Reaction Monitoring (PRM) can help partly address those issues. In DIA, all peptides are fragmented in a set of pre-defined isolation windows, and quantification is refined by integrating signals in the extraction ion chromatogram (EIC) of both precursor and fragment ions [44, 45].

Targeted approaches rely on a predefined inclusion list to selectively fragment peptidoforms based on prior knowledge. DIA and targeted methods enhance the detection of low-abundance peptides by generating more fragmentation spectra and generally reduce the occurrence of missing values.

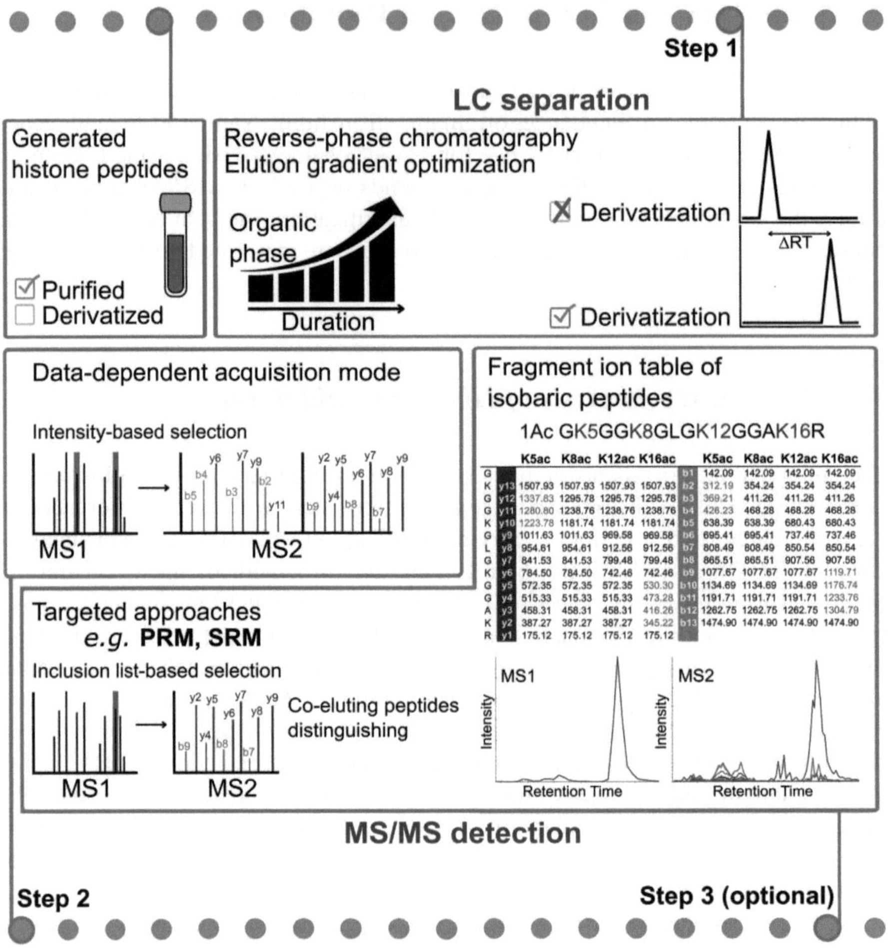

Fig. 4 LC-MS/MS analysis of histone peptidoforms. The LC gradient length and steepness must be adjusted to the increased hydrophobicity of derivatized peptides, which elute later than their non-derivatized counterparts. Any acquisition strategy can be employed to analyze histone peptidoforms, including data-dependent acquisition (DDA) or targeted methods such as selected reaction monitoring (SRM) or parallel reaction monitoring (PRM). Fragment ion table of G4–R17 monoacetylated isobaric peptides demonstrates the necessity of proper LC separation to distinguish positional isomers; the mass of unique fragments is marked in red, while most ion fragments have a common mass across positional isomers

Notably, many fragment ions are common to histone positional isomers, leaving only a limited set of unique—and often low-intensity—ions available for accurate quantification (Fig. 4). Therefore, the quality of LC separation is crucial regardless of the acquisition method employed. With optimized LC conditions and an appropriate derivatization agent, even DDA can achieve a large set of identified histone peptidoforms. This is demonstrated by chemical derivatization with TMA, which enhances the chromatographic retention of short, hydrophilic peptides and enables MS1-level quantification of post-translationally modified forms by increasing the resolution of peptides with similar physicochemical properties [21].

Comparisons between DDA and various DIA methods revealed consistent trends in the relative quantification of histone peptidoforms [46]. Overall, MS-based characterization of histone peptidoforms proves to be a versatile approach, though it demands careful optimization of method settings depending on the available instrumentation.

Here, we present an example LC-MS/MS method used for Arg-C-type peptides, either non-derivatized or labeled with TMA.

Protocol: LC-MS/MS analysis of Histone Peptidoforms using Orbitrap Fusion Lumos Tribrid It is recommended to spike samples with synthetic iRT peptides (Biognosys) to monitor retention time behavior of individual samples during LC separation, especially when analyzing larger sample sets for several days. It is also recommended to analyze the samples in a random order to treat the batch effects.

Samples of both TMA- and non-labeled peptides can be analyzed using an Ultimate 3000 RSLCnano liquid chromatograph coupled to an Orbitrap Fusion Lumos Tribrid mass spectrometer (Thermo Fisher Scientific). Equilibration of the trapping and analytical columns with 99:1 (mobile phase A:B) at a 400 nL/min is done prior to sample injection into the sample loop. Trap and analytical columns are tempered at 25 °C and 50 °C, respectively. Prior to LC separation, tryptic digests are online concentrated and desalted using a trapping column (Acclaim PepMap 100 C18 Neo, dimensions 300 μm ID, 5 mm long, 5 μm particles, Thermo Fisher Scientific; PN 174500). The peptides are separated on an Aurora C18 analytical column (1.6 μm particles, 75 μm ID, 25 mm; Ion Opticks) at a 300 $nL.min^{-1}$ by 120 min gradient using a mobile phase consisting of 0.1% formic acid in water (A) and 0.1% formic acid in 80% ACN (B), with the following proportions of B: 3% (0–4 min), 3–42% (4–107 min), 42–80% (107–115 min), and followed by an isocratic wash of 80% B (115–120 min). The analytical column outlet is directly connected to the Digital PicoView 550 ion source equipped with Active Background Ion Reduction Device (ESI Source Solutions, Woburn, MA, USA). Mass spectra are acquired in data-dependent acquisition (DDA) mode using 350–2000 m/z survey scans at a resolution of 120,000 (at m/z 200) with an automatic gain control target setting of 4×10^5 and maximum injection time of 500 ms. Precursors with charge states from 2+ to 7+ and intensity above 1×10^4 are subjected to higher energy collisional dissociation fragmentation with a normalized collision energy of 30%. Once fragmented, precursors are excluded for 20 s before the next fragmentation. Precursors are isolated by quadrupole with a 1.2 m/z isolation window. Tandem mass spectra are obtained using 30,000 resolution (at m/z 200). Ions are accumulated for a target value of 5×10^4 or 500 ms injection time. The cycle time between master scans is 2.5 s.

3.2 Identification of Histone Peptidoforms

Distinguishing histone sequence variants with high homology and detecting the full range of PTMs at various sites require high sequence coverage in MS database searches while minimizing false-positive identifications. A major complication in the database search process is the large number of PTMs that must be included as variable modifications (Fig. 5). Commonly searched PTMs include lysine acetylation, mono-, di-, and tri-methylation, and phosphorylation at serine, threonine, and tyrosine residues. Other modifications may be considered depending on the specific objectives of the biological study. Importantly, the mass adduct introduced by the derivatization agent must be specified as a variable modification on lysine residues and peptide N-termini and should also be included for serine, threonine, and tyrosine residues to account for potential side reactions on hydroxyl groups during the derivatization process.

An extensive array of PTMs impacts the search speed and the correct assignment of the fragment ion spectrum to a respective peptide sequence. This assignment is performed automatically by search engines and software tools, either through comparison with theoretical spectra generated by search engines based on sequences in a database or via spectral library searching, which matches spectra against libraries of experimentally identified MS/MS spectra from previous studies [47]. Multiple factors can affect histone peptide identification, including the inability of search engines to distinguish closely related sequences, mixed MS/MS spectra from co-eluting precursors, nonspecific protease cleavage, splitting the spectra due to side reactions during derivatization, incomplete databases, and the exclusion of spectra of low-abundance peptidoforms due to low matching scores. A comparative evaluation of several search engines—such as Mascot [48], SEQUEST [49], PEAKS DB [50], ProteinPilot [51], pFind [52], Andromeda [53], OMSSA [54], and X!Tandem [55]—for histone DDA data, identified pFind and Mascot as providing the most reliable results with reasonable processing times [56]. Computational tools suitable for searching histone peptidoforms must be selected with regard to how many PTMs can be added in a single search.

Here we provide an example of the parameters set up for searching raw mass spectrometry data of Arg-C-type peptides, either non-derivatized or labeled with TMA, in the Mascot engine via Proteome Discoverer software. Even though Proteome Discoverer allows a maximum of nine variable modifications to be set, the number can be increased by grouping particular PTM occurring on different amino acids in Mascot (e.g., phosphorylation on serine, threonine, and tyrosine can be configured and searched as a single group).

To enhance the detection of a wide range of histone peptidoforms, it is advisable to create a custom in-house histone database. This can be generated by selecting histone entries specific to the organism of interest from UniProt (https://www.uniprot.org/), exporting them as a FASTA file, and importing the file into Mascot. In addition to this focused database, searches should be conducted against a comprehensive protein database for the organism to identify non-histone proteins that may

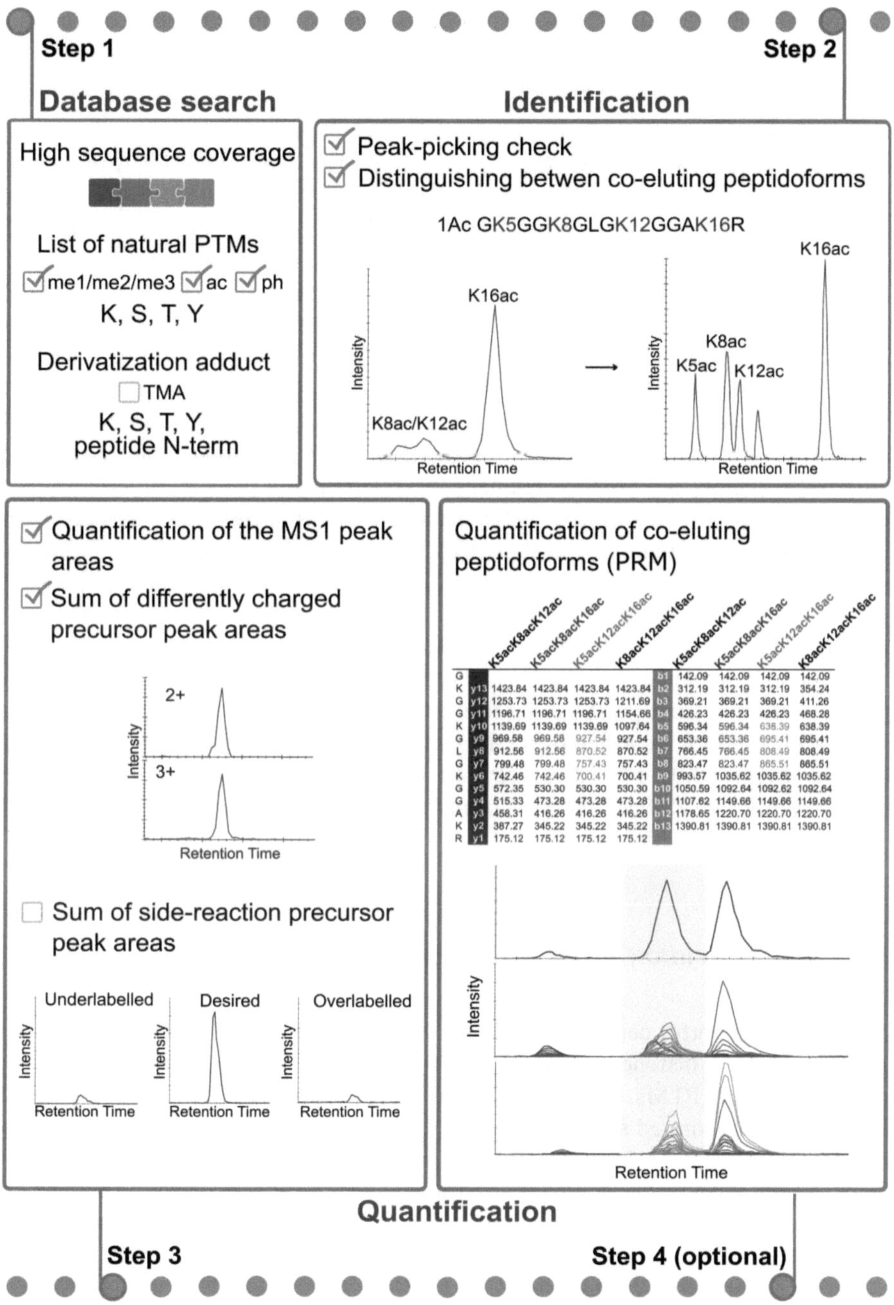

Fig. 5 Identification and quantification of histone peptidoforms. All PTMs relevant to the biological topic have to be set for the database search. Peak-picking has to be performed to check the separation of isobaric peptides. It is recommended to sum the area under the EIC of multiple charged and side reaction precursor peaks with the area of the desired precursor peak to avoid bias in the quantification. The quantity of co-eluting peptides can be determined from MS/MS obtained using a targeted approach (e.g., PRM)

be present in histone extracts. Finally, spectra should be searched against a contaminant database, such as the cRAP database (http://www.thegpm.org/crap/), to account for common laboratory contaminants.

Protocol: Database Search of Human Histone Peptidoforms using Proteome Discoverer software The raw mass spectrometry data acquired in DDA mode can be analyzed using Proteome Discoverer software (Thermo Fisher Scientific) with an in-house Mascot search engine (Matrix Science) to compare acquired spectra with entries in the cRAP contamination database (based on http://www.thegpm.org/crap/), an in-house Histone Human database generated from UniProt, and the UniProt KB Human database. The mass error tolerance for precursor and fragment ions is set at 7 ppm and 0.03 Da, respectively. Enzyme specificity is configured to semi-Arg-C, allowing for two missed cleavages across all databases. Variable modifications for each database is as follows: cRAP—acetylation (protein N-terminal region), deamidation (N and Q), oxidation (M), and trimethylacetylation (peptide N-terminal region, K); Histone Human—acetylation (K), phosphorylation (S, T), methylation (K, R), dimethylation (K), trimethylation (K), and trimethylacetylation (peptide N-terminal region, K, S, T, and Y); UniProt KB Human database—trimethylacetylation (peptide N-terminal region, K). For underivatized sample analyses, the same modifications are used except for trimethylacetylation. Peptides identified using the in-house Human Histone database are refined by fixed-value peptide-spectrum match validator (delta Cn < 0.05), followed by filtering highly confident peptides using Mascot parameters set to Rank 1, expectation value <0.01, and ion score ≥30. Peptides identified using the UniProt KB Human database are refined by Percolator (Cn < 0.05, false discovery rate <0.01). Histone PSMs for the following quantification are selected, exported from Proteome Discoverer to an Excel file, and converted to an SSL file.

3.3 *Quantification of Histone Peptidoforms*

Histone peptidoforms can be quantified at two fundamental levels: (1) by assessing the abundance of histone sequence variants and (2) by assessing the abundance of peptides carrying PTMs. Quantification can be carried out at either the MS1 or MS2 level. Log_2-transformed absolute MS signal values—either from precursor peaks in the MS1 spectrum or fragment ions in the MS2 spectrum—can be normalized across samples to enable comparison of peptide abundance between different biological conditions. Due to the complexity of histone sample preparation for LC-MS/MS, which introduces variability between replicates, relative abundances are often preferred for quantitative analysis.

The relative abundance of post-translationally modified peptides is determined as the percentage of the respective peptide form, calculated by dividing its peak area in EIC by the total EIC area of all forms corresponding to the same peptide sequence [57].

Quantitative analysis of histone sequence variants can be based on the ratios of their unique peptides, since many variants share a substantial part of their sequences [19, 58, 59]. For example, the human H3.3 variant shares 96% sequence identity with H3.1, differing by only five amino acids. Their relative abundance can be determined using the K27 – R40 peptide segment, which contains a distinguishing alanine-to-threonine substitution at position 31 (H3.1 KSAP**A31**TGGVKKPHR; H3.3 KSAP**T31**TGGVKKPHR), resulting in a 30 Da mass difference.

Software Tools Used for Histone Peptidoforms Quantification

Automated peak-picking workflows in commonly used software platforms introduce bias in the quantification of histone peptides due to their highly complex combinatorial patterns of modifications. A complete fragment ion series is a prerequisite for an unambiguous assignment of PTM to the correct amino acid position, especially in the case of a peptide with multiple lysines, which occurs in the form of more positional isomers. Peak-picking, i.e., determining retention time and the boundaries of the chromatographic peaks, has to be edited manually or using specialized software to integrate the correct peak of each peptidoform (Fig. 5).

The probability of the correct detection of certain PTMs can be supported by the presence of diagnostic ions, which are low-mass signature fragment ions generated upon MS/MS fragmentation of a respective precursor ion. These are fragments corresponding to masses of lysine immonium and related ions (m/z 101, 84, and 112; https://www.matrixscience.com/help/fragmentation_help.html) carrying PTM, including N-terminal lysine fragments carrying the derivatization adduct together with PTM. For instance, diagnostic ions of masses 126.0919 (84 + acetyl 42.0106) or 227.175 (101 + trimethylacetyl 84.058 + acetyl 42.0106) are supportive in assigning a given mass spectrum to the acetylated peptide form. However, diagnostic ions cannot be found for all PTMs in MS spectra as their presence depends on the chemical nature of the analyte (e.g., sequence context around the PTM or charge state) and the instrumental settings (e.g., mass resolution, fragmentation technique, scan range, etc.).

Quantification Using Skyline Software

Semi-automatic peak-picking can be performed in Skyline software (MacCoss Lab, University of Washington; https://skyline.ms/project/home/software/Skyline/begin.view), a free, open-source software platform designed primarily for targeted proteomics and quantitative mass spectrometry data analysis [60]. FASTA-based or FASTA-free workflows can be used to import mass spectrometry data to Skyline. In a FASTA-based workflow, either a spectral library built from raw DDA files and search results exported from software platforms (e.g., Proteome Discoverer) is aligned with imported protein sequences in the FASTA file or in-silico digestion is performed from imported protein sequences in the FASTA file and target peptides are defined for monitoring. In a FASTA-free setup, a spectral library, or a custom peptide list (with m/z, retention time, and fragment ions), or external DIA or PRM results are directly imported. There is no restriction regarding the type of modification (either PTM or derivatization adduct) or organism that can be imported to Skyline.

Skyline provides peak alignment across the raw files based on retention time and m/z, which is particularly useful for histone precursor peaks whose MS2 spectra were not generated because the fragmentation was skipped due to close retention time or low intensity. Peak areas, retention time, and mass error comparison can be displayed in Skyline. Due to the complexity of histone peptidoforms, the process is error-prone and thus requires further time-consuming manual inspection of the particular EICs. MS/MS spectra together with annotated fragments are displayed in Skyline in the library match section. However, it is useful to inspect the data in parallel in other software platforms (e.g., Proteome Discoverer), which are more suitable for thorough inspection of fragment ion series related to particular peptidoforms, including low-intensity fragment ions.

Quantification Using EpiProfile Software

A specialized software tool with the ability to accurately discriminate highly modified histone peptides, EpiProfile 2.0, has been developed for the quantitative analysis of histone peptidoforms [61]. It uses pre-defined retention time and m/z values to automatically extract ion chromatograms for each peptidoform, automatically resolves isobaric histone peptides using MS2 fragment ion patterns and elution profiles, and automatically calculates peak areas for both MS1 and MS2 signals to quantify individual peptidoforms. Still, although largely automatic, EpiProfile requires assignment of the peak boundaries or visual inspection of noisy, overlapping, or low-abundance peaks. The software has been adapted to include 13 species used as model systems for epigenetics studies (*Homo sapiens, Mus musculus, Bos taurus, Caenorhabditis elegans, Harpegnathos saltator, Heterocephalus glaber, Neurospora crassa, Oxytricha trifallax, Theileria annulata, Plasmodium falciparum, Saccharomyces cerevisiae, Saccharum officinarum, and Xenopus laevis*). EpiProfile 2.0 is optimized for both DDA and DIA through the use of precursor and fragment extracted ion chromatograms to accurately determine the chromatographic profile and to discriminate isobaric forms of peptides. Quantification of histone data obtained from derivatization with propionic anhydride or phenylisocyanate can be performed. The PTM portfolio is restricted to acetylation, mono-, di-, trimethylation, phosphorylation, ubiquitinylation, and acylation. Nevertheless, EpiProfile is customizable, allowing new PTMs or other organisms to be added via manual editing of configuration files.

Quantification Issues Related to Chemically Derivatized Histone Peptidoforms

Quantification of chemically derivatized samples can be affected by side reactions during derivatization. These reactions may produce multiple precursor peaks in the spectra corresponding to under-labeled forms (e.g., missing adducts at lysine residues or the peptide N-terminus) or over-labeled forms (e.g., unintended adducts at hydroxyl groups of serine, threonine, or tyrosine). Additionally, some peptides may appear in multiple charge states, each generating a distinct EIC. To accurately quantify a given peptide form, the MS signal of all relevant precursors—whether from side reactions or different charge states—should be summed (Fig. 5).

Accurate quantification of isobaric peptides requires sufficient resolution of their peaks, regardless of the acquisition method used. Enhanced LC separation, achieved through optimized instrumentation and method settings, combined with chemical derivatization using TMA, substantially facilitates the quantification of isobaric peptides. This enables the majority of peptides to be quantified from MS1-level spectral information (Fig. 5).

However, some positional isomers may still co-elute. For example, while most of the 16 H4 G4–R17 peptidoforms can be separated, two tri-acetylated forms—G**K5ac**GG**K8ac**GLGK12GGA**K16ac**R and G**K5ac**GGK8GLG**K12ac**GGA**K16ac**R—co-elute even after TMA derivatization. In such cases, targeted acquisition methods (e.g., PRM) can be employed to distinguish and quantify the co-eluting forms using isomer-specific MS2 fragment ions (Fig. 5) [21].

Here, we present an example of a FASTA-based workflow for quantifying MS1 signals of Arg-C-type peptides using Skyline software.

Protocol: Quantification of Histone Peptidoforms using Skyline Software The SSL file (prepared as described in Sect. 3.2) has to be stored in the same folder as raw DDA files. Peptide and transition settings have to be defined.

Peptide settings are accommodated to database search settings. Digestion: Arg-C(semi), Max missed cleavages: 2, Prediction: Retention time predictor, none; Filter: Peptide length, 6–25; Library: Pick peptides matching; Library: spectral library built from SSL file; Modifications: Structural modifications, the same as used for histone database search; Maximum of variable modifications: 8. For N-terminal amino acid containing two amine residues, one from side chain and the second from peptide bond, modification of corresponding mass has to be defined in Skyline (e.g., for N-terminal lysine, which is naturally acetylated at side chain and chemically derivatized at peptide N-terminus, monoisotopic mass for acetyl-trimethylacetyl (N-term) corresponding to 126.0681 Da has to defined).

Transition settings are accommodated to the LC-MS/MS and database search settings. Filter—Peptides: Precursor charges, 2–4; Ion charges, 1; Ion types, p; Product ion selection: from m/z > precursor to 3 ions; Library: Ion match tolerance, 0.05 m/z; Instrument: m/z 250–1500; Method match tolerance m/z, 0.055; Full scan—MS1 filtering: Isotope peaks included, count; Precursor mass analyzer,

Orbitrap; Peaks, 3; resolving power 120,000 at m/z 400; Full scan—MS/MS filtering: Acquisition method, none; Retention time filtering, Include all matching scans.

The peptide target list is created by importing a FASTA file containing histone proteins of interest. Then, raw DDA files are imported, and peak-picking is manually edited. Quantities determined from peak area in the respective precursor's EIC are exported to an Excel file for the following statistical analysis and data interpretation.

3.4 Statistical Analysis of Between-Sample Differences of Histone Peptidoforms

Careful experimental design is essential in proteomic studies, including those analyzing epigenetic marks, as various factors can influence perceived changes in peptide abundance. Histone characterization by proteomics involves multiple steps, such as cell disruption, histone extraction, protein digestion, chemical derivatization of amine groups, LC-MS/MS analysis, and data processing. This complex and time-consuming workflow can introduce variability into the data, especially when processing a large sample set. To ensure reproducibility and comparability of findings, reporting statistically significant results is critical.

Robust experimental design, high reproducibility, and accurate quantification are fundamental to sound statistical analysis. In differential proteomic studies, statistical power should complement traditional metrics such as fold change and p-value [62]. Both p-value and statistical power are influenced by variability and the number of replicates. Indeed, detection and exclusion of outliers using appropriate statistical methods (e.g., the Grubbs test) require a sufficient number of replicates. Therefore, carefully determining the number of biological replicates and identifying both biological and technical sources of variation are crucial for obtaining meaningful and reproducible results in the study [62, 63].

In studies of human histone peptidoforms, more than three biological replicates are recommended, with higher numbers needed for samples exhibiting substantial variability. However, it is important to note that obtaining such a large amount of starting material from certain types of samples may be difficult, which may compromise the quality and reliability of the results. Although differences between samples are typically evaluated using statistical tests such as t-tests or ANOVA, non-significant changes in histone peptidoform abundances can still be informative in certain contexts—particularly when they reveal meaningful trends in the balance of epigenetic marks, such as during time-course studies.

Here, we present an example of statistical evaluation of Arg-C-type peptidoforms using the KNIME platform.

Protocol: Statistical Analysis of Histone Peptidoforms in KNIME Platform The KNIME Analytics Platform can be used for quantitative statistical analysis (http://www.knime.com). The peak areas corresponding to individual histone peptidoforms are treated as compositions, and Aitchison's methodology based on log-ratios is applied in the statistical evaluation [64]. The missing values are imputed by iterative least trimmed squares regression [65], and areas are transformed to relative abundances (percentages). Then, acetylated and non-acetylated forms of each peptide are amalgamated, and all forms of acetylated/non-acetylated histones are processed as follows. Technical replicates are aggregated by geometric mean, followed by closure. The log2-ratio of amalgamated acetylated and non-acetylated forms is calculated (alr-transformation of 2-parts composition), and t-test or ANOVA is used to compare means of the ratios between corresponding pairs of experimental and control groups. The data analysis is performed in R version 3.6.3 (http://www.R-project.org, accessed on November 17, 2023) using the composition R package (version 1.40–5; http://CRAN.R-project.org/package=compositions, accessed on November 17, 2023) for alr-transformation. To uncover patterns within the data, hierarchical clustering can be performed. Percentage data are transformed using the pivot coordinate (pivotCoord) method (pivot coordinates as a special case of isometric logratio, or ilr coordinates) in order to convert the compositional dataset into a suitable format for subsequent clustering. The "dist" function is used to calculate pairwise distances between rows in the "ilr" matrix, and the "average" linkage method is chosen to construct the hierarchical tree (dendrogram) of the data.

4 Example Application: Mapping of Histones in Human Sperm for Improved Insight into Male Fertility Disorders

We illustrate the clinical potential of the above-described methodology through a case study focused on male infertility (Fig. 6). Infertility affects approximately 10–15% of couples worldwide [66], with male factors contributing to around 50% of all cases [67]. While a comprehensive physical examination and standard semen analysis remain the cornerstone of male infertility evaluation [68], over one-third of cases are classified as idiopathic, as conventional diagnostic tests often fail to identify the underlying cause [69, 70]. Although genetic factors have been implicated in a significant portion of idiopathic infertility cases [71, 72], substantial number remains unexplained.

This pilot study highlights the inter-individual variability among normozoospermic samples and demonstrates the potential of proteomic-based approaches to identify individuals with outlier molecular features. A total of 26 semen samples from normozoospermic men undergoing routine semen analysis at the REPROMEDA clinic due to infertility concerns (www.repromeda.com) were analyzed. Cluster analysis incorporating mixed data—qualitative questionnaire responses, spermiogram results, and quantitative histone measurements—identified three distinct patient groups. Notably, mass spectrometry–based histone profiles emerged as the primary factor driving the clustering. For instance, group 2 was characterized by altered levels of H3K23ac and the testis-specific histone variant H3.1T.

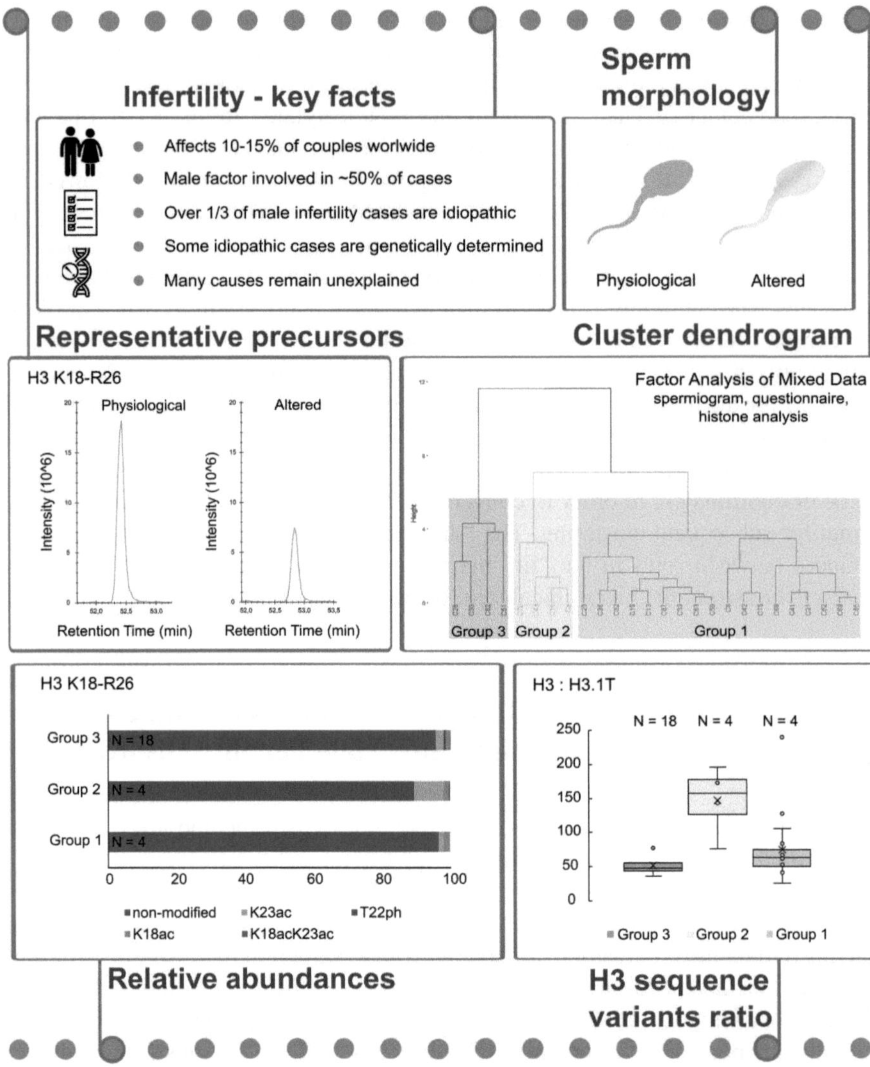

Fig. 6 Representative study of histone peptidoforms quantification in sperm cells in the context of male fertility disorders. The quantity of each modified peptide was determined from the peak area of the respective precursor's EIC in Skyline software. Here, the area of representative precursors' EICs of K18-R26 post-translationally modified forms of histone H3 was found to differ between samples. Cluster analysis revealed three groups of 26 normozoospermic samples. Each group exhibited unique molecular outlier features based on the relative abundance of specific PTMs and the levels of histone sequence variants

This example illustrates how a deeper understanding of histone epigenetics can contribute to more accurate diagnosis and improved treatment of certain diseases.

5 Conclusions

LC-MS/MS stands as the most powerful technique for investigating histone peptidoforms, offering unparalleled sensitivity and resolution for decoding the complexity of the epigenetic landscape. As epigenetics has become a central pillar of molecular biology, the precise mapping and quantification of histone modifications are now essential for elucidating mechanisms of gene regulation, cellular plasticity, and disease pathology.

Beyond the identification of both established and novel PTMs, MS enables precise localization of these modifications to specific amino acid residues and quantification based on their relative abundance. This capability reveals dynamic changes in chromatin structure across diverse biological contexts. Through comparative analyses across different cell types, tissues, and experimental conditions, proteomic analysis delivers critical insights into how chromatin states respond to internal and external stimuli—firmly establishing its role as a leading tool in modern epigenetic research.

Acknowledgments This work was supported by the Czech Science Foundation (project No. 22-28190S). We acknowledge CEITEC Proteomics Core Facility of CIISB, Instruct-CZ Centre, supported by MEYS CR (LM2023042, CZ.02.01.01/00/23_015/0008175, e-INFRA CZ (ID:90254)).

We gratefully acknowledge Repromeda clinic (www.repromeda.com) for kindly providing clinical samples used to demonstrate the application of the described methodology.

References

1. Allfrey V, Faulkner R, Mirsky A (1964) Acetylation and methylation of histones and their possible role in the regulation of RNA synthesis. Proc Natl Acad Sci USA 51. https://doi.org/10.1073/pnas.51.5.786
2. Jenuwein T, Allis CD (2001) Translating the histone code. Science 293:1074–1080. https://doi.org/10.1126/science.1063127
3. Sun L, Zhang H, Gao P (2022) Metabolic reprogramming and epigenetic modifications on the path to cancer. Protein Cell 13. https://doi.org/10.1007/s13238-021-00846-7
4. Schueller E, Paiva I, Blanc F, Wang X, Cassel J, Boutillier A, Bousiges O (2020) Dysregulation of histone acetylation pathways in hippocampus and frontal cortex of Alzheimer's disease patients. Eur Neuropsychopharmacol 33. https://doi.org/10.1016/j.euroneuro.2020.01.015
5. Hosseini M-S, Sanaat Z, Akbarzadeh MA, Vaez-Gharamaleki Y, Akbarzadeh M (2024) Histone deacetylase inhibitors for leukemia treatment: current status and future directions. Eur J Med Res 29:514. https://doi.org/10.1186/s40001-024-02108-8
6. El Omari N, Bakrim S, Elhrech H, Aanniz T, Balahbib A, Lee L-H, Al Abdulmonem W, Bouyahya A (2025) Clinical efficacy and mechanistic insights of FDA-approved HDAC

inhibitors in the treatment of lymphoma. Eur J Pharm Sci 208:107057. https://doi.org/10.1016/j.ejps.2025.107057
7. O'Doherty AM, McGettigan PA (2015) Epigenetic processes in the male germline. Reprod Fertil Dev 27:725–738. https://doi.org/10.1071/RD14167
8. Laurentino S, Beygo J, Nordhoff V, Kliesch S, Wistuba J, Borgmann J, Buiting K, Horsthemke B, Gromoll J (2015) Epigenetic germline mosaicism in infertile men. Hum Mol Genet 24:1295–1304. https://doi.org/10.1093/hmg/ddu540
9. Stuppia L, Franzago M, Ballerini P, Gatta V, Antonucci I (2015) Epigenetics and male reproduction: the consequences of paternal lifestyle on fertility, embryo development, and children lifetime health. Clin Epigenetics 7:120. https://doi.org/10.1186/s13148-015-0155-4
10. Krejčí J, Stixová L, Pagáčová E, Legartová S, Kozubek S, Lochmanová G, Zdráhal Z, Sehnalová P, Dabravolski S, Hejátko J, Bártová E (2015) Post-translational modifications of histones in human sperm. J Cell Biochem 116:2195–2209. https://doi.org/10.1002/jcb.25170
11. Agarwal A, Panner Selvam MK, Baskaran S (2020) Proteomic analyses of human sperm cells: understanding the role of proteins and molecular pathways affecting male reproductive health. Int J Mol Sci 21:1621. https://doi.org/10.3390/ijms21051621
12. Schon SB, Luense LJ, Wang X, Bartolomei MS, Coutifaris C, Garcia BA, Berger SL (2019) Histone modification signatures in human sperm distinguish clinical abnormalities. J Assist Reprod Genet 36:267–275. https://doi.org/10.1007/s10815-018-1354-7
13. Ueda J, Harada A, Urahama T, Machida S, Maehara K, Hada M, Makino Y, Nogami J, Horikoshi N, Osakabe A, Taguchi H, Tanaka H, Tachiwana H, Yao T, Yamada M, Iwamoto T, Isotani A, Ikawa M, Tachibana T, Okada Y, Kimura H, Ohkawa Y, Kurumizaka H, Yamagata K (2017) Testis-specific histone variant H3t gene is essential for entry into spermatogenesis. Cell Rep 18:593–600. https://doi.org/10.1016/j.celrep.2016.12.065
14. Martianov I, Brancorsini S, Catena R, Gansmuller A, Kotaja N, Parvinen M, Sassone-Corsi P, Davidson I (2005) Polar nuclear localization of H1T2, a histone H1 variant, required for spermatid elongation and DNA condensation during spermiogenesis. Proc Natl Acad Sci 102:2808–2813. https://doi.org/10.1073/pnas.0406060102
15. Montellier E, Boussouar F, Rousseaux S, Zhang K, Buchou T, Fenaille F, Shiota H, Debernardi A, Héry P, Curtet S, Jamshidikia M, Barral S, Holota H, Bergon A, Lopez F, Guardiola P, Pernet K, Imbert J, Petosa C, Tan M, Zhao Y, Gérard M, Khochbin S (2013) Chromatin-to-nucleoprotamine transition is controlled by the histone H2B variant TH2B. Genes Dev 27:1680–1692. https://doi.org/10.1101/gad.220095.113
16. Zhang X, San Gabriel M, Zini A (2006) Sperm nuclear histone to protamine ratio in fertile and infertile men: evidence of heterogeneous subpopulations of spermatozoa in the ejaculate. J Androl 27:414–420. https://doi.org/10.2164/jandrol.05171
17. Guo J-E, Wang H, Yang Y, Li J, Zhu Z (2023) Histone deacetylase gene SlHDA3 is involved in drought and salt response in tomato. Plant Growth Regul 99:359–372. https://doi.org/10.1007/s10725-022-00913-x
18. Xue Y, Cao X, Chen X, Deng X, Deng XW, Ding Y, Dong A, Duan C-G, Fang X, Gong L, Gong Z, Gu X, He C, He H, He S, He X-J, He Y, He Y, Jia G, Jiang D, Jiang J, Lai J, Lang Z, Li C, Li Q, Li X, Liu B, Liu B, Luo X, Qi Y, Qian W, Ren G, Song Q, Song X, Tian Z, Wang J-W, Wang Y, Wu L, Wu Z, Xia R, Xiao J, Xu L, Xu Z-Y, Yan W, Yang H, Zhai J, Zhang Y, Zhao Y, Zhong X, Zhou D-X, Zhou M, Zhou Y, Zhu B, Zhu J-K, Liu Q (2025) Epigenetics in the modern era of crop improvements. Sci China Life Sci 68:1570–1609. https://doi.org/10.1007/s11427-024-2784-3
19. Pírek P, Kryštofová K, Kováčová I, Kromerová A, Zachová D, Helia O, Panzarová K, Fajkus J, Zdráhal Z, Lochmanová G, Fojtová M (2023) Unraveling epigenetic changes in a. thaliana Calli: impact of HDAC inhibitors. Plants (Basel) 12:4177. https://doi.org/10.3390/plants12244177
20. Karch KR, DeNizio JE, Black BE, Garcia BA (2013) Identification and interrogation of combinatorial histone modifications. Front Genet 4. https://doi.org/10.3389/fgene.2013.00264
21. Kuchaříková H, Dobrovolná P, Lochmanová G, Zdráhal Z (2021) Trimethylacetic anhydride-based derivatization facilitates quantification of histone marks at the MS1 level. Mol Cell Proteomics 20:100114. https://doi.org/10.1016/j.mcpro.2021.100114

22. Ryzhaya P, Pírek P, Zdráhal Z, Lochmanová G (2025) Arg-C ultra simplifies histone preparation for LC-MS/MS. Anal Chem. https://doi.org/10.1021/acs.analchem.5c02238
23. Garcia BA, Mollah S, Ueberheide BM, Busby SA, Muratore TL, Shabanowitz J, Hunt DF (2007) Chemical derivatization of histones for facilitated analysis by mass spectrometry. Nat Protoc 2:933–938. https://doi.org/10.1038/nprot.2007.106
24. Činčárová L, Lochmanová G, Nováková K, Šultesová P, Konečná H, Fajkusová L, Fajkus J, Zdráhal Z (2012) A combined approach for the study of histone deacetylase inhibitors. Mol BioSyst 8:2937–2945. https://doi.org/10.1039/c2mb25136a
25. Shechter D, Dormann HL, Allis CD, Hake SB (2007) Extraction, purification and analysis of histones. Nat Protoc 2:1445–1457. https://doi.org/10.1038/nprot.2007.202
26. Rodriguez-Collazo P, Leuba SH, Zlatanova J (2009) Robust methods for purification of histones from cultured mammalian cells with the preservation of their native modifications. Nucleic Acids Res 37:e81. https://doi.org/10.1093/nar/gkp273
27. Homsi C, Rajan RE, Minati R, St-Hilaire E, Bonneil E, Dufresne SF, Wurtele H, Verreault A, Thibault P (2023) A rapid and efficient method for the extraction of histone proteins. J Proteome Res 22:2765–2773. https://doi.org/10.1021/acs.jproteome.3c00266
28. Ledvinová D, Mikulášek K, Kuchaříková H, Brabencová S, Fojtová M, Zdráhal Z, Lochmanová G (2018) Filter-aided sample preparation procedure for mass spectrometric analysis of plant histones. Front Plant Sci 9. https://doi.org/10.3389/fpls.2018.01373
29. Scheid R, Dowell JA, Sanders D, Jiang J, Denu JM, Zhong X (2022) Histone acid extraction and high throughput mass spectrometry to profile histone modifications in Arabidopsis thaliana. Curr Protoc 2:e527. https://doi.org/10.1002/cpz1.527
30. Frohlich J, Liorni N, Mangoni M, Lochmanová G, Pírek P, Kaštánková N, Pata P, Kucera J, Chaldakov GN, Tonchev AB, Pata I, Gorbunova V, Leire E, Zdráhal Z, Mazza T, Vinciguerra M (2024) Epigenetic and transcriptional control of adipocyte function by centenarian-associated SIRT6 N308K/A313S mutant. Clin Epigenetics 16:96. https://doi.org/10.1186/s13148-024-01710-1
31. Mao S, Goodrich RJ, Hauser R, Schrader SM, Chen Z, Krawetz SA (2013) Evaluation of the effectiveness of semen storage and sperm purification methods for spermatozoa transcript profiling. Syst Biol Reprod Med 59:287–295. https://doi.org/10.3109/19396368.2013.817626
32. Green GR, Do DP (2009) Purification and analysis of variant and modified histones using 2D PAGE. Methods Mol Biol 464:285–302. https://doi.org/10.1007/978-1-60327-461-6_16
33. Abshiru N, Caron-Lizotte O, Rajan RE, Jamai A, Pomies C, Verreault A, Thibault P (2015) Discovery of protein acetylation patterns by deconvolution of peptide isomer mass spectra. Nat Commun 6:8648. https://doi.org/10.1038/ncomms9648
34. Giansanti P, Tsiatsiani L, Low TY, Heck AJR (2016) Six alternative proteases for mass spectrometry–based proteomics beyond trypsin. Nat Protoc 11:993–1006. https://doi.org/10.1038/nprot.2016.057
35. Dupree EJ, Jayathirtha M, Yorkey H, Mihasan M, Petre BA, Darie CC (2020) A critical review of bottom-up proteomics: the good, the bad, and the future of this field. Proteomes 8:14. https://doi.org/10.3390/proteomes8030014
36. Noberini R, Robusti G, Bonaldi T (2022) Mass spectrometry-based characterization of histones in clinical samples: applications, progress, and challenges. FEBS J 289:1191–1213. https://doi.org/10.1111/febs.15707
37. Maile TM, Izrael-Tomasevic A, Cheung T, Guler GD, Tindell C, Masselot A, Liang J, Zhao F, Trojer P, Classon M, Arnott D (2015) Mass spectrometric quantification of histone post-translational modifications by a hybrid chemical labeling method. Mol Cell Proteomics 14:1148–1158. https://doi.org/10.1074/mcp.O114.046573
38. Sidoli S, Yuan Z-F, Lin S, Karch K, Wang X, Bhanu N, Arnaudo AM, Britton L-M, Cao X-J, Gonzales-Cope M, Han Y, Liu S, Molden RC, Wein S, Afjehi-Sadat L, Garcia BA (2015) Drawbacks in the use of unconventional hydrophobic anhydrides for histone derivatization in bottom-up proteomics PTM analysis. Proteomics 15:1459–1469. https://doi.org/10.1002/pmic.201400483
39. Liao R, Wu H, Deng H, Yu Y, Hu M, Zhai H, Yang P, Zhou S, Yi W (2013) Specific and efficient N-propionylation of histones with propionic acid N-hydroxysuccinimide ester for

histone marks characterization by LC-MS. Anal Chem 85:2253–2259. https://doi.org/10.1021/ac303171h
40. Chen J, Gao J, Peng M, Wang Y, Yu Y, Yang P, Jin H (2015) *In-gel* NHS-propionate derivatization for histone post-translational modifications analysis in *Arabidopsis thaliana*. Anal Chim Acta 886:107–113. https://doi.org/10.1016/j.aca.2015.06.019
41. Herwig-Carl MC, Sharma A, Tischler V, Pelusi N, Loeffler KU, Holz FG, Zeschnigk M, Landreville S, Auw-Haedrich C, Noberini R, Bonaldi T (2024) Mass spectrometry-based profiling of histone post-translational modifications in uveal melanoma tissues, human melanocytes, and uveal melanoma cell lines—a pilot study. Invest Ophthalmol Vis Sci 65:27. https://doi.org/10.1167/iovs.65.2.27
42. Noberini R, Savoia EO, Brandini S, Greco F, Marra F, Bertalot G, Pruneri G, McDonnell LA, Bonaldi T (2021) Spatial epi-proteomics enabled by histone post-translational modification analysis from low-abundance clinical samples. Clin Epigenetics 13:145. https://doi.org/10.1186/s13148-021-01120-7
43. Jung HR, Pasini D, Helin K, Jensen ON (2010) Quantitative mass spectrometry of histones H3.2 and H3.3 in Suz12-deficient mouse embryonic stem cells reveals distinct, dynamic post-translational modifications at Lys-27 and Lys-36 *. Mol Cell Proteomics 9:838–850. https://doi.org/10.1074/mcp.M900489-MCP200
44. Searfoss RM, Karki R, Lin Z, Robison F, Garcia BA (2023) An optimized and high-throughput method for histone propionylation and data-independent acquisition analysis for the identification and quantification of histone post-translational modifications. J Am Soc Mass Spectrom 34:2508–2517. https://doi.org/10.1021/jasms.3c00223
45. Krautkramer KA, Reiter L, Denu JM, Dowell JA (2015) Quantification of SAHA-dependent changes in histone modifications using data-independent acquisition mass spectrometry. J Proteome Res 14:3252–3262. https://doi.org/10.1021/acs.jproteome.5b00245
46. Cole J, Hanson EJ, James DC, Dockrell DH, Dickman MJ (2019) Comparison of data-acquisition methods for the identification and quantification of histone post-translational modifications on a Q Exactive HF hybrid quadrupole Orbitrap mass spectrometer. Rapid Commun Mass Spectrom 33:897–906. https://doi.org/10.1002/rcm.8401
47. Nesvizhskii AI, Vitek O, Aebersold R (2007) Analysis and validation of proteomic data generated by tandem mass spectrometry. Nat Methods 4:787–797. https://doi.org/10.1038/nmeth1088
48. Perkins DN, Pappin DJ, Creasy DM, Cottrell JS (1999) Probability-based protein identification by searching sequence databases using mass spectrometry data. Electrophoresis 20:3551–3567. https://doi.org/10.1002/(SICI)1522-2683(19991201)20:18<3551::AID-ELPS3551>3.0.CO;2-2
49. Eng JK, McCormack AL, Yates JR (1994) An approach to correlate tandem mass spectral data of peptides with amino acid sequences in a protein database. J Am Soc Mass Spectrom 5:976–989. https://doi.org/10.1016/1044-0305(94)80016-2
50. Zhang J, Xin L, Shan B, Chen W, Xie M, Yuen D, Zhang W, Zhang Z, Lajoie GA, Ma B (2012) PEAKS DB: de novo sequencing assisted database search for sensitive and accurate peptide identification. Mol Cell Proteomics 11(M111):010587. https://doi.org/10.1074/mcp.M111.010587
51. Shilov IV, Seymour SL, Patel AA, Loboda A, Tang WH, Keating SP, Hunter CL, Nuwaysir LM, Schaeffer DA (2007) The Paragon Algorithm, a next generation search engine that uses sequence temperature values and feature probabilities to identify peptides from tandem mass spectra. Mol Cell Proteomics 6:1638–1655. https://doi.org/10.1074/mcp.T600050-MCP200
52. Wang L-H, Li D-Q, Fu Y, Wang H-P, Zhang J-F, Yuan Z-F, Sun R-X, Zeng R, He S-M, Gao W (2007) pFind 2.0: a software package for peptide and protein identification via tandem mass spectrometry. Rapid Commun Mass Spectrom 21:2985–2991. https://doi.org/10.1002/rcm.3173
53. Andromeda: a peptide search engine integrated into the MaxQuant environment | Journal of Proteome Research. https://pubs.acs.org/doi/10.1021/pr101065j. Accessed 23 Jul 2025

54. Geer LY, Markey SP, Kowalak JA, Wagner L, Xu M, Maynard DM, Yang X, Shi W, Bryant SH (2004) Open mass spectrometry search algorithm. J Proteome Res 3:958–964. https://doi.org/10.1021/pr0499491
55. Fenyö D, Beavis RC (2003) A method for assessing the statistical significance of mass spectrometry-based protein identifications using general scoring schemes. Anal Chem 75:768–774. https://doi.org/10.1021/ac0258709
56. Yuan Z-F, Lin S, Molden RC, Garcia BA (2014) Evaluation of proteomic search engines for the analysis of histone modifications. J Proteome Res 13:4470–4478. https://doi.org/10.1021/pr5008015
57. Brabencová S, Ihnatová I, Potěšil D, Fojtová M, Fajkus J, Zdráhal Z, Lochmanová G (2017) Variations of histone modification patterns: contributions of inter-plant variability and technical factors. Front Plant Sci 8:2084. https://doi.org/10.3389/fpls.2017.02084
58. Franek M, Nešpor Dadejová M, Pírek P, Kryštofová K, Dobisová T, Zdráhal Z, Dvořáčková M, Lochmanová G (2024) Histone chaperone deficiency in Arabidopsis plants triggers adaptive epigenetic changes in histone variants and modifications. Mol Cell Proteomics 23:100795. https://doi.org/10.1016/j.mcpro.2024.100795
59. Kolářová K, Nešpor Dadejová M, Loja T, Lochmanová G, Sýkorová E, Dvořáčková M (2021) Disruption of NAP1 genes in Arabidopsis thaliana suppresses the fas1 mutant phenotype, enhances genome stability and changes chromatin compaction. Plant J 106:56–73. https://doi.org/10.1111/tpj.15145
60. MacLean B, Tomazela DM, Shulman N, Chambers M, Finney GL, Frewen B, Kern R, Tabb DL, Liebler DC, MacCoss MJ (2010) Skyline: an open source document editor for creating and analyzing targeted proteomics experiments. Bioinformatics 26:966–968. https://doi.org/10.1093/bioinformatics/btq054
61. Yuan Z-F, Sidoli S, Marchione DM, Simithy J, Janssen KA, Szurgot MR, Garcia BA (2018) EpiProfile 2.0: a computational platform for processing epi-proteomics mass spectrometry data. J Proteome Res 17:2533–2541. https://doi.org/10.1021/acs.jproteome.8b00133
62. Levin Y (2011) The role of statistical power analysis in quantitative proteomics. Proteomics 11:2565–2567. https://doi.org/10.1002/pmic.201100033
63. Al Shweiki MR, Mönchgesang S, Majovsky P, Thieme D, Trutschel D, Hoehenwarter W (2017) Assessment of label-free quantification in discovery proteomics and impact of technological factors and natural variability of protein abundance. J Proteome Res 16:1410–1424. https://doi.org/10.1021/acs.jproteome.6b00645
64. Aitchison J (1982) The statistical analysis of compositional data. J R Stat Soc Ser B (Methodol) 44:139–160. https://doi.org/10.1111/j.2517-6161.1982.tb01195.x
65. Hron K, Templ M, Filzmoser P (2010) Imputation of missing values for compositional data using classical and robust methods. Comput Stat Data Anal 54:3095–3107. https://doi.org/10.1016/j.csda.2009.11.023
66. Zegers-Hochschild F, Adamson GD, de Mouzon J, Ishihara O, Mansour R, Nygren K, Sullivan E, Vanderpoel S, International Committee for Monitoring Assisted Reproductive Technology, World Health Organization (2009) International Committee for Monitoring Assisted Reproductive Technology (ICMART) and the World Health Organization (WHO) revised glossary of ART terminology, 2009. Fertil Steril 92:1520–1524. https://doi.org/10.1016/j.fertnstert.2009.09.009
67. Jarow JP (2007) Diagnostic approach to the infertile male patient. Endocrinol Metab Clin N Am 36:297–311. https://doi.org/10.1016/j.ecl.2007.03.012
68. Esteves SC, Miyaoka R, Agarwal A (2011) An update on the clinical assessment of the infertile male. [corrected]. Clinics (Sao Paulo) 66:691–700. https://doi.org/10.1590/s1807-59322011000400026
69. Hamada A, Esteves SC, Nizza M, Agarwal A (2012) Unexplained male infertility: diagnosis and management. Int Braz J Urol 38:576–594. https://doi.org/10.1590/s1677-55382012000500002

70. Naz M, Kamal M (2017) Classification, causes, diagnosis and treatment of male infertility: a review. Orient Pharm Exp Med 17:89–109. https://doi.org/10.1007/s13596-017-0269-7
71. Guerri G, Maniscalchi T, Barati S, Busetto GM, Del Giudice F, De Berardinis E, Cannarella R, Calogero AE, Bertelli M (2019) Non-syndromic monogenic male infertility. Acta Bio Medica Atenei Parmensis 90:62–67. https://doi.org/10.23750/abm.v90i10-S.8762
72. Houston BJ, Riera-Escamilla A, Wyrwoll MJ, Salas-Huetos A, Xavier MJ, Nagirnaja L, Friedrich C, Conrad DF, Aston KI, Krausz C, Tüttelmann F, O'Bryan MK, Veltman JA, Oud MS (2021) A systematic review of the validated monogenic causes of human male infertility: 2020 update and a discussion of emerging gene-disease relationships. Hum Reprod Update 28:15–29. https://doi.org/10.1093/humupd/dmab030

Proteomic and Functional Comparison of Extracellular Vesicles from Wild-Type and Lyn-Deficient Stromal Cells

Thaís Dolzany de Oliveira, Prerana Wagle, and Hinrich P. Hansen

Abstract Chronic lymphocytic leukemia (CLL) progression critically depends on bidirectional communication between malignant B cells and non-malignant bystander cells within the tumor microenvironment, including stromal cells, macrophages, and T lymphocytes. Among the key regulators of this cross talk is the Src-family kinase Lyn, whose activity in stromal cells controls extracellular matrix remodeling, fibrotic niche formation, and the release of tumor-supportive extracellular vesicles (EVs). This chapter presents a comprehensive proteomics-based strategy to identify Lyn-dependent effector molecules conveyed by stromal-cell–derived EVs and to evaluate their functional relevance for leukemic cell survival. State-of-the-art EV purification methods compliant with the Minimal information for studies of extracellular vesicles (MISEV) guidelines, nanoparticle tracking analysis, and high-resolution Orbitrap mass spectrometry are integrated with CRISPR-mediated Lyn knockout models and functional co-culture assays using primary CLL cells. Comparative metaproteomic profiling of EVs from Lyn-proficient and Lyn-deficient stromal cells reveals profound alterations in extracellular matrix proteins, adhesion molecules, redox regulators, and signaling mediators, highlighting downstream pathways through which Lyn programs a tumor-supportive microenvironment. Bioinformatic network analysis and pathway enrichment further identify druggable candidate proteins, including matricellular components and integrin-associated

T. D. de Oliveira · H. P. Hansen (✉)
Department I of Internal Medicine, Center for Integrated Oncology Aachen Bonn Cologne Düsseldorf, University of Cologne, Cologne, Germany

CECAD Center of Excellence on Cellular Stress Responses in Aging-Associated Diseases, University of Cologne, Cologne, Germany
e-mail: h.hansen@uni-koeln.de

P. Wagle
CECAD Center of Excellence on Cellular Stress Responses in Aging-Associated Diseases, University of Cologne, Cologne, Germany

Proteomics Facility, University of Cologne, Cologne, Germany
e-mail: pwagle@uni-koeln.de

J.-L. Capelo-Martínez (ed.), *Emerging Sample Treatments in Proteomics*, Advances in Experimental Medicine and Biology 1510,
https://doi.org/10.1007/978-3-032-21638-0_5

anchors such as CD248, as potential therapeutic nodes. Collectively, this chapter illustrates how EV-focused proteomics enables the systematic dissection of microenvironment-driven oncogenic signaling and provides a translational framework for discovering actionable stromal targets in CLL and other Lyn-dependent malignancies.

Keywords Chronic lymphocytic leukemia · Tumor microenvironment · Extracellular vesicles · Metaproteomics · Lyn kinase · Cancer-associated fibroblasts · Src family kinases · Mass spectrometry · Protein–protein interactions · Extracellular matrix · Precision oncology · Drug target discovery

1 Introduction

Cancer cells need the support of non-malignant bystander cells within a so-called tumor microenvironmental niche. Chronic lymphocytic leukemia (CLL) is a good example. This disease is characterized by malignant monoclonal CD5$^+$, CD19$^+$, CD20$^+$, CD23$^+$ B cells, which are concentrated in the bone marrow and lymphoid tissue and expanded in the peripheral blood [1]. CLL cell survival is dependent on the support from non-malignant cells within the tumor microenvironment (TME) of the lymphoid tissue. They receive support mainly from stromal, nurse-like macrophages and T cells (Fig. 1). Isolation of CLL cells from the peripheral blood of patients causes their spontaneous apoptosis unless they are co-cultivated with non-malignant supporter cells such as bone marrow stromal cells or monocyte-derived nurse-like cells [2–4]. The interaction between tumor and supporter cells is bidirectional. CLL cells stimulate macrophages and stromal cells to differentiate to cells with elevated supporter capacity, such as nurse-like or cancer-associated fibroblast (CAF)-like phenotype, respectively [5]. Vice versa, nurse-like macrophages and CAF-like activated mesenchymal stromal cells play an essential role in creating a tumor-promoting niche and mitigating the efficacy of anticancer treatment [6]. Thus, it is crucial to study these interactions in detail to identify possible options to block this environmental tumor support.

It has been shown already that CLL cells experience an upregulation of the T-cell leukemia/lymphoma 1 (TCL1) oncogene and a drop of TCL1-regulatory microRNAs (miR-29b, miR-181b, and miR-34b) upon contact-dependent coculture with bone marrow stromal cells [7]. In addition to soluble factors from stromal cells, like structural extracellular matrix proteins, transforming growth factor beta 1 (TGFB1), and fibroblast growth factor (FGF), the direct membrane-associated contact plays a more prominent role in CLL stimulation. Adhesion molecules such as ICAM1, integrins, and CD248 are able to stabilize the contact. The transmembrane glycoprotein CD248 (endosialin, tumor endothelial marker 1/TEM1) belongs to a family of group XIV C-type lectins [8]. CD248 is an activation marker of fibroblasts and myofibroblasts but is also expressed in pericytes and malignant cells of a few tumor entities [9]. Like integrins, CD248 binds to several ECM molecules, such as

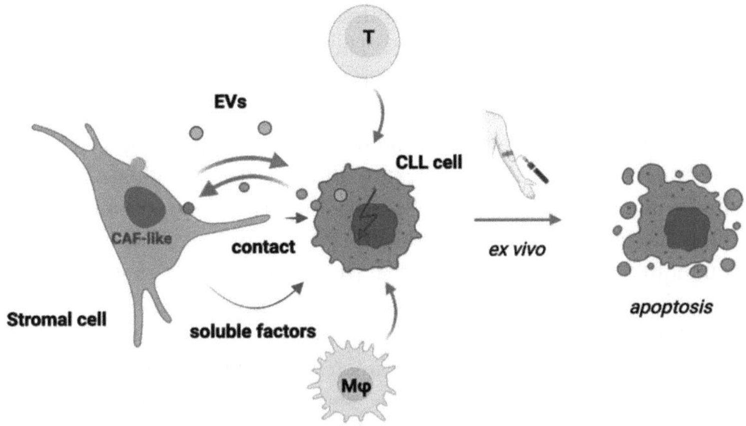

Fig. 1 Tumor microenvironment in chronic lymphocytic leukemia. CLL is characterized by malignant B cells, which are concentrated in the bone marrow, lymphoid tissue, and peripheral blood. For their survival, they need the support of non-malignant bystander cells, such as macrophages, lymphocytes, and stromal cells. Isolated CLL cells undergo rapid apoptosis when cultivated in vitro. In particular, cancer-associated fibroblasts (CAFs) support CLL cells, implicating direct cell contact, soluble factors, and extracellular vesicles (EVs). Tumor cell EVs stimulate stromal cells to acquire a CAF-like phenotype, and stromal cell EVs support the survival of CLL cells

collagen, fibronectin, the metastasis-related protein LGALS3BP (Mac-2 BP/90K) as well as the ECM glycoprotein multimerin 2 (MMRN2), and might serve as an anchor and cross-linker of structural ECM proteins [10]. These interactions of CD248 contribute to cancer progression, mainly by enhancing the deposition of collagen and fibronectin. Such ECM proteins create a microenvironment that fosters tumor cell migration, invasion, and metastasis. As an example, in colorectal and breast cancer, CD248 expression correlates with a worse outcome [11].

Stromal cells interact with malignant CLL cells not only through direct cell contact and soluble factors but also through extracellular vesicles (EVs) [12]. According to the latest guidelines of the International Society of Extracellular Vesicles (ISEV), EVs are regarded as a group of lipid bilayer-enclosed vesicles but cannot replicate on their own [13]. They are released by virtually all cell types and transport lipids, nucleic acids as well as structural and functional proteins [14]. These vesicles transport traits of the donor cell and emerged as an important platform for intercellular communication, including tumor development [15]. Dependent on their biogenesis and diameter, they were subdivided into three groups. The groups with the largest EVs are apoptotic blebs that are released from dying cells. Because of their large diameter up to 5 μm, they are even detectable by conventional microscopy. However, most scientists focus on EVs from viable cells, which are smaller than the apoptotic blebs. Viable cells release EVs predominantly by two mechanisms. The second group of EVs is formed by outward budding of the plasma membrane from viable cells. Such EVs are heterogeneous in diameter (approximately 100–700 nm) but smaller than apoptotic blebs. Because of their origin and diameter, they were often referred to as microvesicles or ectosomes. The third group derives from intraluminal endosomal vesicles, and these particles are released into the cell environment by exocytosis of multivesicular endosomes (MVEs) (Figs. 1 and 2). The latter, which is occasionally called "exosomes," contains predominantly small vesicles with a diameter between 40 and 120 nm. Because all EV types are released and dispersed in body fluids and difficult to identify and distinguish, the International Society of Extracellular Vesicles (ISEV) recommends to use the term extracellular vesicle (EV) throughout for all EV types. If clustering is necessary, the society suggests to only differentiate between large EVs (>200 nm) and small EVs (<200 nm), independent of the origin [16]. A variety of physical and immunological means are available to enrich EV fractions by diameter or density (Fig. 2). The methods include different centrifugation, filtration, and size-exclusion techniques. The last two groups contain EVs that are too small for identification in light microscopes. Transmission or cryo-electron microscopy is recommended for direct visualization.

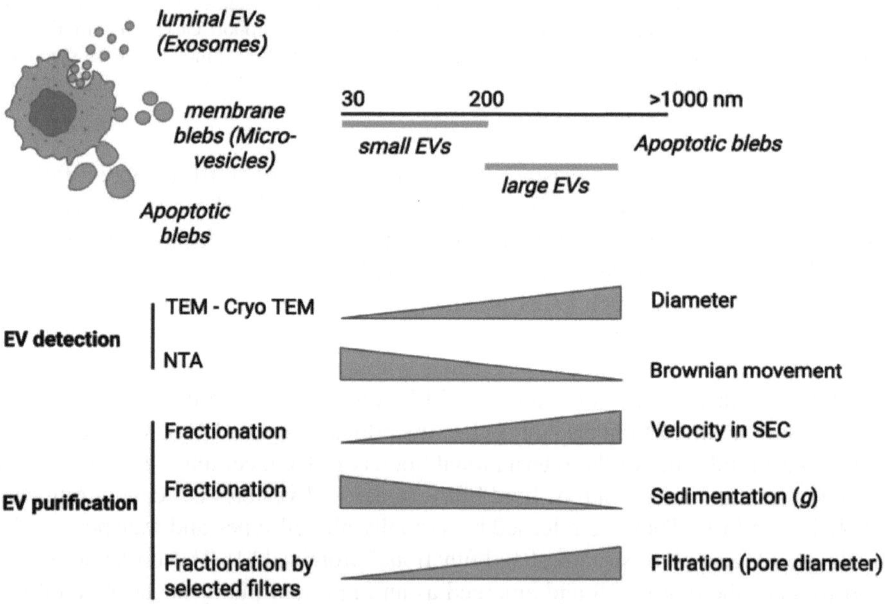

Fig. 2 Selected methods to detect and purify extracellular vesicles. Cells release multivesicular endosome-derived (luminal), membrane-derived, and apoptotic vesicles into the environment where they are found in a heterogeneous mixture. Independent of their origin, they are collectively referred to as extracellular vesicles (EVs). Because the cell supernatant generally contains a mixture of luminal and membrane-derived vesicles with overlapping diameters, recent publications prefer to distinguish only between small and large EVs. EVs are small and difficult to detect by conventional microscopy. Therefore, electron microscopy and nanoparticle tracking analysis (NTA) are utilized to detect EVs. There are different methods to enrich EVs by their diameter, by size-exclusion chromatography (SEC), differential centrifugation, and filtration. Then, according to the MISEV suggestions and independent of their origin, they are grouped as small or large EVs

Because these methods are time-consuming, EVs are often indirectly detected by tracking the scattered light of a laser beam (nanoparticle tracking analysis, NTA, see below).

EVs carry typical traits of the donor cells, including nucleic acids, receptors, ligands as well as luminal proteins, and transport them in the environment of the donor cell to interact with distant cells. This functionality opens an alternative avenue of intercellular communication. Like cell contact, EVs are able to perform multiple receptor–ligand interactions within a membrane-associated context, but due to their mobility, this interaction is possible over distance (*in trans*). Moreover, target cells might take up EVs along with their entire payload. EVs play a role in physiological and pathological conditions, like inflammation and cancer [17]. In addition to their primary composition, they are able to acquire soluble molecules, which enrich the original payload. These molecules bind to the surface and form a so-called corona, which can participate in the functionality of EVs [18, 19]. In chronic lymphocytic leukemia, the malignant cells communicate with different non-malignant supporter cells such as immune and stromal cells [12]. Similar to the direct cell contact, the EV-based interaction between stromal and CLL cells is bidirectional (Fig. 1). EVs from the malignant CLL cells induce the transition of stromal cells to tumor-supporting cancer-associated fibroblasts (CAF) and vice versa, EVs from activated stromal cells bind and are taken up by CLL cells [18]. Then, the stromal cell EVs stimulate migration and survival in the malignant cells [20]. Thus, EVs are powerful signaling particles. They are able to bind to ECM proteins like collagen, laminin, and fibronectin [21, 22] and influence target cells in the microenvironment by presenting membrane-anchored and attached molecules in a membrane-associated context. Moreover, they might communicate the internalizing functional proteins and nucleic acids into the target cell lumen.

The supporting communication between tumor and tumor supporter cells is a hallmark in cancer. CLL cells that in cell culture die soon after isolation from the blood of patients are strongly supported when co-cultivated with non-malignant supporter cells, such as stromal cells, lymphocytes, and activated macrophages [3, 23]. In CLL, the non-receptor tyrosine kinase Lyn plays an essential role in this communication. This proto-oncogene Lyn is a non-receptor-type tyrosine kinase and member of the Src family kinases (SFK). Other family members are the structurally related kinases, Abl, Csk, Src, Lyn, Fyn, Yes, Fgr, Hck, Lck, and Blk [24]. Lyn participates in signal transduction of various cell surface receptors, and in CLL, it is expressed in malignant B-CLL cells as well as bystander cells of the CLL-affected tissue [24]. Its expression correlates with the progression of the disease [25]. Although the malignant and most non-malignant cells of the tumor microenvironment express Lyn, loss- and gain-of-function experiments indicate that the disease is predominantly driven by Lyn in bystander cells, particularly in non-hematopoietic bystander cells, such stromal cells [26]. Stromal cells are prominent producers of ECM proteins, and activated Lyn can induce ECM stiffness by activating the twist 1 pathway [27]. In addition, Lyn plays a major role in fibrotic disorders, and constitutively active Lyn in myofibroblasts correlates with enhanced fibrosis and worse outcome, whereas Lyn inhibition reduces the synthesis of procollagen [28]. These symptoms were also demonstrated in two patients with an inherited activating truncation of Lyn. They suffered from small vessel vasculitis and liver fibrosis [29]. A dense collagenous stroma is a strong indicator of tumor aggressiveness and therapeutic resistance in solid tumors [30]. Thus, dysregulated Lyn and its downstream effectors play a decisive role in tumor progression.

Recently, it was shown that in the absence of stromal cells themselves, the EVs thereof are sufficient to stimulate the survival of CLL cells. Like direct stromal/cancer cell co-culture, this stimulation is dependent on active Lyn in the supporter cell [31–33]. It was also shown that Lyn significantly influences the protein payload and the tumor-supporting functionality of EVs from stromal cells [29]. Currently, there is no Lyn-specific inhibitor available for the clinical use, and broad-spectrum inhibitors have severe side effects in CLL [34]. Therefore, it is of great importance to identify downstream effectors of Lyn, not only in cells but also in EVs. Therapeutic inhibition of Lyn's downstream effector proteins might provide an alternative to the inhibition of Lyn or part of a combination treatment strategy in cancer or other Lyn-driven diseases. Using the search for Lyn's downstream effectors as a model, we herein explain the strategy of how to identify and test novel therapeutic targets in EVs with the help of mass spectrometry (Fig. 4).

2 Deletion of Lyn in Stromal Cell Line and Isolation of Primary CLL Cells

The influence of tumor supporter cells on the activation of tumor cells is a hallmark in the tumor microenvironment of cancer in general, and it is a strategy in cancer therapy to block this support by inhibition of key pathway members in this activation cascade. Lyn is such a pathway member, but good therapeutic inhibitors are not available. Because Lyn-dependent support is conveyed by downstream effector molecules in EVs and EVs have a limited protein and nucleic acid payload as compared to their donor cell, the EVs are well suited as material for a mass spectrometry–based search for functional downstream molecules of Lyn. To test the functionality of downstream molecules of Lyn in EVs, it is necessary to generate Lyn-deficient donor cells. This is possible by generating a strain of Lyn-deficient sister cells from Lyn-proficient supporter cells, e.g., by means of CRISPR Cas9 deletion of Lyn. In the case of CLL, there are several candidate supporter cells published. In addition to stromal cells, T lymphocytes and M2-polarized macrophages are regarded as drivers of malignant CLL cells in tissue [35], all of which express Lyn and release EVs. Therefore, the study of Lyn and EVs in stromal cells might serve as a template for other tumor-supporting interactions. Here, we describe methods to identify Lyn-dependent effector proteins in EVs, including the production, purification of EVs, their identification and evaluation by mass spectrometry and proteomics. As shown in Fig. 4, we need to compare EVs from cells, which are typically found in the tumor microenvironment and, apart from Lyn, are as related as possible. In our example, we use mesenchymal stromal cells with characteristics of cancer-associated fibroblasts (CAFs), and the HS-5 cell best replicates this capacity to influence tumor biology and molecular mechanisms that allow the tumor immune escape mediated by stroma cells [36]. To study the influence of Lyn in EVs, it is necessary to deplete or delete the driver protein. A transient or stable knockdown of the target protein is necessary. Both the wild-type and the defective cells should be viable and have comparable cell growth.

The tumor-supporting functionality of the EVs from wild-type and Lyn-defective cells needs to be tested in co-culture experiments with primary tumor cells. Alike other primary tumor cells, the cells from CLL patients are different from patient to patient because the cells have different supporting mutations such as TP53 dysfunction, IGHV status, or a complex karyotype [37]. Therefore, to generate reliable results, the influence of EVs should be tested in individual experiments, on cells from a large number of patients. In the case of CLL, we preferred to test cells from nine different patients in co-culture experiments with EVs. Here, the viability means ± SEM of nine patients were necessary to generate reliable results. Viable primary CLL cells are circulating in abundance in the peripheral blood of affected patients. However, CLL cells that are removed from affected patients and in-vitro purified by immunological means are difficult to cultivate because they quickly start the process of apoptosis due to a lack of support (Fig. 1). Therefore, cells need to be quickly purified and used for subsequent experiments. Direct conservation at −150 °C after isolation is possible and associated with an acceptable loss of viability. Blood of CLL patients was obtained from the CLL-Biobank after written informed consent and according to the Declaration of Helsinki and with Institutional Review Board approvals at the University of Cologne. Although the malignant B cells are the predominant fraction in the blood of CLL patients, the malignant B cells are needed from the whole blood by negative selection using the RosetteSepTM human B-cell enrichment cocktail (Stemcell Technologies) and a subsequent purification on top of the density medium (density: 1.081 g/ml), which is centrifuged for 20 min at 1200× *g* at room temperature. The enriched B-CLL cells are collected from the interphase and washed twice with PBS + 10% FBS. The harvested cells are maintained in RPMI 1640 medium with 10% FBS for direct use or directly frozen with culture medium containing 10% dimethyl sulfoxide (DMSO). Frozen cell aliquots were stored at −150 °C and thawed freshly for every test.

3 Purification and Quality Control of EVs

The tyrosine kinase Lyn is upregulated in tumor and bystander cells and plays a major role in the progression of CLL [25]. After successful co-culture experiments confirming the model that malignant cells in CLL-affected tissue are at least in part stimulated by stromal cell–released extracellular vesicles in a Lyn-dependent manner, it is necessary to identify functional EV proteins [20]. Proteins that are responsible for the tumor-supporting functionality of EVs are best detected and studied by mass spectrometry and proteomics. Most publications describe to purify and study the functionality of small EVs. However, it is not always granted that small EVs are responsible for the effect and that the tested EVs were originally generated as small

EVs. Large EVs, such as apoptotic blebs, might be degraded to smaller EVs after artificial shear force–generated fission. To avoid this degradation, all purification steps demand great care. Contaminating proteins, such as serum proteins and other non-particular soluble factors from the donor cells, are another problem. An accurate purification protocol is needed, best according to the MISEV guidelines from the international society of extracellular vesicles to purify the EVs [16]. There is a debate ongoing whether corona proteins belong to the EV payload because they are not originally present but acquired from the environment [13]. Because they influence the function of EVs, they might be regarded as EV proteins. When EVs are used for mass spectrometry, bovine serum proteins should be avoided if possible. Although mass spectrometry is generally adjusted to only validate human peptides of seven amino acids or more, contaminating peptides dilute the wanted ones and reduce the sensitivity of the analysis. Considering this issue, stromal cells are only cultivated for 12 h in serum to get attached to the culture plate. For this cell type, this period of time allows complete cell adherence to the culture flasks. But then, to reduce the amount of contaminating protein from the serum, the adherent cells are carefully washed and cultivated for another 18 h with serum-free medium. In our case, an incubation period of 18 h in serum-free medium was possible without any loss of cell viability and is sufficient to generate enough EVs for subsequent tests. The described conditions should be verified in experiments with other cells.

Most EVs are small and difficult to visualize by standard light microscopy. The most common techniques to directly detect EVs are standard electron microscopy or cryo-electron microscopy. These techniques are complicated and time-consuming; therefore, an indirect measurement of EVs by nanoparticle tracking analysis (NTA) is often used when testing many different conditions. This method detects the Brownian movement of vesicles in a cuvette and calculates their diameter and their concentration per ml (Fig. 2).

3.1 Method: Nanoparticle Tracking Analysis

EVs are small and difficult to detect by conventional methods. Nanoparticle tracking analysis is a quick method to validate the EV quality at different steps of the EV purification protocol. EV samples were suspended in filtered phosphate-buffered saline (PBS) and diluted to a suitable working volume. It is of great importance to use particle-free PBS for the suspension of EV-containing samples. Although standard 0.2 μm filters do not retain all particles, the filtration of reagents with these filters dramatically reduces particle contaminations from the reagents. Measurement of PBS alone might help to determine the particle background. To quantify the concentration of EVs and assess their size profiles, a nanoparticle tracking analysis (NTA) was performed on a Nanosight NS300 instrument (Malvern Panalytical). All procedures were carried out following standard protocols to ensure accuracy and reproducibility. Prior to NTA measurement, control samples with PBS alone were measured to estimate the background particle concentration. Each sample (including technical replicates) was then loaded into the Nanosight NS300 sample chamber using a syringe pump to maintain constant flow. Five or seven 60-second video recordings were captured for each sample, and particle movement was analyzed frame by frame using NTA software (NanoSight NTA 3.4). The Brownian motion of individual particles was tracked, and their hydrodynamic diameter was calculated using the Stokes–Einstein equation. The results were presented as size distribution histograms and concentration profiles expressed in particles per milliliter (particles/ml).

To purify EVs or EV fractions from cell supernatants or body fluids that contain mixtures of EVs, a variety of methods are available that consider different physical and immunological properties of EVs, such as density, diameter, or the presence of certain biomarkers on the outer surface (Fig. 2). The most commonly used methods are the purification by a combination of low speed and ultracentrifugation. Large EVs sediment already at 3000–10,000× *g* and can be depleted before the remaining EVs are pelleted in a subsequent ultracentrifugation step at 100,000 *g*. Ultrafiltration is another method to purify EVs. Here, the hydrodynamic diameter of the EVs is used to generate EV fractions by ultrafiltration using filters with different pore sizes or size-exclusion chromatography (SEC) techniques. A third strategy is to use immobilized ligands (antibodies, annexin V) to detect and bind markers on the surface of EV's plasma membrane, such as tetraspanins (CD 9, CD63, CD81) or the lipid phosphatidyl serine. Phosphatidyl serine is exposed on the outer leaflet of the plasma membrane of larger EVs [38].

To purify small EVs for mass spectrometry, it is preferred to avoid proteins for enrichment procedures that might influence the sensitivity of mass spectrometry. A typical purification strategy is shown in Fig. 3. This is to deplete larger particles and soluble proteins from the preparations. Several purification steps generate shear forces that are an important obstacle in the purification methods because they might destroy and split larger particles such as apoptotic blebs and artificially generate smaller ones. For this reason, most protocols start with mild purification steps to deplete larger particles without destroying them. Thus, the cell supernatant (SN) is collected and cleared by three consecutive centrifugation steps, i.e., 5 min at 300× *g*, 10 min at 3000× *g*, and 20 min at 10,000× *g* (Fig. 3). Most apoptotic blebs and other large EVs sediment at 3000× *g* or 10,000× *g*. Then, the EV-containing supernatants are concentrated using Amicon Ultra −15 centrifugal filters (Millipore) at 3000× *g* until a final volume of 500 μl is reached. Filters of regenerated cellulose and 50,000 Dalton molecular weight cutoff can be used. The resulting 500 μl with enriched EVs, which still contain contaminations, such as soluble molecules as well as large protein and vesicle aggregates, needs further purification. Size-exclusion chromatography is used to purify vesicles with a defined diameter. Different qEV columns are available from IZON (Lyon, France). The 35 nm and the 70 nm series are generally preferred for the purification of EV samples. While the 35 nm columns have an optimal recovery of EVs in the diameter range of 35–400 nm, the 70 nm series is optimal between 70 and 2000 nm. Before sample loading, the columns are washed with particle-free PBS. 0.2 μm filters remove most particles from the buffer. After complete loading of the 0.5 ml sample, 3 ml of PBS is applied in 0.5 ml steps, and the flow-through is discarded (void volume). Then, 2.5 ml in 0.5 ml aliquots is applied, and the flow-through is collected and pooled. The EV-containing flow-through fractions are pooled (5 × 500 μl) and concentrated by ultracentrifugation for 2 h at 100,000× *g* and finally suspended in 50 μl of particle-free PBS. EVs in this preparation can be identified, characterized, and validated by nanoparticle tracking analysis (NTA) and electron microscopy, quantified by NanoDrop and bicinchoninic acid (BCA) protein assays, and used for mass-spectrometry and biological tests (Fig. 3b).

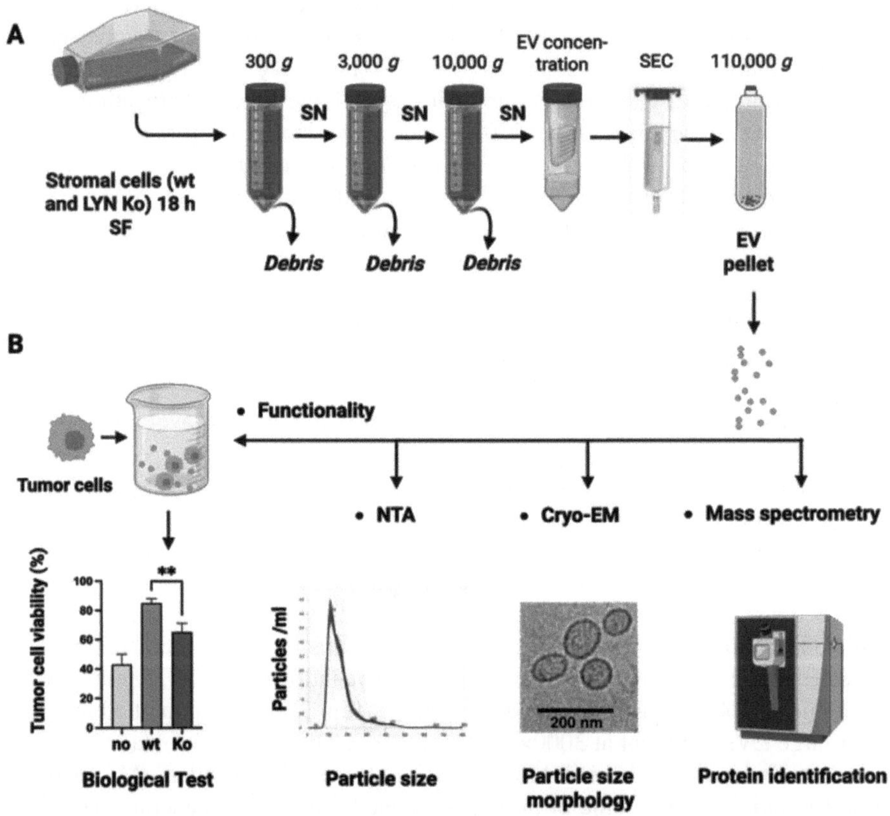

Fig. 3 Purification and evaluation of EVs. (**a**) As an example, the enrichment and evaluation of EVs from stromal cell supernatants are shown. To enrich small EVs from EV-containing material, the samples are typically purified in three consecutive centrifugation steps with increasing acceleration of gravity, while saving the supernatant for the subsequent centrifugation step. Then, the supernatant is concentrated in Amicon Ultra −15 centrifugal filters and applied to size-exclusion chromatography using qEVoriginal, 35 nm columns. The EV-containing fractions are then concentrated by ultracentrifugation. (**b**) The purified EV preparations are evaluated and processed by NTA, protein concentration, mass spectrometry, and biological activity. To determine the biologic activity, the supportive effect of EVs on CLL cells is tested with samples from nine different donors by co-incubating them with EVs. Then, the viability of untreated and EV-treated primary CLL cells is determined

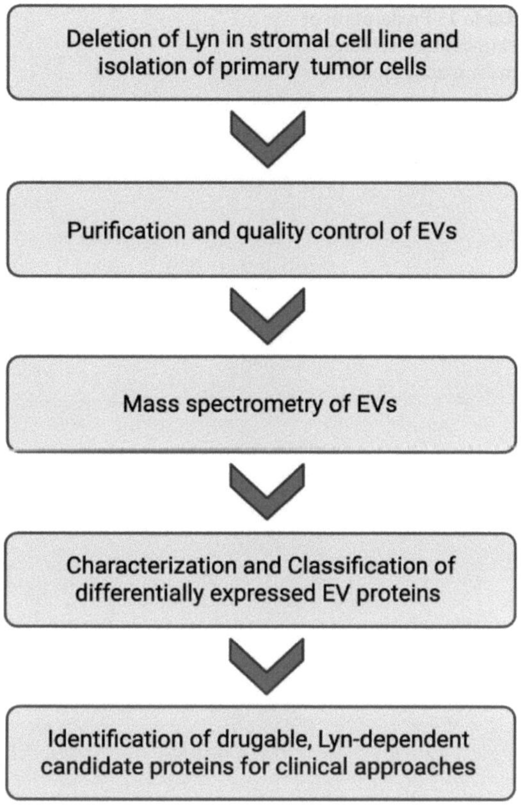

Fig. 4 The principal steps in the purification of EVs from Lyn-proficient and deficient stromal cells for mass spectrometry and subsequent biological validation of detected Lyn-dependent downstream proteins for the suitability in clinical approaches

4 Mass Spectrometry of EVs

The above enriched EV suspension was used for mass spectrometry. Therefore, 2 μg aliquots were lysed in 10 μl of SP3 buffer (5% SDS in PBS) (Table 1). Lysis was performed by thoroughly pipetting the EV suspension to disrupt the vesicle membranes and release their protein contents. To enhance lysis and ensure a complete denaturation of vesicular proteins, the samples were heated at 95 °C for 10 min. Although large chromatin contaminations are unlikely in EV samples, we degraded residual nucleic acids by sonication with a Bioruptor Plus sonication device (Diagenode) through 30/30-second cycles over 10 min. After confirming proper viscosity, the protein concentration was measured and adjusted to 0.5–2 μg/μl. Then, proteins were reduced with dithiothreitol (DTT) to a final concentration of 5 mM. The samples were vortexed and incubated at 55 °C for 30 min. Afterward, alkylation was performed by adding chloroacetamide (CAA) to a final concentration of 40 mM, followed by incubation in the dark at room temperature for 30 min to prevent reformation of disulfide bonds. Then, trypsin was added at an enzyme/substrate ratio of 1:75 and incubated at 25 °C overnight before formic acid was added to a final concentration of 1% to stop the enzyme activity. Samples were

Table 1 Preparation of extracellular vehicles for mass spectrometry

EV sample preparation
– EVs (3 μg) were lysed in 5% SDS/PBS (SP3)
– 10 min 95 °C
– Sonication
– Protein conc. adjusted to 0.5–2 μg/μl
+ DTT ad 10 mM and 30 min incubation at 55 °C
+ Chloroacetamid (CAA) ad 40 mM, 30 min incubation at RT
+ Trypsin to 1:75 enzyme/protein ratio and incubated for 4 h at 37 °C
+ 1% formic acid
– Centrifugation: 5 min at 11000 g
– Supernatant was placed on preconditioned SDB-RP-loaded stage tip
– Centrifugation: 5 min at 800 g
– Loaded stage tips were washed with
– 0.1% formic acid
– 0.1% formic acid + 80% acetonitril (twice)
– Stage tips were dried and stored at 4 °C
Before mass spectrometry:
– Sample reconstitution in 30 μl of 1% NH_4OH/60% acetonitril for 30 min
– Buffer depletion in SpeedVac
0.1% formic acid

centrifuged at 11,000× *g* for 5 min, and the supernatant was loaded onto a preconditioned SDB-RP-loaded Stage Tip (Attract SPE, Affisep). Before application of the samples, the Stage Tips were consecutively equilibrated with 20 μl 100% methanol, 20 μl 0.1% formic acid in 80% acetonitrile, and twice with 20 μl 0.1% formic acid in water. Between the conditioning steps, the reagents were removed from the Stage Tips by centrifugation for 1 min at 800× *g*. Then, the formic-acid-containing samples were loaded on preconditioned Stage Tips and centrifuged for 5 min at 800× *g*. The loaded Stage Tips were washed three times, first with 0.1% formic acid, then twice with 0.1% formic acid in 80% acetonitrile. Afterward, the stage tips were dried and stored at 4 °C. Before mass spectrometry, the dried stage tips were treated for 15 min with 30 μl 1% NH_4OH in 60% acetonitrile to dissolve the attached peptides. To concentrate the peptides and remove the NH_4OH, the samples were dried in a SpeedVac. Then, the peptides were dissolved in 20 μl of a solution of standardization peptides in 2% acetonitrile and 5% formic acid.

4.1 Mass Spectrometry

All samples were analyzed by the CECAD Proteomics Facility on a Q Exactive Plus Orbitrap (Thermo Scientific) mass spectrometer that was coupled to an EASY nLC (Thermo Scientific, Waltham, MA, USA). Peptides were loaded with solvent A (0.1% formic acid in water) onto an in-house packed analytical column (50 cm −75 μm I.D., filled with 2.7 μm Poroshell 120 EC-C18, Agilent). The peptides were chromatographically separated at a constant flow rate of 250 nl/minutes using the following gradient: 3–5% solvent B (0.1% formic acid in 80% acetonitrile) within 1.0 min, 5–30% solvent B within 121.0 min, 30 –40% solvent B within 19.0 min, 40–95% solvent B within 1.0 min followed by washing and column equilibration. The mass spectrometer was operated in data-dependent acquisition mode. The MS1 survey scan was acquired from 300 to 1750 m/z at a resolution of 70,000. The top 10 most abundant peptides were isolated within a 1.8 Th window and subjected to higher-energy collisional dissociation (HCD) fragmentation at a normalized collision energy of 27%. The automatic gain control (AGC) target was set to 5e5 charges, allowing a maximum injection time of 55 ms. Product ions were detected in the Orbitrap at a resolution of 17,500. Precursors were dynamically excluded for 30.0 s.

All mass spectrometric raw data were processed with Maxquant (version 1.5.3.8) using default parameters [39]. Briefly, MS2 spectra were searched against a canonical human Uniprot fasta file (UP000005640, downloaded at: 26.08.2020) database, including a list of common contaminants. False discovery rates (FDR) on protein and peptide spectrum match (PSM) level were estimated by the target-decoy approach to 1% (Protein FDR) and 1% (PSM FDR), respectively. The minimal peptide length was set to seven amino acids, and carbamido-methylation at cysteine residues was considered as a fixed modification. Oxidation (M) and Acetyl (Protein N-term) were included as variable modifications for whole proteome searches. The match-between runs option was enabled. Initial data filter was done with Perseus (version 1.6.1.1) [39].

5 Characterization and Classification of Differentially Expressed EV Proteins from Wild-Type and Lyn-Deficient Stromal Cells

EVs from Lyn-proficient stromal cells support tumor cells significantly more than EVs from Lyn-deficient ones. EV proteins that are differentially expressed might be responsible for the support. Therefore, using this study as an example, we demonstrate how to identify such proteins in EVs by means of mass spectrometry. Then, the main aim is to identify one or more candidate proteins that are differentially expressed and might function as tumor-driving, downstream effectors of Lyn in EVs. To reach this aim, the particular raw files are assigned to two respective groups. Differentially expressed proteins are identified under the following conditions. A student *t*-test with S0 = 0.2 and a permutation-based FDR of 0.05 with 250 randomizations was performed to obtain differentially regulated proteins between the two groups. The resulting proteins are identified and visualized using Instant Clue software [40]. For the identification of proteins, protein groups or families that are principally influenced by cellular Lyn, it is useful to initially evaluate hits with the largest negative log student *t*-test p-value, like the largest 100. Such Lyn-sensitive hits, up- or downregulated, can be applied separately to gene ontology (GO) evaluation. This is for receiving an overview of Lyn-dependent proteins or protein groups released in EVs. If necessary and possible, the number can be extended to >100 or to all FDR significant hits.

Enrichment analyses are conveniently performed with Cytoscape software. Cytoscape is an open-source software to visualize molecular interactions and test different gene ontology groups, e.g., cellular component, molecular function, reactome pathways, and KEGG pathways. Stromal cells generally release soluble ECM proteins, and the EVs thereof also transport such proteins. However, there is a difference between EVs from Lyn-proficient and Lyn-deficient cells because Lyn depletion in cells causes a depletion in ECM proteins and ECM anchoring proteins in EVs, most of which clustered in extracellular matrix–associated GO terms such as extracellular matrix and extracellular matrix organization [31]. Many ECM proteins including elastin microfibril interfacer 1/2, collagens, fibronectin, and laminins as well as ECM anchoring proteins, like integrins, heparan sulfate proteoglycan, and CD248, are dependent on cellular Lyn. In particular, these Lyn-dependent protein hits are candidates for targeted cancer therapy. An exception might be the matricellular proteins (MCPs) thrombospondin 1 (THBS1). It is not depleted in Lyn-deficient cells and EVs thereof [31, 33]. THBS1 is a glycoprotein that binds structural proteins like laminin, fibronectin, and collagens as well as proteoglycans and receptors like integrins, CD36, and CD47 [41]. The functionality of this molecule in the interaction with primary CLL cells was recently studied in detail [33]. CD47 was found on primary CLL cells and CLL cell lines [42]. In stromal/CLL cell interaction, THBS1 binds to CD47 on CLL cells and significantly reduces the CLL cell viability [33]. This is an example of a reverse effect because Lyn depletion causes an upregulation of a defined EV proteins that inhibits the tumor support.

6 Identification of Drugable, Lyn-Dependent Candidate Proteins for Clinical Approaches

Among the Lyn-dependent proteins in EVs from stromal cells, CD248 is an example of a possibly clinically relevant hit. It is downregulated in EVs from Lyn-deficient cells, and both, Lyn-deficient and CD248-deficient cells show significantly reduced tumor cell support as compared to wild-type stromal cells [31]. Its expression is limited in healthy adult tissue, but it is typically expressed in activated stromal cells, particularly in CAFs and in EVs thereof [43]. In cells, CD248 is expressed as a transmembrane protein and binds structural proteins of the ECM such as collagen I, collagen IV, and fibronectin as well as the tumor cell protein galectin 3 binding protein (LGAL3BP or MAC-2 BP) [10]. All structural proteins are reduced in EVs from cells with defective Lyn suggesting a correlation between Lyn and the above molecules. In addition, by binding to LGAL3BP on tumor cells, CD248 mediates adhesion and migration, directly or indirectly [44, 45]. Therefore, on the one hand, it might stabilize the ECM as an ECM anchor or cross-linker and, on the other hand, stimulate the tumor cells directly [45, 46]. Recent data suggest that CD248 not only functions by binding to ECM molecules but also stimulates another anchor (integrin β1, ITGB1) to activate the ECM binding capacity of this integrin [9, 47]. The tumor-supporting mechanism of CD248 in CLL is not completely unveiled. However, it is interacting with typical CLL-supporting ECM proteins, able to activate the ECM anchor integrin β1 and stimulate tumor cells directly. Therefore, it might both coordinate and stimulate directly the support of CLL cells.

7 Conclusion

Lyn is an important driver for the progress in CLL. However, no specific inhibitors are available for the clinical use. The overall aim of the project is to identify downstream effectors of Lyn in EVs by mass spectrometry and test them as possible targets for the treatment of CLL-affected patients. Here, we describe the methods to generate and purify EVs for mass spectrometry and biological tests. We also show how to identify downstream proteins of Lyn by proteomics and select promising hit for further preclinical and clinical testing.

Acknowledgments This research was supported by grants from the German Research Council (DFG) to HH (HA2432/5-2) and the large instrument grant INST 216/1163-1 FUGG (DFG Großgeräteantrag).

Software GraphPad Prism version 10.4.2, Cytoscape version 3.10.3, Instant Clue version 0.12.1, and BioRender Software (Toronto, Canada).

References

1. Hallek M, Shanafelt TD, Eichhorst B (2018) Chronic lymphocytic leukaemia. Lancet 391:1524–1537
2. Kurtova AV et al (2009) Diverse marrow stromal cells protect CLL cells from spontaneous and drug-induced apoptosis: development of a reliable and reproducible system to assess stromal cell adhesion-mediated drug resistance. Blood 114:4441–4450
3. Burger JA et al (2000) Blood-derived nurse-like cells protect chronic lymphocytic leukemia B cells from spontaneous apoptosis through stromal cell-derived factor-1. Blood 96:2655–2663

4. Panayiotidis P, Jones D, Ganeshaguru K, Foroni L, Hoffbrand AV (1996) Human bone marrow stromal cells prevent apoptosis and support the survival of chronic lymphocytic leukaemia cells in vitro. Br J Haematol 92:97–103
5. Dumontet E, Mancini SJC, Tarte K (2021) Bone marrow lymphoid niche adaptation to mature B cell neoplasms. Front Immunol 12:784691
6. Sarkar M et al (2023) Cancer-associated fibroblasts: the chief architect in the tumor microenvironment. Front Cell Dev Biol 11:1089068
7. Sivina M et al (2012) Stromal cells modulate TCL1 expression, interacting AP-1 components and TCL1-targeting micro-RNAs in chronic lymphocytic leukemia. Leukemia 26:1812–1820
8. Yee EJ et al (2024) Group XIV C-type lectins: emerging targets in tumor angiogenesis. Angiogenesis 27:173–192
9. Lu S et al (2024) Endosialin in cancer: expression patterns, mechanistic insights, and therapeutic approaches. Theranostics 14:379–391
10. Khan KA, McMurray JL, Mohammed F, Bicknell R (2019) C-type lectin domain group 14 proteins in vascular biology, cancer and inflammation. FEBS J 286:3299–3332
11. O'Shannessy DJ et al (2014) Influence of tumor microenvironment on prognosis in colorectal cancer: tissue architecture-dependent signature of endosialin (TEM-1) and associated proteins. Oncotarget 5:3983–3995
12. Dubois N et al (2020) Importance of crosstalk between chronic lymphocytic leukemia cells and the stromal microenvironment: direct contact, soluble factors, and extracellular vesicles. Front Oncol 10:1422
13. Welsh JA et al (2024) Minimal information for studies of extracellular vesicles (MISEV2023): from basic to advanced approaches. J Extracell Vesicles 13:e12404
14. Colombo M, Raposo G, Thery C (2014) Biogenesis, secretion, and intercellular interactions of exosomes and other extracellular vesicles. Annu Rev Cell Dev Biol 30:255–289
15. Kalluri R, McAndrews KM (2023) The role of extracellular vesicles in cancer. Cell 186:1610–1626
16. Thery C et al (2018) Minimal information for studies of extracellular vesicles 2018 (MISEV2018): a position statement of the International Society for Extracellular Vesicles and update of the MISEV2014 guidelines. J Extracell Vesicles 7:1535750
17. Shahbazi R, Kalishwaralal K, Paul MK, Anto RJ (2023) Editorial: role of extracellular vesicles (EVs) in pathogenesis, diagnosis, therapeutic delivery, treatment and theranostic applications in cancer. Front Bioeng Biotechnol 11:1288806
18. Toth EA et al (2021) Formation of a protein corona on the surface of extracellular vesicles in blood plasma. J Extracell Vesicles 10:e12140
19. Liam-Or R et al (2024) Cellular uptake and in vivo distribution of mesenchymal-stem-cell-derived extracellular vesicles are protein corona dependent. Nat Nanotechnol 19:846–855
20. Crompot E et al (2017) Extracellular vesicles of bone marrow stromal cells rescue chronic lymphocytic leukemia B cells from apoptosis, enhance their migration and induce gene expression modifications. Haematologica 102:1594–1604
21. Huleihel L et al (2016) Matrix-bound nanovesicles within ECM bioscaffolds. Sci Adv 2:e1600502
22. Lewin S, Hunt S, Lambert DW (2020) Extracellular vesicles and the extracellular matrix: a new paradigm or old news? Biochem Soc Trans 48:2335–2345
23. Ten Hacken E, Burger JA (2016) Microenvironment interactions and B-cell receptor signaling in chronic lymphocytic leukemia: implications for disease pathogenesis and treatment. Biochim Biophys Acta 1863:401–413
24. Robinson DR, Wu YM, Lin SF (2000) The protein tyrosine kinase family of the human genome. Oncogene 19:5548–5557
25. Nguyen PH et al (2016) LYN kinase in the tumor microenvironment is essential for the progression of chronic lymphocytic leukemia. Cancer Cell 30:610–622
26. Nguyen PH, Niesen E, Hallek M (2019) New roles for B cell receptor associated kinases: when the B cell is not the target. Leukemia 33:576–587

27. Fattet L et al (2020) Matrix rigidity controls epithelial-mesenchymal plasticity and tumor metastasis via a mechanoresponsive EPHA2/LYN complex. Dev Cell 54:302–316.e7
28. Pham H et al (2016) Essential role of Lyn in fibrosis. Front Physiol 7:387
29. de Jesus AA et al (2023) Constitutively active Lyn kinase causes a cutaneous small vessel vasculitis and liver fibrosis syndrome. Nat Commun 14:1502
30. Sharma V, Letson J, Furuta S (2022) Fibrous stroma: driver and passenger in cancer development. Sci Signal 15:eabg3449
31. de Oliveira TD et al (2023) Stromal cells support the survival of human primary chronic lymphocytic leukemia (CLL) cells through Lyn-driven extracellular vesicles. Front Med (Lausanne) 9
32. Vom Stein AF, Hallek M, Nguyen PH (2024) Role of the tumor microenvironment in CLL pathogenesis. Semin Hematol 61:142–154
33. Vom Stein AF et al (2023) LYN kinase programs stromal fibroblasts to facilitate leukemic survival via regulation of c-JUN and THBS1. Nat Commun 14:1330
34. Amrein PC et al (2011) Phase II study of dasatinib in relapsed or refractory chronic lymphocytic leukemia. Clin Cancer Res 17:2977–2986
35. Koehrer S, Burger JA (2024) Chronic lymphocytic leukemia: disease biology. Acta Haematol 147:8–21
36. Adamo A et al (2020) HS-5 and HS-27A stromal cell lines to study bone marrow mesenchymal stromal cell-mediated support to cancer development. Front Cell Dev Biol 8:584232
37. Eichhorst B, Hallek M (2016) Prognostication of chronic lymphocytic leukemia in the era of new agents. Hematol Am Soc Hematol Educ Program 2016:149–155
38. Jeppesen DK, Zhang Q, Franklin JL, Coffey RJ (2023) Extracellular vesicles and nanoparticles: emerging complexities. Trends Cell Biol 33:667–681
39. Tyanova S et al (2016) The Perseus computational platform for comprehensive analysis of (prote)omics data. Nat Methods 13:731–740
40. Nolte H, MacVicar TD, Tellkamp F, Kruger M (2018) Instant clue: a software suite for interactive data visualization and analysis. Sci Rep 8:12648
41. Tabary M, Gheware A, Penaloza HF, Lee JS (2022) The matricellular protein thrombospondin-1 in lung inflammation and injury. Am J Physiol Cell Physiol 323:C857–C865
42. Paggetti J et al (2015) Exosomes released by chronic lymphocytic leukemia cells induce the transition of stromal cells into cancer-associated fibroblasts. Blood 126:1106–1117
43. Kim HS et al (2012) Proteomic analysis of microvesicles derived from human mesenchymal stem cells. J Proteome Res 11:839–849
44. Tomkowicz B et al (2007) Interaction of endosialin/TEM1 with extracellular matrix proteins mediates cell adhesion and migration. Proc Natl Acad Sci USA 104:17965–17970
45. Zhang L et al (2024) CD248 interacts with ECM to promote hypertrophic scar formation and development. Gene 927:148730
46. Di Benedetto P et al (2019) Linking myofibroblast generation and microvascular alteration: the role of CD248 from pathogenesis to therapeutic target (review). Mol Med Rep 20:1488–1498
47. Lu T et al (2023) Endosialin-positive tumor-derived pericytes promote tumor progression through impeding the infiltration of CD8(+) T cells in clear cell renal cell carcinoma. Cancer Immunol Immunother 72:1739–1750

Proteomic Sample Preparation for the Petroleum Industry: A Biocorrosion Case Study

Icoquih Zapata-Peñasco and Jorge Herrera-Díaz

Abstract Petroleum-associated environments are among the most chemically complex and biologically extreme systems encountered in the field of industrial biotechnology. Here, microbial activity plays a pivotal role in hydrocarbon biodegradation, reservoir souring, and microbiologically influenced corrosion (MIC). In these systems, proteins constitute the functional interface between microbial metabolism and physicochemical processes affecting infrastructure integrity and environmental impact. This chapter presents an integrated proteomics-based workflow for the characterization of microbial communities inhabiting oil pipeline sludges, with particular emphasis on sample preparation strategies tailored to hydrocarbon-rich, metal-laden, and saline matrices. Optimized phenol-based extraction, electrochemical in vitro corrosion assays, two-dimensional gel electrophoresis, and high-resolution mass spectrometry are combined with metagenomic information to enable robust identification and functional interpretation of proteins involved in redox metabolism, biofilm formation, extracellular electron transfer, sulfur and nitrogen cycling, and stress adaptation. The approach is illustrated through a biocorrosion case study of marine pipeline sludge, revealing key enzymatic systems, including oxidoreductases, hydrolases, cytochromes, ABC transporters, and biofilm-associated structural proteins that mediate metal dissolution and microbial energy conservation. By integrating proteomics with electrochemical measurements and systems-level analysis, this chapter highlights how tailored sample preparation and functional protein profiling can overcome the limitations of culture-dependent methods, providing mechanistic insight into complex petroleum microbiomes. These advances establish proteomics as a critical tool for monitoring,

I. Zapata-Peñasco (✉)
Instituto Mexicano del Petróleo, Eje Central Lázaro Cárdenas, Ciudad de México, Mexico
e-mail: izapata@imp.mx

J. Herrera-Díaz
Facultad de Química, Universidad Nacional Autónoma de México,
Ciudad de México, Mexico

J.-L. Capelo-Martínez (ed.), *Emerging Sample Treatments in Proteomics*,
Advances in Experimental Medicine and Biology 1510,
https://doi.org/10.1007/978-3-032-21638-0_6

predicting, and ultimately mitigating biocorrosion, as well as for guiding the development of biotechnology-based strategies in the oil and gas industry.

Keywords Proteomics · Oil and gas industry · Biocorrosion · Microbiologically Influenced Corrosion (MIC) · Pipeline sludge · Metaproteomics · Hydrocarbon-degrading microorganisms · Redox enzymes · Biofilms · Mass spectrometry · Functional microbial ecology · Systems biology

1 Introduction

The most abundant hazardous organic pollutants from crude oil are hydrocarbon compounds, including the n-alkanes, branched aliphatics, monoaromatics, polyaromatics, and heterocyclic aromatics. They are released at all stages of oil life cycles including oil production, drilling, transportation, and storage, leading to their accumulation in nearby environments with infrastructures and dependent ecosystems [1]. Petroleum hydrocarbons are structurally complex, relatively hydrophobic, and prone to weathering effects; hence, remediation of these contaminants is a serious challenge.

Another important problem in the petroleum industry is corrosion, one of the causes of which constitute between 3% and 4% of a country's gross domestic product (GDP) [2]. Microbiologically influenced corrosion (MIC) accounts for 10–20% of the total corrosion costs, with microorganisms responsible for approximately 40% of internal corrosion in pipelines and 20–30% of external pipeline corrosion [3].

A large percentage of microbial diversity—80–99% of environmental microorganisms—are non-culturable in conventional laboratory environments [4], constituting an unexploited source of metabolic and genetic diversity. Culture-independent techniques have transformed the field of microbial ecology, allowing for the comprehensive characterization of culturable and unculturable microbiota in a variety of habitats. As microorganisms dynamically alter their protein expression with environmental stressors, proteomics analysis could overcome the limitations of metagenomics and metatranscriptomics, directly providing insights into functional metabolic pathways [5].

1.1 Proteomics in the Oil Industry: Groundbreaking Studies and Early Challenges

The basis of petroleum proteomics was established through preliminary environmental studies, particularly by the pioneering study of Venter et al. [6], which established that large-scale metagenomic and proteomic analyses would be informative. While not specific to petroleum, this work opened the way for engineering and industrial applications by uncovering the enormous diversity of microbial communities in natural systems. At the time, it was difficult to apply these techniques to the petroleum environment due to the complexity of oilfield samples and the technological limitations in those days [7].

One of the major takeaways from these early endeavors was an understanding of how microorganisms are involved and influence oil reservoirs, including biodegradation, reservoir souring, pore plugging, and microbial-enhanced oil recovery (MEOR) [5, 8–10]. To comprehend these microbial activities at a deeper level, insights into protein functions were needed, and these paved the way for petroleum proteomics.

1.2 Early Era: Technological Constraints

2D gel electrophoresis and mass spectrometry dominated protein analysis at the time [11–13]. However, the limited amount of microbial genomic data associated with petroleum sources hindered the integration of genomics and proteomics. Nevertheless, these preliminary studies established the path for present-day petroleum proteomics and demonstrated that proteins contained answers to some of the industry's most important biological questions.

1.3 Emergence of Metagenomics and Community Analysis

Improvements in DNA sequencing technology allowed researchers to investigate the genetic composition of oilfield microbiomes and offered a broader understanding of microbial diversity and function in subsurface ecosystems [14, 15]. This genomic framework was crucial, but only half the story; the real engines of microbial activity were proteins, the molecular machines that performed metabolic processes. From these findings, three major research themes took shape:

Microbial Proteomics. Scientists began to decipher the proteomes of extremophiles and hydrocarbon-degrading bacteria (e.g., *Geobacillus, Pseudomonas*), revealing how their enzymes and metabolic pathways enable them to succeed in extreme, oil-rich environments [12, 13, 16].

Biocorrosion Studies. The MIC of pipelines, equipment, and steel infrastructure, mainly by sulfate-reducing bacteria (SRB) such as *Desulfovibrio*, and their corrosive proteins and secreted materials have been characterized [17, 18]. Important discoveries were made about the molecules responsible for direct electron uptake from iron, involving outer membrane redox proteins such as c-type cytochromes in SRB [19].

Bioremediation Strategies. By identifying the main proteins involved in hydrocarbon degradation, scientists devised focused strategies to remediate oil-contaminated sites, from terrestrial spills to marine ecosystems [11, 13, 20].

1.4 Technological Breakthroughs and Applications in Petroleum Proteomics for Deeper Insights

The 2010s were a decade of great strides in proteomic technology, enabled and accompanied by advances in high-resolution mass spectrometry, bioinformatic tools, and integrative genomics, transcriptomics, and metabolomics. These advances gave us a more in-depth understanding of the within-systems interactions occurring in these petroleum-enriched environments [21–24]. Consequently, several major research areas are poised to greatly benefit from this technology, including metaproteomics, enzyme discovery, stress response characterization, and biomarker development.

Metaproteomics has emerged as a valuable strategy to investigate the collective proteome of microbial communities in oil reservoirs and relies on the knowledge of metabolic pathways used in the degradation of hydrocarbons [25]. Meanwhile, similar endeavors in enzyme discovery were directed at finding new biocatalysts such as hydrolases and oxidoreductases, which would be beneficial for biofuel generation as well as industrial biocatalysis [26]. Moreover, studying microbial stress responses has enabled understanding of the reasons behind protein expression changes under extreme conditions of reservoirs, which include high-temperature, high-pressure, and salinity conditions, providing invaluable information for the optimization of MEOR strategies [27–29].

Protein biomarkers for tracking microbial activity and prediction of oil reservoir performance are another important application of proteomics. A prominent example of enzymatic application in the energy field is Enzyme-Enhanced Oil Recovery (EEOR), which operates on the principle of using enzymes such as proteases, hydrolases, and dehydrogenases to enhance the efficiency of oil extraction. They alter reservoir wettability, lower interfacial tension, and catalyze heavy hydrocarbons breakdown to lighter fractions [30]. Combining proteomics with other omics disciplines has heightened the capacity to link protein expression to genomic and metabolic potential and has provided a more integrated view of microbial processes in petroleum systems [31]. The role of extracellular polymeric substances (EPSs) and biofilm-associated proteins in microbial adhesion, corrosion, and hydrocarbon degradation has also been investigated by means of a proteomic strategy [32].

The 2020s saw proteomics becoming fixed in place as an essential tool of the petroleum industry, fueled by developments in high-throughput mass spectrometry and advanced data analysis methodologies. The Deepwater Horizon (DWH) oil spill represented a watershed event in the form of the first wide-scale deployment of omics techniques for environmental monitoring. The data collected from such an event aided better models to use for spill response, improving tracking, and mitigation strategies in future oil contamination events [33].

The integration of proteomics with flux balance analysis and metabolic network reconstruction has further improved the predictive modeling of microbial behavior in several environments, such as the oil reservoirs with systems biology approaches [34]. In addition, through proteomics, it has contributed to the pursuit of a sustainable practice by enabling green innovations in oil production and remediation in coherence with global sustainability goals [35]. Likewise, the large-scale proteomic datasets can provide insights into significant trends in microbial responses to changing environmental conditions for better predictions on industrial and ecological utilization [36].

New tools, such as metagenomics techniques combined with stable isotope probing, single-cell sequencing, metaproteomics, and metabolomics, will yield insights into microbial ecophysiology of subsurface and oil ecosystems [37]. At the same time, in order to achieve efficient biocatalysts through proteins, synthetic biology plays a key role in the optimization of the assembly of selected functions into engineered microorganisms or consortia [26, 30, 37]. Extensive collaboration between academia and the fossil fuel industry is also crucial for developing and applying biotechnologies to address a wide range of operational problems. The importance of petroleum has a long history in microbiological study; it is evident that the topic continues to be vital to industry and the ecosystems. It will remain a working research area for decades to come [33, 37].

Proteomics in industrial environments, specifically the oil industry, offers critical insights into microbial activity, biodegradation processes, and biocorrosion mechanisms. However, sample preparation is uniquely challenging due to complex matrices such as hydrocarbons, oily sludges, water with hydrocarbons, heavy metals, and salts, which interfere with protein extraction and its analysis. Traditional proteomic workflows often fail to address these complexities and require tailored protocols to guarantee accurate detection and identification of proteins (Fig. 1).

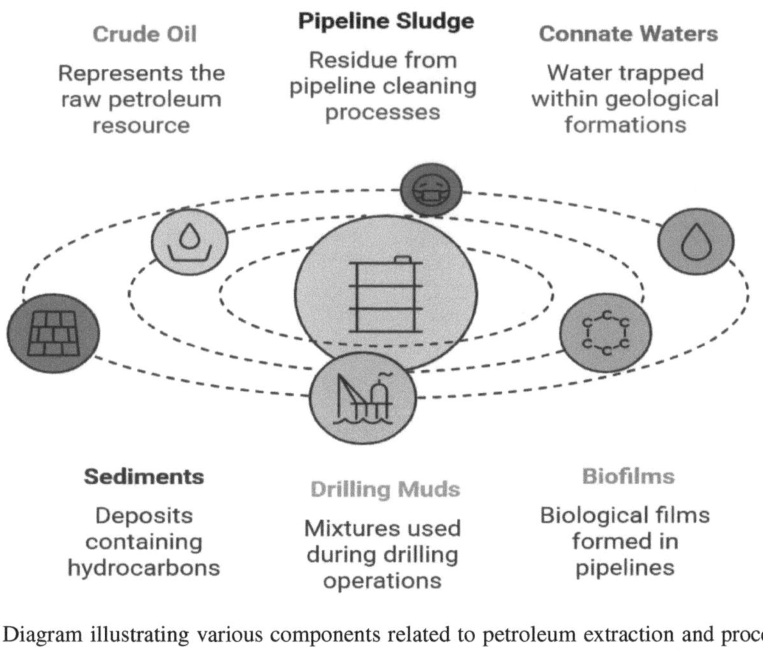

Fig. 1 Diagram illustrating various components related to petroleum extraction and processing. Central icon represents a storage tank, surrounded by six labeled elements: "Crude Oil" (raw petroleum resource), "Pipeline Sludge" (residue from cleaning processes), "Connate Waters" (water in geological formations), "Sediments" (hydrocarbon deposits), "Drilling Muds" (mixtures for drilling), and "Biofilms" (biological films in pipelines). Each element is depicted with a unique icon and color, connected by dashed lines, indicating their relationship to the central theme

This chapter outlines some optimized strategies for sample preparation in oil processing proteomics, focusing on robustness, reproducibility, and compatibility with downstream analytical techniques.

Herein, we present methods for rapid extraction of proteins from crude oil or other complex hydrocarbon-containing samples, including sludges from pipeline cleaning, connate waters containing hydrocarbons, and sediment with hydrocarbons. It thus allows for quicker methods of accessing native proteins from complex hydrocarbon sources that are applicable for downstream analysis in proteomic workflows, which gives results with less bias in sample preparation and helps drive the understanding of biological steps. The description of complex hydrocarbon samples relates to those entered the unconventional category: from sediments deposited within oil pipelines; from tanks for storage, production lines to distribute produced waters, connate water, and injection water from processes to recover oil; from waters carrying hydrocarbons, drill muds, and cuttings containing hydrocarbons; and from biofilms and biofouling deposits within pipelines in petroleum installations, resulting from oil-field exploitation, whether this is at shallow or deep levels (Fig. 2).

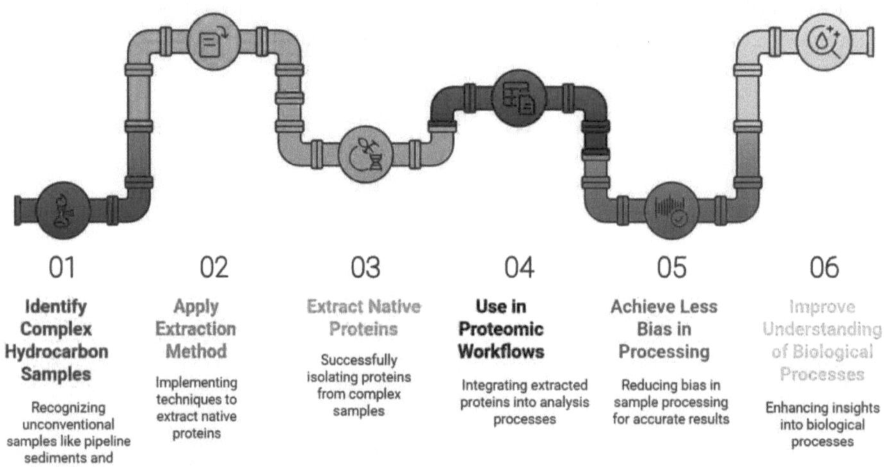

Fig. 2 Flow chart illustrating a six-step process for protein extraction and analysis. Step 1: Identify complex hydrocarbon samples, focusing on unconventional samples like pipeline sediments. Step 2: Apply extraction method, implementing techniques to extract native proteins. Step 3: Extract native proteins, successfully isolating them from complex samples. Step 4: Use in proteomic workflows, integrating extracted proteins into analysis processes. Step 5: Achieve less bias in processing, reducing bias for accurate results. Step 6: Improve understanding of biological processes, enhancing insights into biological processes. Each step is represented by a colored pipe segment with an icon

2 Sampling of Hydrocarbon Sludge

When evaluating biotechnologies within the engineering setting of the oil industry, many factors must be considered, including sample itself, matrix type, chemical composition, and location. Sample acquisitions are sometimes found to be more methodologically complex and expensive; they not only require special types of equipment to obtain the necessary amount of sample but also to have appropriate authorization to access oil industry facilities, which have very strict regulations and safety codes. In addition, the meteorological conditions are a significant factor of the sampling process. The flowchart of the methodology used, from sampling the inside of the pipeline on the offshore platform to the final proteomic analysis, is shown in Fig. 3.

The samples are taken after the process of internal cleaning of pipeline by means of instruments that run several kilometers inside the pipeline. In this case, the sludge (1.5 kg) was collected from a cleaning program of a X52 oil pipeline (24″ Ø × 5.388 km) in Marine Region-PEMEX in the Gulf of Mexico by applying the standard NACE RP0102 [38]. The oil pipeline operates using heavy crude oil (15.3°API) with 11% water and 0.9% sediment, evaluated in accordance with NACE RP0102 [38], ASTM D4007–11 [39], and ASTM D1928–12b [40]. Following the standard, the sulfur content is 3.5% by weight, and this makes the crude oil a sour one [41]. The sediment consisted of 48.5% inorganic matter, 33.5% organic matter, and 18% water as determined by the UOP 46–85 standard method [42]. It also measured 19.5% liquid hydrocarbons, 11.4% asphaltenes, and 2.6% paraffin [43]. The sludge was stored in a cryogenic container with N_2 before being transported to the laboratory for analysis.

3 Assessment of the Hydrocarbon Sludge Samples

3.1 Sample Preprocessing

The sampling preprocessing is determined depending on the source, matrix, and chemical composition of the sample. In this case, the sediment had a high content of hydrocarbons and organic compounds; it was therefore washed vigorously with three to five vortexing cycles of 5 min each and centrifuged at 4 °C/14,940× *g* with sterile synthetic seawater (35 g/L) prepared following ASTM D1141–98 [44, 45], with a final step of washing with protect buffer reagent, then immediately preserved the in vivo profile of DNA, RNA, and protein (RNAprotect® Cell Reagent. QIAGEN) by centrifugation of the cell mixture at 5000× *g* for 5 min.

There are two separate steps after the pretreatment carried out in parallel: the first step is protein extraction and purification of the extracted microbiota, and in the second step, this entire oil pipeline microbiota is inoculated into electrochemical corrosion rate tests using metal coupons made from the same type of material as the

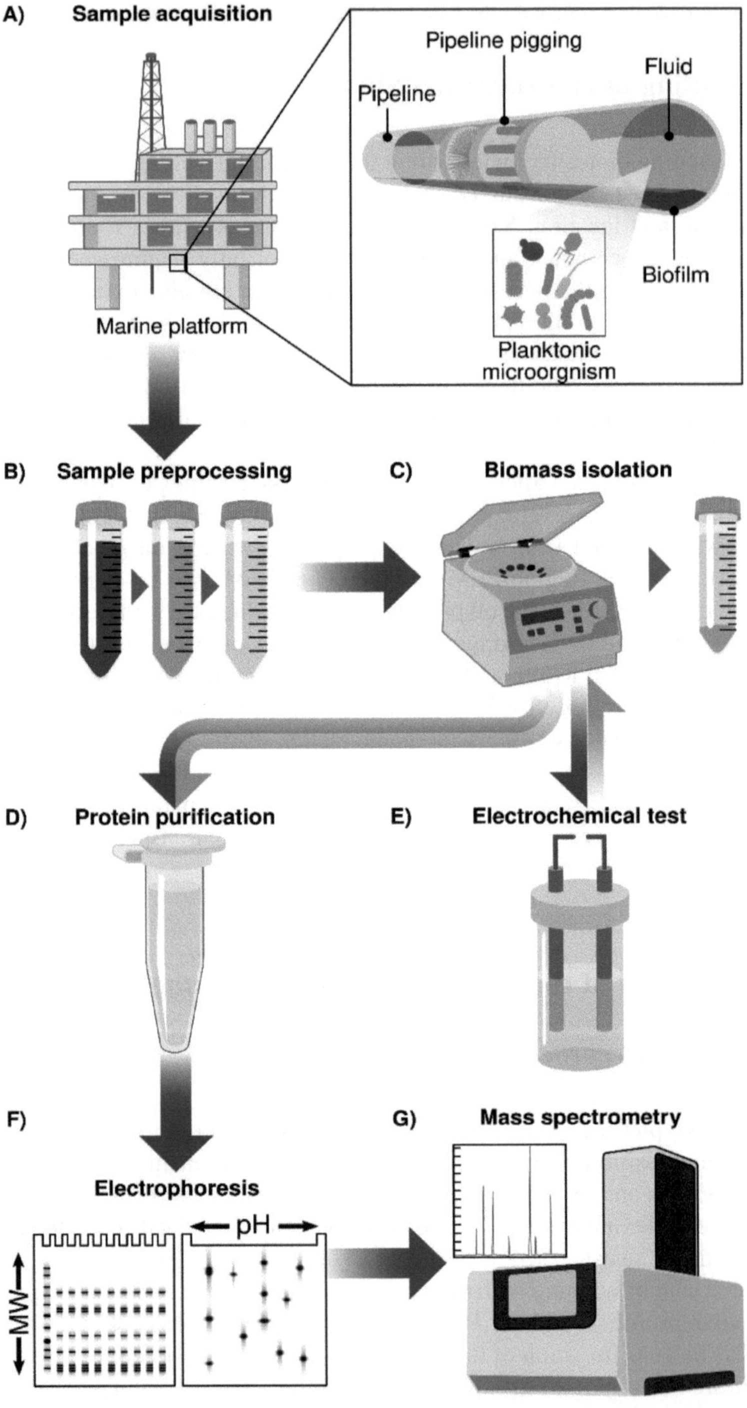

Fig. 3 Flow chart illustrating a scientific process. **a**) Sample acquisition from a marine platform, showing a pipeline with fluid, biofilm, and planktonic microorganisms. **b**) Sample preprocessing with test tubes. **c**) Biomass isolation using a centrifuge. **d**) Protein purification depicted with a filtration setup. **e**) Electrochemical test using an electrode setup. **f**) Electrophoresis showing separation by molecular weight and pH. **g**) Mass spectrometry with a machine displaying a spectrum. Arrows indicate the flow of the process

pipeline. Biomass from the electrochemical tests will be used for DNA, RNA, and protein extractions to study the corrosive behavior of the microbial communities inhabiting the pipeline derived from these electrochemical systems.

3.2 *In Vitro Electrochemical Assays*

All kinetic and electrochemical measurements were conducted in sterile environmental conditions in an appropriate medium M1: $MgSO_4$ (0.2 g/L), $CaCl_2$ (0.002 g/L), KH_2PO_4 (1.0 g/L), K_2HPO_4 (1.0 g/L), NH_4NO_3 (1.0 g/L), yeast extract (1.0 g/L), and artificial seawater (15 g/L) according to ASTM D1141–98 [44, 45] and were adjusted with a YSI-Multiparameter to get the optimal environmental settings for the microbiota. Electrochemical experiments in standard three-electrode cell with 450 mL of medium solution (Fig. 4) at 40 °C were performed following ASTM G3 and ASTM G5–14 [46, 47]. This temperature was selected because of the oil pipeline condition. 10^6 cells/mL of microorganisms were added into each electrochemical system. Silver/silver chloride (Ag/AgCl) reference electrode and an auxiliary electrode as a graphite bar were used. The working electrode was prepared using an API 5 L X52 steel coupon (exposure area = 4.85 cm^2), the same alloy as the oil pipeline, with the following composition (wt%): C (0.08), Mn

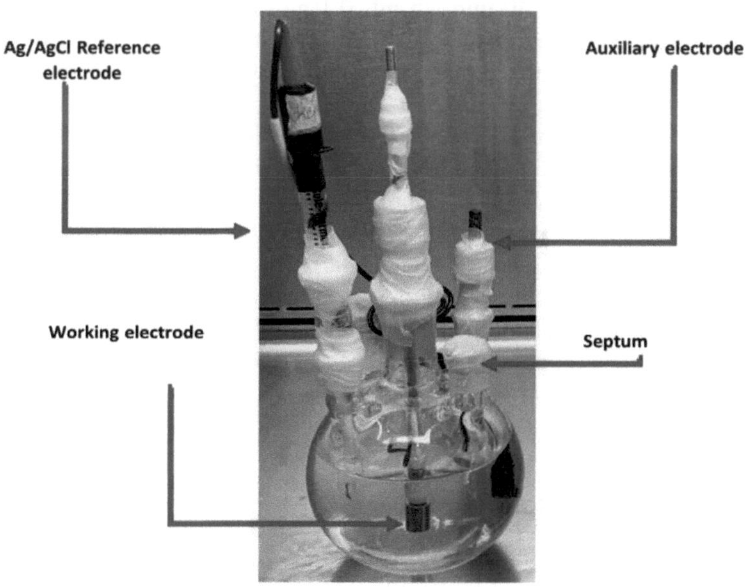

Fig. 4 Conventional electrochemical tests in a standard three-electrode cell with marine solution at room temperature, using an Ag/AgCl reference electrode, a graphite rod counter electrode, and a X52 steel working electrode (exposed area 1.32 cm^2). Microorganisms are inoculated with 10^6 cells/mL. Electrochemical measurements are taken hourly

(1.06), Si (0.26), P (0.019), S (0.003), Al (0.0041), Nb (0.0041), Cu (0.0018), Cr (0.002), Ni (0.0019), and Fe (Bal). Steel coupons were treated according to ASTM E3 [48] prior to the electrochemical tests. ATCC 51192 strain *Shewanella* (SW) was used as a microbial control to assess the deterioration and electrochemical process on X52 steel due to its proven biocorrosive role on metallic surfaces.

4 Protein Extraction from Hydrocarbon Sludge and Electrochemical Systems

Following the hydrocarbon sludge sample pretreatment (preserved at the final step of the preprocessing protocol using the protect buffer reagent RNAprotect® Cell Reagent, QIAGEN, which can be frozen at −20 °C), the protein extraction is performed according to the protocol reported below for the biomass from electrochemical systems.

Biomass collected from the electrochemical cell solution was centrifuged for 10 min at 5975 *g*, and the supernatant decanted, the pellet resuspended in 10 mL of buffer (50% [v/v] phenol pH 8.8% and 50% 0.1 mol/L ethylenediaminetetraacetic acid (EDTA), 0.4% [v/v] 2-betamercaptoethanol, and protease inhibitors (without EDTA) (Complete™; Roche Molecular Diagnostics). The samples were homogenized using an Ultra-turrax T-25 polytron, incubated on ice for 5 min, and then centrifuged at 5975 *g* for 40 min. 25 mL 0.1 mol/L ammonium acetate in 100% methanol was added in the organic phase and, after standing at −20 °C for 16 h, able to precipitate proteins for 1 week. The samples were centrifuged for 30 min at 5975 *g*, then washed twice in 5 mL of 0.1 mol/L ammonium acetate in 100% methanol, twice in 80% acetone (0.1 mol/L), and once in 70% ethanol, dried at room temperature, and finally dissolved in 400 µl of IEF (isoelectric focusing buffer) that was composed of 2 mol/L thiourea, 8 mol/L urea, 2% (v/v) Triton X-100, 0.05 mol/L dithiothreitol (DTT), 0.4% (w/v) CHAPS, and Complete™ protease inhibitors (without EDTA, Roche Molecular Diagnostic). This mixture was later centrifuged (20,913 *g*/15 min) to acquire crystalline supernatant. Final protein concentrations were assessed using a Nanodrop (Thermo Scientific), and the integrity of proteins was assessed using 12% one-dimensional SDS-PAGE for at least 16 h followed by colloidal Coomassie staining [49].

5 2D Gel Electrophoresis

After the Isoelectric focusing (IEF), the total protein samples (350 ug) from oil pipeline microbiota and biomass from electrochemical cells were applied to a 10 cm immobilized pH gradient polyacrylamide gel strip (IPG BIORAD) and rehydrated for 10 min at room temperature at a range of 3–10 on a rehydration tray (Immobiline DryStrip Reswelling Tray, BIORAD). The tray was subsequently inserted into a

PROTEAN i12 IEF isoelectric focusing unit (BioRad) and was incubated for 12 h at 20 °C to allow rehydration; this was accomplished using a concentric of five steps composed of 500 V for 30 min, 1000 V for 1 h, 3000 XV for 1 h, 5000× for 2 h, and finally 800 for 12 h and 8000 V for 3 h, for a total voltage of 30–35 kVh. After isoelectric focusing, the IPG strips were removed from the tray in and equilibrated in sodium dodecyl sulfate (SDS) equilibration buffer (1.5 mol/L Tris-HCl, 6 mol/L urea, 30% (v/v) glycerol, and 5% (w/v) SDS) containing 2% (w/v) dithiothreitol [DTT] for 15 min at room temperature on a shaker. This step was repeated twice. The samples were added with an SDS alkylation buffer (1.5 mol/L Tris-HCl pH 8.8, 6 mol/L urea, 30% (v/v) glycerol, 5% (w/v) SDS, and 2.5% iodoacetamide) and incubated for 15 min under stirring at room temperature. This was repeated twice, after which they were washed with 1x SDS-PAGE buffer. To accomplish second-dimension separation, washed strips were applied on to denaturing 12% acrylamide gels (SE 600 Ruby System, GE Healthcare) at 25 °C and separated at constant voltage of 50 V for 22 h. Gels were fixed for 30 min in 50% methanol and stained with Coomassie colloidal [20% (v/v) ethanol, 1.6% (v/v) phosphoric acid, 1.6% (v/v) ammonium sulfate, and 0.08% (w/v) Coomassie Brilliant Blue G-250] for ≥16 h. Two-dimensional gels were washed with double distilled water, scanned using ChemiDoc MP system and analyzed with PDQuest 2-D version 7.0 analysis software (BioRad) to detect, quantify, and join spots. The regions with reproducibly altered relative abundance were then chosen for subsequent mass spectrometric analysis [49].

6 Mass Spectrometry Identification of Proteins

The representative spots with obvious changes in electrophoresis were selected for identification by mass spectroscopy. Then they were washed with distilled water and incubated twice for 15 min in 100 mM ammonium bicarbonate. Reduction was performed with 50 mM dithiothreitol during 45 min. Alkylation was performed using 30 mM iodoacetamide for 2 h, samples were rehydrated with 100% acetonitrile and dried to completion. Protein digestion was performed with the addition of 30 μl porcine trypsin (20 ng/μl), followed by incubation for 18 h at 37 °C. Peptides were extracted with 50% [v/v] acetonitrile and 5% [v/v] formic acid by sonication and lyophilized. The samples were resuspended in 20 μl of 1% [v/v] formic acid and forward concentrated by Ziptip C14 and finally eluted with 12 μl of high-performance liquid chromatography (HPLC) mobile phase. Protein searches were performed in the protein sequence and functional information database UniProt, with trypsin specified as a protease. Peptide matches with PLGS (ProteinLynx Global Server™) scores >95% confidence were considered correct [49].

Taxonomic profiling based on metaproteomic datasets is tightly associated with a protein inference, which means assigning a given peptide to a respective protein. While peptide sequences can be unique to a single protein, from a practical point of view and particularly when dealing with short peptides, the peptide sequence can match several proteins having homologous sequence. A peptide uniqueness also has some relationship to length, where long peptides are more likely to track as unique, whereas short peptides tend to get too many hits across proteins to be useful [50]. In this regard, the taxonomic data obtained from the metagenomic analysis of the oil pipeline microbial community were used to perform a better peptide assignment (for details on the metagenomic analysis, see [31]).

7 Main Results of the Proteomic Analysis

During the biocorrosion process with the oil pipeline microbiota, proteomics revealed 32,284 proteins. On the other hand, 15 proteins in *Shewanella* were discovered (Table 1. Detailed list of proteins in [31]).

Proteins expressed during corrosion electrochemical behavior of sludge from the pipeline were oxidoreductase, lyases, nitrogenases, metalloprotease, and hydrolase associated with the biocorrosion process such as electron transport, sulfur oxidation, nitrate reduction, biofilm formation, and oxygen reduction. The effect of the

Table 1 Some proteins found in pipeline microbiota and *Shewanella* SW by mass spectrometry analysis

Accession	Description	mW (Da)	pI (pH)	PLGS score	Peptides
C2SNB4	Aconitate hydratase OS Bacillus cereus BDRD ST196	99,022	4.7622	2844.032	26
A0A5E9INQ10	Fe S cluster assembly protein SufB	52,490	4.9775	176.8422	4
A0A073KSG3	Enolase	46,213	4.5059	1015.493	10
A0A1S9T1S1	Aminodexychorismate	21,844	5.3291	519.5247	5
A0A0K0QFJ3	Glutamine synthetase	50,144	4.8691	1375.762	14
A0A243K4W9	Malonate semialdehyde dehydrogenase	52,890	5.1226	15618.13	23
SW					
Accession	Description	mW (Da)	pI (pH)	PLGS score	Peptides
A0A252ERF1	50S ribosomal protein	20,190	9.6533	1292.67	3
A0A2K9A364	Antibiotic biosynthesis monooxygenase	12,327	4.541	753.0068	1
A0A565DCL5	IncF plasmid conjugative transfer pilus assembly protein TraF	35,172	4.3828	707.8275	2
A0A2K8ZK77	ATP dependent Clp protease ATP binding subunit ClpX	46,402	5.1167	624.5455	6
A0A380C7A1	Redox sensitive transcriptional activator SoxR	17,551	9.3691	561.8464	11
A0A1S2TY19	Sulfite reductase NADPH hemoprotein beta component	63,563	6.1421	546.8323	9
A0A117V0U7	Basal body rod modification protein FlgD	9597	3.2651	982.0353	3

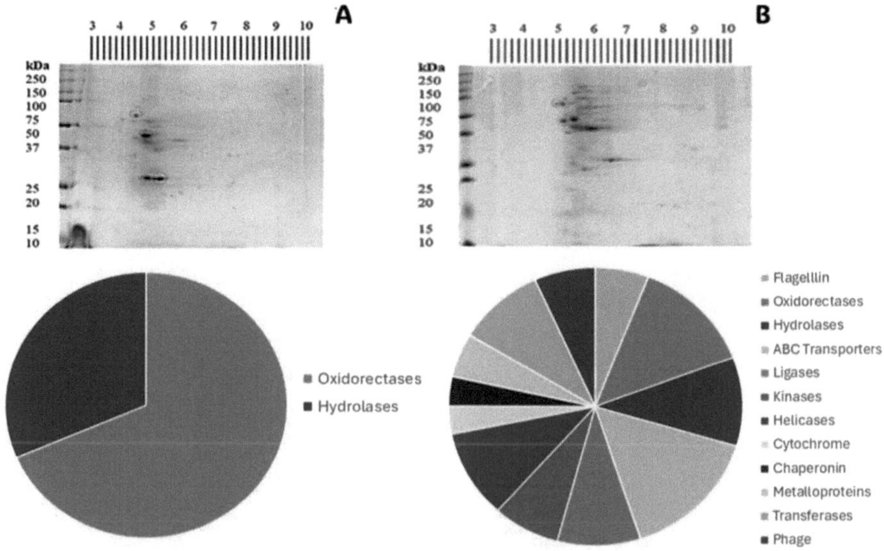

Fig. 5 2D protein profiling. Expression of proteins and some of the functional categories on each system during the biocorrosion process. (**a**) *Shewanella* SW; (**b**) oil pipeline microbiota

biofilm on the metal coupon during the electrochemical tests was assessed with the ATCC strain *Shewanella* and the entire microbiota inhabiting the oil pipeline (Fig. 5). In this last case, the protein profiles found in the original sludge sample did not differ in the type of proteins found in the electrochemical tests, but in terms of the abundance or concentration (for more details on the electrochemical profiles and results, see [31]).

In the biofilm (Fig. 6), extracellular polymeric substances (EPSs) are composed mainly of proteins, polysaccharides, and other small molecules, such as nucleic acids, humic acids, enzymes, and cytochromes. EPS consists of proteins, including heme-binding proteins, that serve as a unique matrix for mediated electron transfer between cells and solids. Electron transfer involving electrodes mediated by molecular structures can also be performed by humic-like substances. Under EPS, iron is oxidized into Fe(II) and loses an electron, which is transferred by mediators and transported across the cell membrane by electron transport chain [51].

Adhesion in cells takes place in non-specific physicochemical interactions, where appendages and proteins bind to specific chemical fractions on the surface through non-covalent interactions. Specifically, they use the generally attractive van der Waals forces, the repulsive electrostatic charges, or the dual behavior of acid–base interactions. Properties are influenced by the composition of the medium, temperature, pressure, pH, availability of oxygen and nutrients, and surface conditions. Certain interactions result in unique adhesion processes, as special appendages can adhere to chemical species on particular surfaces [52]. The junctions of coaggregation are very specialized, as the adhesins of the partner cell recognize receptors from only one cell type. For instance, the ability of a given bacterium to adhere to the surface of another bacterium is based on the presence of polysaccharides on the

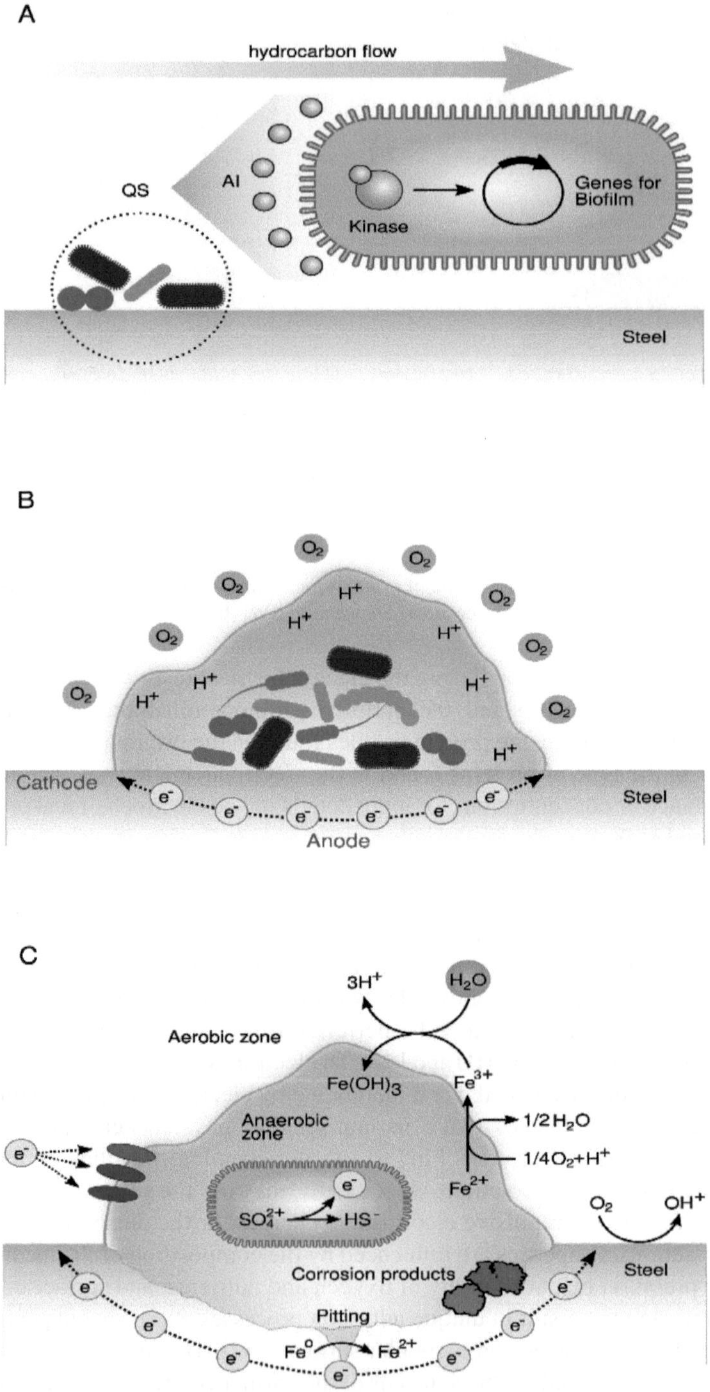

Fig. 6 Diagram showing the process of biofilm formation on the metal surface of the pipeline and the resulting process of biocorrosion and pitting. (**a**) Attachment of planktonic bacteria to the surface; (**b**) Electrochemical reaction on the pipeline surface: metal acts as the anode and the surrounding dissolved oxygen as the cathode; (**c**) Mature biofilm and complete biotic and abiotic assembly of a pipeline scale

surface of the latter, consisting of mannose, glucose, and galactose [52, 53]. However, surface mechanosensing describes the scenario in which a bacterium passively probes a surface upon contact with it. Flagellin as well as tension in the retraction of pili mediates adhesion. Signal transduction and the microorganism response are signature features of adhesion [53]. Adhesin (accession number A0A243KCZ4) and flagellin (A0A1B1L3Q0; A0A7M3FPJ7; A0A158RQU8; A0A1W6A6D7; A0A5K1KY26; A0A243KRE6) expressed during the electrochemical process of biocorrosion on metal were abundant proteins found in the biofilms of the oil sludge microbiota.

The biofilm forming process with select proteins identified in this study is shown in Fig. 6. It begins with the attachment of planktonic bacteria to the surface, cell–cell and cell–protein interactions leading to the elaborated three-dimensional architecture of growing colonies. Kinase also plays an important role in biofilm development, quorum sensing (QS), and bacteria virulence [54]. Such a protein was detected in this proteomics study during electrochemical monitoring of biocorrosion with the pipeline microbiota (some accession numbers: C2X5M9; A0A2B5YKH7; A0A511F234; A0A6I7XCB8; A0A090YXU0; B7KQX5; A0A090YCW3). Highly differentiated and columnar biofilm structures, comprising predominantly of mucoid exopolysaccharide, are bathed in water-filled channels. Intercellular communication through diffusible signaling molecules shapes the final three-dimensional architecture of mature biofilms. At a population level, bacteria utilize QS for signaling its cell density to coordinate the expression of certain genes. Homoserine lactones (AIs) are examples of these types of compounds [57].

The electrochemical reaction predominates on the pipeline wall, where metal serves as the anode and dissolved oxygen acts as the cathode, which is why the corresponding pit can be the first sign of corrosion [31, 51]. This pipeline stickiness has reversible microbial adherence and high mineral deposition [51]. Fine adherent scales deposited on a pipeline wall by chemogenic processes increase the surface roughness. Mineral precipitation is another robust phenomenon that leads to microbiota local concentration [31, 51], as well as chemocatalysis and microbial-mediated biocatalysis of several eukaryotes, such as fungi, microalgae, and amoebae. Pathways of microbial corrosion of nitrogen metabolism on the biofilm on the metal surface, which is then oxidized to nitrate, promote nitrate reproduction metabolism and lead to a bioavailable microenvironment; anaerobic iron reproduction metabolism can stabilize the scale on the pipeline [51]. While the pipeline is corroded by pitting, the electrochemical reactions between the metal surface and the surrounding fluid are still taking place (See all proteins reported during oil pipeline biocorrosion [31]). This is complex and multipath microbial corrosion behavior with mature biofilms, microbial metabolic guilds, and complete pipeline-scale assemblies. Both pipeline surface and scale components serve as substrates to support microbial metabolism, and pipelines are currently in the most severe corrosion phase and are likely to leak after hydrodynamic transformation or external failure [31, 51]. Accumulating sulfur promotes the sulfate-respiring bacteria that produce HS^-, which can interact with semiconducting minerals (FeSx) on the pipeline surface. Moreover, acid-producing bacteria (APB) grow rapidly as organic matter is biodegraded, generating acids and CO_2 [51].

Oxidoreductases were the other proteins significantly expressed in the oil pipeline microbiota during the biocorrosion process, as well as in *Shewanella* (A0A150AXT4; A0A1S1NYR2; A0A640MUM8; A0A1P8QP69; A0A1P8QQA1). Metals are most utilized as redox centers by oxidoreductases, particularly when dioxygen activation is necessary, with iron being the selective metal of choice. In redox catalysis, metals are often bound to organic cofactors (for example, heme for Fe, B12 for Co, pterin for Mo), whereas there is no such scenario in non-redox catalysis. Such cofactors can be used in order to tweak the redox potential of metals, revealing that biological systems take special care in dealing with processes that could lead to toxic redox-active species [55].

Another type of proteins expressed during the biocorrosion of the oil pipeline steel by the microbiota and the SW strain are the hydrolases (A0A243K649; A0A2B5X812; A0A2B6CSH5; A0A2A8U2N0; A0A090YBD4). These are enzymes that cleave chemical bonds by using water and are found in bacteria and fungi. Cell wall maturation involves the installation of the tailoring changes, the regulation of the cross-linking degree, trimming glycan strands to their final lengths after synthesis and remodeling during the growth phases. As cells grow, hydrolases allow the peptidoglycan polymer that surrounds the bacterium to stretch to allow for cell expansion. Hydrolases cleave peptidoglycan to assemble structures such as flagella, fimbriae, and secretion systems. For almost all cross-links in peptidoglycan, there is a hydrolase capable of breaking it. Based on their chemical specificity of cleavage, peptidoglycan hydrolases can broadly be classified into two groups: glycosidases and peptidases (C2NEP2; A0A2C1LJN1; A0A242WMV8; A0A2C3NV74; A0A2H2X3B3; A0A1J9US76; A0A2C2UTM0; A0A073KE65) [56]. There is no metal-dependent redox reaction for hydrolases, and the functions of the metal can be separated into substrate activation and electrostatic stabilization. For the case of metal-dependent hydrolytic reaction mechanisms, a water molecule is used as a substrate and two main stages of water activation take place: first, the formation of reactive hydroxyl species followed by the nucleophilic attack of this molecule on the substrate. Hydroxyl group formation requires metal ions and sometimes to increase the electrophilicity of the substrate [55].

The c-type cytochromes (C2R5U7; A0A2C5NTS2) were detected in the oil pipeline microbiota during electrochemical tests but not for *Shewanella*. The microbial c-type cytochromes in electron transport (reduction of sulfate to hydrogen sulfide in sulfate reduction, conversion of nitrogen to ammonia in nitrogen fixation, or conversion of nitrate to dinitrogen in denitrification) are terminal electron donors in sulfate-reducing bacteria. These taxa use sulfur or oxidized sulfur compounds as terminal electron acceptors in their respiratory chain. Among protein members in this class, one of the best-studied is based on *Desulfovibrio*, which serves as a natural electron acceptor and donor in hydrogenases and ferredoxins. These proteins are involved in electron transport to elemental sulfur and in energy-generating associated acetate oxidation and dissimilatory reduction of Fe (III) and Mn (IV) in these taxa [57]. Importantly, pipeline microbiota is also structured by sulfate-reducing bacteria carrying such proteins (for metagenomic analysis of oil pipeline community, see [31]).

ABC Transporters were revealed with proteomics in the oil pipeline microbiota during biocorrosion (some accession numbers: A0A063CMB1; A0A2C2A6A8; D8H3L5; A0A7V7V4P9; A0A120EE64; A0A1J9TAZ1; A0A2B1FJB0; A0A1D3PK85). ATP-binding cassette (ABC) transporters are present in all lifeforms. ABC exporters have two nucleotide-binding domains (NBDs) and two transmembrane domains (TMDs). The NBDs interact with and hydrolyze ATP, and the TMDs promote substrate translocation. ATP binding/hydrolysis induces conformational changes in the NBDs that are communicated to the TMDs by coupling helices. Although NBDs are highly conserved across all ABC transporters, TMDs differ quite extensively, which mirrors the diversity of ABC transporter substrates [58].

Since nitrogen metabolism on the biofilm on the metal surface led to a bioavailable corrosion environment, specific proteins involved in this metabolism pathway were detected in the pipeline microbiota, including nitrate reductase quinone (A0A2C3HWD3), nitrate reductase beta subunit (A0A2B4LUC1), and nitrite reductase (A0A2B4LVD0). Oilfield-associated proteins are linked to competition for organic electron donors between sulfate- and nitrate-reducing bacteria. Previous metatranscriptomic analyses show that expressions of genes encoding NAD proteins, ABC transporters, and cytochromes P450 are related to hydrocarbon degradation, sulfur metabolism, and metal and amino acid transport, suggesting nutritional and metabolic processes of microbial communities in oil-impacted sites [59].

8 Exploration of Proteomic Methodologies in Petroleum Industry Scenarios

The petroleum industry generates a diversity of complex samples requiring methodological adaptations. The methodological protocols employed depend on the physicochemical properties of the materials and matrices obtained from different operations and processes within the oil industry. Table 2 shows the challenges posed by environmental oil samples for the study and treatment of proteomics, based on our experience in different processes of the oil industry, both in the field and in the laboratory. To learn more about the complexities of samples handled in the petroleum industry and how they are approached in omics sciences laboratories, particularly in proteomics, we recommend the following references, which provide broader scope for correct methodological analysis [60–63].

The protocol outlined in this chapter is based on phenolic extraction, followed by ammonium acetate/methanol precipitation. This method is comparable to classical techniques employed in the analysis of soils and sediments [63, 64, 67–69]. However, it differs from other approaches. The table below (Table 3) shows the distinctions between various protocols, as well as their respective strengths and weaknesses.

Table 2 Samples in the petroleum industry and their challenges

Sample type	Origin	Main interferents	Typical proteomic target
Pipeline sludge	Internal pipeline cleaning	Heavy hydrocarbons, asphaltenes, paraffins, salts, metals, inorganic solids	Biocorrosion proteins, biofilm, biofouling, aerobic/anaerobic biodegradation proteins (Chapter's case)
Produced water	Wells, separators	Water-oil emulsions, high salinity, H S, trace metals, residual biocides	SRB/NRB proteins, sulfate/nitrate metabolisms
Biofilm from metal surfaces	Tanks, heat exchangers, pipe walls	EPS matrix (exopolysac charides), embedded cells, mineral salts	Adhesins, extracellular enzymes, electron transfer proteins
Contaminated soils/sediments	Spill sites, bioremediation	Non-petrogenic organic matter, clays, variable moisture	Hydrocarbon degradation enzymes, oxidative stress
Drilling fluids	Drilling muds	Barite, synthetic polymers, high viscosity, high microbial load	Extremophile proteins, adaptation to high pressure/temperature

Table 3 Different techniques for analysis of soils and sediments from oil industry

Protocol	Advantages	Disadvantages	Typical application	Reference
Phenolic extraction	High efficiency in lipid/hydrocarbon-rich samples, reduces polymeric interferents	Use of phenol (toxic), multiple steps, loss of hydrophilic proteins	Oily sludges, biofilms	[67, 68]
Direct SDS/Detergent extraction	Fast, good for membrane proteins, compatible with shotgun proteomics	Interference from hydrocarbons, difficulty removing detergents for MS	Produced water, pure cultures	[65, 66]
Sonication/Chemical lysis extraction	Low cost, scalable	Low efficiency in high-mineral samples, protein degradation.	Lightly contaminated sediments	[68, 69]
Pressure/Temperature-assisted extraction (e.g., SpeedMill)	High efficiency, reproducible, automatable	High cost, requires specialized equipment. The extraction of protein from actual oil reservoir cores is an emerging niche application related to the study of enhanced oil recovery and microbial activity	Reservoir core samples	[70, 71]
Supercritical fluid extraction (CO)	Efficiently removes hydrocarbons, green	Underdeveloped for proteomics, risk of denaturation	Preliminary research Biorefinery	[72, 73]

The following reasons were behind the choice of the phenolic protocol in this work:

1. The matrix complexity. Pipeline sludge contains asphaltenes and paraffins that interfere with aqueous extractions; phenol allows separation of organic (proteins) and aqueous (polar interferents) phases.
2. Compatibility with 2D electrophoresis. This protocol produces clean proteins for isoelectric focusing technique (IEF), minimizing residual ionic charges affecting the isoelectric point.
3. Prior experience with marine sediments. Similar methods have been used in proteomics of contaminated sediments, validating their efficacy in hydrocarbon-rich matrices [62–64].
4. Preservation of post-translational modifications. Phenolic extraction reduces the activity of proteases and phosphatases, which is crucial for studying redox modifications, that are relevant in biocorrosion.

This chapter introduces key improvements compared to generic protocols. These include the pre-wash with synthetic seawater (ASTM D1141-98), which has the advantage of removing salts and metals without lysing cells and preserving the native profile of extracellular and membrane proteins. Other protocols include washing with phosphate buffer solution in Milli-Q water, which can alter osmolarity and cause undesirable cell lysis. Using RNAprotect® before extraction stabilizes proteins, DNA, and RNA simultaneously, enabling multi-omic studies from the same sample. In classical environmental proteomics, samples are frozen directly at −80 °C, which can cause enzymatic degradation during thawing. Another important improvement is the parallel extraction of biomass from electrochemical assays, which allows the direct correlation of corrosive activity (electrochemically measured) with the protein profile. This is rare in biocorrosion studies, as these typically analyze static biofilms.

Taking a balanced view of the pros and cons of the current way of working revealed the following advantages and limitations.

Advantages

1. Robustness against interferents. Phenol efficiently removes hydrocarbons, pigments, and humic acids, which can inhibit tryptic digestion and MS ionization or mass spectrometry.
2. High protein purity. This is essential for shotgun proteomics and quantitative studies.
3. Integration with metagenomic data. Prior microbiome sequencing enables more accurate annotation and reduces the protein inference problem.
4. Real conditions can be simulated. Electrochemical assays at 40 °C replicate the temperature of the pipeline, capturing heat shock proteins.

Limitations

1. Moderate yield. Loss of low-abundance proteins (e.g., *quorum sensing* and transcriptional regulators).
2. Incompatibility with in vivo labelling. Phenol interferes with metabolic labeling (e.g., the use of 15 N stable amino acids as powerful tracers).
3. Limited scalability. Filter-based and in-solution digestion methods are faster for large sample numbers as they are highly labor-intensive.

In-depth reflection on the treatment of extreme environmental samples reveals that pipeline sediment is one of the most challenging matrices in environmental proteomics.

The following factors make processing petroleum industry samples a difficult process:

(a) Physical heterogeneity. A mixture of solid, liquid, and biological phases. The chapter's solution involves homogenization and vortex-centrifugation cycles.
(b) Massive MS inhibitors. Asphaltenes, sulfides and corrosive metals (Fe, Mn). In the current work, precipitation occurs with ammonium acetate in methanol, co-precipitating metals and sulfides.
(c) Low microbial biomass. Despite the high organic content, the cells are embedded in an inorganic matrix. The proposed strategy was to concentrate biomass from large volumes (1.5 kg sludge) by staged centrifugation.
(d) Recalcitrant proteins. Biofilm EPS and outer membrane proteins (C-type cytochromes) are difficult to extract. The solution was to use beta-mercaptoethanol and thiourea/urea in the rehydration buffer, as this denatures protein aggregates.

9 Conclusion

This chapter presents a specialized protocol for one of the most complex scenarios in industrial proteomics: oil pipeline sludge. Its value lies in its dual electrochemistry and proteomics validation and integration with metagenomics. The standards for biocorrosion diagnosis, monitoring, and control of microbial communities are based on electrochemical techniques and classical microbiological methods, but these neglect the vast universe of uncultivable microbes (bacteria, archaea, fungi, and viruses) that play a decisive role in the functions responsible for biofouling and corrosion processes. Furthermore, these conventional cultivation methods are inherently biased when assessing the efficacy of biocides in the oil production process. The effect of oxide layers and biofilms on the passivation (decreased electrical conductivity) of X52 steel has been observed, but the surface of the metal may still be susceptible to the action of metabolites, proteins, and extracellular electron transfer from these microbial formations, particularly during pipeline cleaning programs. *Shewanella* produces more biocorrosion products, which are important as standards for assessing corrosion rates in pipelines; however, in the services provided to the oil industry, the microbiota as a whole is not considered when the inner walls of gas and oil pipelines are monitored. This may underestimate the action of microbes. Proteome mapping enabled the description of a broad and thorough protein spectrum with diverse functions expressed during the temporal frame of the electrochemical process, which may provide reasons for the dissimilar biocorrosion mechanisms inside the oil pipelines. However, further investigation is needed to understand how the proteome expression evolves and space in the bioprocess. Finding relevant protein biomarkers to specifically detect corrosive bioprocess remains an achievable goal.

Acknowledgments The authors acknowledge financial support from the Mexican Petroleum Institute through Project D.62009.

References

1. Laczi K, Erdeiné Kis Á, Szilágyi Á, Bounedjoum N, Bodor A, Vincze GE, Kovács T, Rákhely G, Perei K (2020) New frontiers of anaerobic hydrocarbon biodegradation in the multi-omics era. Front Microbiol 11:590049. https://doi.org/10.3389/fmicb.2020.590049
2. Shi X, Zhang R, Sand W, Mathivanan K, Zhang Y, Wang N, Duan J, Hou B (2023) Comprehensive review on the use of biocides in microbiologically influenced corrosion. Microorganisms 11(9):2194. https://doi.org/10.3390/microorganisms11092194
3. Hashemi SJ, Bak N, Khan F, Hawboldt K, Lefsrud L, Wolodko J (2018) Bibliometric analysis of microbiologically influenced corrosion (MIC) of oil and gas engineering systems. Corrosion 74(4):468–486. https://doi.org/10.5006/2620
4. Mocali S, Benedetti A (2010) Exploring research frontiers in microbiology: the challenge of metagenomics in soil microbiology. Res Microbiol 161(6):497–505. https://doi.org/10.1016/j.resmic.2010.04.010
5. Pan H, Wattiez R, Gillan D (2024) Soil metaproteomics for microbial community profiling: methodologies and challenges. Curr Microbiol 81:257. https://doi.org/10.1007/s00284-024-03781-y
6. Venter JC, Remington K, Heidelberg JF, Halpern AL, Rusch D, Eisen JA, Wu D, Paulsen I, Nelson KE, Nelson W, Fouts DE, Levy S, Knap AH, Lomas MW, Nealson K, White O, Peterson J, Hoffman J, Parsons R, Baden-Tillson H, Pfannkoch C, Rogers YH, Smith HO (2004) Environmental genome shotgun sequencing of the Sargasso Sea. Science 304(5667):66–74. https://doi.org/10.1126/science.1093857
7. Bargiela R, Mapelli F, Rojo D, Chouaia B, Tornés J, Borin S, Kalogerakis N, Daffonchio D, Yakimov M, Golyshin PN (2015) Bacterial population and biodegradation potential in chronically crude oil-contaminated marine sediments are strongly linked to temperature. Sci Rep 5:11651. https://doi.org/10.1038/srep11651
8. Youssef N, Elshahed MS, McInerney MJ (2009) Microbial processes in oil fields: culprits, problems, and opportunities. Adv Appl Microbiol 66:141–251. https://doi.org/10.1016/S0065-2164(08)00806-X
9. Rabus R, Boll M, Heider J, Meckenstock RU, Buckel W, Einsle O, Ermler U, Golding BT, Gunsalus RP, Kroneck PM, Krüger M, Lueders T, Martins BM, Musat F, Richnow HH, Schink B, Seifert J, Szaleniec M, Treude T, Ullmann GM, Vogt C, von Bergen M, Wilkes H (2016) Anaerobic microbial degradation of hydrocarbons: from enzymatic reactions to the environment. J Mol Microbiol Biotechnol 26(1–3):5–28. https://doi.org/10.1159/000443997

10. Nikolova C, Gutierrez T (2020) Use of microorganisms in the recovery of oil from recalcitrant oil reservoirs: current state of knowledge, technological advances and future perspectives. Front Microbiol 10:2996. https://doi.org/10.3389/fmicb.2019.02996
11. Singh OV, Nagaraj NS (2006) Transcriptomics, proteomics and interactomics: unique approaches to track the insights of bioremediation. Brief Funct Genomic Proteomic 4(4):355–362. https://doi.org/10.1093/bfgp/eli006
12. Graham RL, Pollock CE, Ternan NG, McMullan G (2006) Top-down proteomic analysis of the soluble sub-proteome of the obligate thermophile, Geobacillus thermoleovorans T80: insights into its cellular processes. J Proteome Res 5(4):822–828. https://doi.org/10.1021/pr0504642
13. Sabirova JS, Ferrer M, Regenhardt D, Timmis KN, Golyshin PN (2006) Proteomic insights into metabolic adaptations in Alcanivorax borkumensis induced by alkane utilization. J Bacteriol 188(11):3763–3773. https://doi.org/10.1128/JB.00072-06
14. Joshi MN, Dhebar SV, Dhebar SV, Bhargava P, Pandit AS, Patel RP (2014) Metagenomic approach for understanding microbial population from petroleum muck. Genome Announc 2:e00533–e00514. https://doi.org/10.1128/genomeA.00533-14
15. Tan B, Fowler SJ, Abu Laban N, Dong X, Sensen CW, Foght J (2015) Comparative analysis of metagenomes from three methanogenic hydrocarbon-degrading enrichment cultures with 41 environmental samples. ISME J 9:2028–2045. https://doi.org/10.1038/ismej.2015.22
16. Li Y, Wu J, Wang W, Ding P, Feng L (2012) Proteomics analysis of aromatic catabolic pathways in thermophilic Geobacillus thermodenitrificans NG80-2. J Proteome 75(4):1201–1210. https://doi.org/10.1016/j.jprot.2011.10.035
17. Clark ME, He Z, Redding AM, Chen J, Hatzenpichler R, Chivian D, Simmons BA, Yelton AP, Thorgersen MP, Verberkmoes NC, Mouser PJ, Shah M, Zhou J, Arkin AP, Lovley DR (2012) Transcriptomic and proteomic analyses of Desulfovibrio vulgaris biofilms: carbon and energy flow contribute to the distinct biofilm growth state. BMC Genomics 13:138. https://doi.org/10.1186/1471-2164-13-138
18. Chatterjee M, Fan Y, Cao F, Jones AA, Pilloni G, Zhang X (2021) Proteomic study of Desulfovibrio ferrophilus IS5 reveals overexpressed extracellular multi-heme cytochrome associated with severe microbiologically influenced corrosion. Sci Rep 11:15458. https://doi.org/10.1038/s41598-021-95060-0
19. Dinh HT, Kuever J, Mussmann M, Hassel AW, Stratmann M, Widdel F (2004) Iron corrosion by novel anaerobic microorganisms. Nature 427(6977):829–832. https://doi.org/10.1038/nature02321
20. Kim SJ, Kweon O, Cerniglia CE (2009) Proteomic applications to elucidate bacterial aromatic hydrocarbon metabolic pathways. Curr Opin Microbiol 12(3):301–309. https://doi.org/10.1016/j.mib.2009.03.006
21. Vílchez-Vargas R, Junca H, Pieper DH (2010) Metabolic networks, microbial ecology and 'omics' technologies: towards understanding in situ biodegradation processes. Environ Microbiol 15(12):3089–3108. https://doi.org/10.1111/j.1462-2920.2010.02340.x
22. Herbst FA, Bahr A, Duarte M, Pieper DH, Richnow HH, von Bergen M, Seifert J, Bombach P (2013) Elucidation of in situ polycyclic aromatic hydrocarbon degradation by functional metaproteomics (protein-SIP). Proteomics 13(18–19):2910–2920. https://doi.org/10.1002/pmic.201200569
23. Bargiela R, Herbst FA, Martinez M, Seifert J, Rojo D, Cappello S, Genovese M, Crisafi F, Denaro R, Chernikova TN, Barbas C, von Bergen M, Yakimov M, Ferrer M, Golyshin PN (2015) Metaproteomics and metabolomics analyses of chronically petroleum-polluted sites reveal the importance of general anaerobic processes uncoupled with degradation. Proteomics 15:3508–3520. https://doi.org/10.1002/pmic.201400614
24. Pandey A, Tripathi PH, Tripathi AK, Pandey SC, Gangola S (2019) Omics technology to study bioremediation and respective enzymes. In: Bhatt P (ed) Smart bioremediation technologies. Academic Press, pp 23–43. https://doi.org/10.1016/B978-0-12-818307-6.00002-0
25. Xu J, Wang L, Lv W, Song X, Nie Y, Wu XL (2023) Metabolic profiling of petroleum-degrading microbial communities incubated under high-pressure conditions. Front Microbiol 14:1305731. https://doi.org/10.3389/fmicb.2023.1305731

26. Kotlar HK, Lewin A, Johansen J, Throne-Holst M, Haverkamp T, Markussen S, Winnberg A, Ringrose P, Aakvik T, Ryeng E, Jakobsen K, Drabløs F, Valla S (2011) High coverage sequencing of DNA from microorganisms living in an oil reservoir 2.5 kilometres subsurface. Environ Microbiol Rep 3(6):674–681. https://doi.org/10.1111/j.1758-2229.2011.00279.x
27. Sen R (2008) Biotechnology in petroleum recovery: the microbial EOR. Prog Energy Combust Sci 34(6):1–16. https://doi.org/10.1016/j.pecs.2008.05.001
28. Lin JH, Zhang KC, Tao WY, Shaheen SM, Rensing C, Chen H, Rinklebe J (2019) Geobacillus strains that have potential value in microbial enhanced oil recovery. Appl Microbiol Biotechnol 103:8339–8350. https://doi.org/10.1007/s00253-019-10115-7
29. Li M, Yu J, Cao L, Yin Y, Su Z, Chen S, Li G, Ma T (2023) Facultative anaerobic conversion of lignocellulose biomass to new bioemulsifier by thermophilic Geobacillus thermodenitrificans NG80-2. J Hazard Mater 443:130210. https://doi.org/10.1016/j.jhazmat.2022.130210
30. Yernazarova A, Shaimerdenova U, Akimbekov N, Kaiyrmanova G, Shaken M, Izmailova A (2024) Exploring the use of microbial enhanced oil recovery in Kazakhstan: a review. Front Microbiol 15:1394838. https://doi.org/10.3389/fmicb.2024.1394838
31. Avelino-Jiménez IA, Hernández-Maya L, Larios-Serrato V, Quej-Ake L, Castelán-Sánchez H, Herrera-Díaz J, Garibay-Febles V, Rivera-Olvera JN, Zavala-Olivares G, Zapata-Peñasco I (2023) Biofouling and biocorrosion by microbiota from a marine oil pipeline: a metagenomic and proteomic approach. J Environ Chem Eng 11(2):109413. https://doi.org/10.1016/j.jece.2023.109413
32. Wang Y, Zhang R, Mathivanan K, Zhang Y, Yang L, Guan F, Duan J (2024) Proteomics and EPS compositional analysis reveals Desulfovibrio bisertensis SY-1 induced corrosion on Q235 steel by biofilm formation. Materials 17(20):5060. https://doi.org/10.3390/ma17205060
33. Hazen TC (2020) Lessons from the 2010 Deepwater Horizon accident in the Gulf of Mexico. In: Wilkes H (ed) Hydrocarbons, oils and lipids: diversity, origin, chemistry and fate. Handbook of hydrocarbon and lipid microbiology. Springer, Cham, pp 847–864. https://doi.org/10.1007/978-3-319-90569-3_31
34. Hansen ASL, Lennen RM, Sonnenschein N, Herrgård MJ (2017) Systems biology solutions for biochemical production challenges. Curr Opin Biotechnol 45:85–91. https://doi.org/10.1016/j.copbio.2016.11.018
35. Sharma P, Singh SP, Iqbal HMN, Tong YW (2022) Omics approaches in bioremediation of environmental contaminants: an integrated approach for environmental safety and sustainability. Environ Res 211:113102. https://doi.org/10.1016/j.envres.2022.113102
36. Jiang Y, Luo J, Huang D, Liu Y, Li DD (2022) Machine learning advances in microbiology: a review of methods and applications. Front Microbiol 13:925454. https://doi.org/10.3389/fmicb.2022.925454
37. Ismail WA, Van Hamme JD, Kilbane JJ, Gu JD (2017) Editorial: petroleum microbial biotechnology: challenges and prospects. Front Microbiol 8:833. https://doi.org/10.3389/fmicb.2017.00833
38. NACE International (2017) NACE standard RP0102: standard recommended practice in-line inspection of pipelines. NACE International, Houston
39. ASTM International (2016) ASTM D4007–11(2016)e1: standard test method for water and sediment in crude oil by the centrifuge method (laboratory procedure). ASTM International, West Conshohocken
40. ASTM International (2017) ASTM D1928-12b(2017): standard test method for density, relative density, or API gravity of crude petroleum and liquid petroleum products by hydrometer method. ASTM International, West Conshohocken
41. ASTM International (1998) ASTM D4294–98: standard test method for sulfur in petroleum products by energy-dispersive X-ray fluorescence spectroscopy. ASTM International, West Conshohocken
42. UOP LLC (1985) UOP 46–85: paraffin wax content of petroleum oils and asphalts. UOP LLC, A Honeywell Company

43. Hernández-Gayosso MJ, Nava N, Zavala Olivares G (2019) Characterisation and comparison of corrosion products originated in steel pipelines transporting sour gas and crude oil. Corros Eng Sci Technol 51(8):249–256. https://doi.org/10.1080/1478422X.2016.1173421
44. Salgar-Chaparro SJ, Tarazona J, Machuca LL (2022) Corrosion of carbon steel by Shewanella chilikensis DC57 under thiosulphate and nitrate reducing conditions. Front Bioeng Biotechnol 10:825776. https://doi.org/10.3389/fbioe.2022.825776
45. ASTM International (2013) ASTM D1141–98: standard practice for the preparation of substitute ocean water. ASTM International, West Conshohocken
46. ASTM International (2014) Standard practice for conventions applicable to electrochemical measurements in corrosion testing (ASTM G3). ASTM International, West Conshohocken
47. ASTM International (2021) ASTM G5–14: standard reference test method for making potentiodynamic anodic polarization measurements. ASTM International, West Conshohocken
48. ASTM International (2011) ASTM E3–11: standard guide for preparation of metallographic specimens. ASTM International, West Conshohocken
49. Herrera-Díaz J, Jelezova MK, Cruz-García F, Dinkova TD (2018) Protein disulfide isomerase (PDI1-1) differential expression and modification in Mexican malting barley cultivars. PLoS One 13(11):e0206470. https://doi.org/10.1371/journal.pone.0206470
50. Nesvizhskii AI, Aebersold R (2005) Interpretation of shotgun proteomic data: the protein inference problem. Mol Cell Proteomics 4(10):1419–1440. https://doi.org/10.1074/mcp.R500012-MCP200
51. Song X, Zhang G, Zhou Y, Li W (2023) Behaviors and mechanisms of microbially-induced corrosion in metal-based water supply pipelines: a review. Sci Total Environ 895:165034. https://doi.org/10.1016/j.scitotenv.2023.165034
52. Pal MK, Lavanya M (2022) Microbial influenced corrosion: understanding bioadhesion and biofilm formation. J Bio Tribo Corros 8:76. https://doi.org/10.1007/s40735-022-00677-x
53. Kreve S, Reis ACD (2021) Bacterial adhesion to biomaterials: what regulates this attachment? A review. Jpn Dent Sci Rev 57:85–96. https://doi.org/10.1016/j.jdsr.2021.05.003
54. Rashid MH, Rumbaugh K, Passador L, Davies DG, Hamood AN, Iglewski BH, Kornberg A (2000) Polyphosphate kinase is essential for biofilm development, quorum sensing, and virulence of Pseudomonas aeruginosa. Proc Natl Acad Sci USA 97(17):9636–9641. https://doi.org/10.1073/pnas.170283397
55. Andreini C, Bertini I, Cavallaro G, Holliday GL, Thornton JM (2008) Metal ions in biological catalysis: from enzyme databases to general principles. J Biol Inorg Chem 13(8):1205–1218. https://doi.org/10.1007/s00775-008-0404-5
56. Do T, Page JE, Walker S (2008) Uncovering the activities, biological roles, and regulation of bacterial cell wall hydrolases and tailoring enzymes. J Biol Chem 295(10):3347–3361. https://doi.org/10.1074/jbc.REV119.010155
57. Liu J, Chakraborty S, Hosseinzadeh P, Yu Y, Tian S, Petrik I, Bhagi A, Lu Y (2014) Metalloproteins containing cytochrome, iron-sulfur, or copper redox centers. Chem Rev 114(8):4366–4469. https://doi.org/10.1021/cr400479b
58. Bilsing FL, Anlauf MT, Hachani E, Khosa S, Schmitt L (2023) ABC transporters in bacterial nanomachineries. Int J Mol Sci 24(7):6227. https://doi.org/10.3390/ijms24076227
59. Song Z, Li S, Li Z, Zhang Q, Fu S, Wang H, Yang T, Wang Y (2021) Functional metagenomic and enrichment metatranscriptomic analysis of marine microbial activities within a marine oil spill area. Environ Pollut 274:116555. https://doi.org/10.1016/j.envpol.2021.116555
60. Das N, Kumar V, Chaurec K, Pandey P (2025) Environmental restoration of polyaromatic hydrocarbon-contaminated soil through sustainable rhizoremediation: insights into bioeconomy and high-throughput systematic analysis. Environ Sci: Adv 4:842
61. Rajbongshi A, Gogoi SB (2024) A review on oilfield produced wáter and its treatment technologies. Pet Res 9(4):640–656. https://doi.org/10.1016/j.ptlrs.2024.06.003
62. Lindenbah Bonfá, Pereira RM, Piubeli FA, Durrant LR (2022) Microbial bioremediation of petroleum hydrocarbons in moderate to extreme environments and application of "omics". Techniques to evaluate bioremediation approaches and efficiency. In: Mukherjee G, Dhiman S,

Engineering & Technology, Environment and Sustainability (eds) Waste management opportunities and challenges for sustainable development. CRC Press. ISBN9780429341106, Boca Raton. https://doi.org/10.1201/9780429341106
63. Mishra S, Lin Z, Pang S, Zhang W, Bhatt P, Chen S (2021) Recent advanced technologies for the characterization of xenobiotic-degrading microorganisms and microbial communities. Front Bioeng Biotechnol 9:632059. https://doi.org/10.3389/fbioe.2021.632059
64. Bastida F, Jehmlich N, Lima K, Morris BEL, Richnow HH, Hernández T, von Bergen M, García C (2016) The ecological and physiological responses of the microbial community from a semiarid soil to hydrocarbon contamination and its bioremediation using compost amendment. J Proteome 135:162–169. https://doi.org/10.1016/j.jprot.2015.07.023
65. Gan G, Xu X, Chen X, Zhang XF, Wang J, Zhong ZQ (2021) SCASP: a simple and robust SDS-aided sample preparation method for proteomic research. Mol Cell Proteomics 20. https://doi.org/10.1016/j.mcpro.2021.100051
66. Iranpour T, Mirimba M, Shenouda C, Lynch A, Doucette AA (2025) SDS depletion from intact membrane proteins by KCl precipitation ahead of mass spectrometry analysis. Proteomes 13(3):30. https://doi.org/10.3390/proteomes13030030
67. Benndorf D, Balcke GU, Harms H, von Bergen M (2007) Functional metaproteome analysis of protein extracts from contaminated soil and groundwater. ISME J 1(3):224–234. https://doi.org/10.1038/ismej.2007.39
68. Keiblinger KM, Wilhartitz IC, Schneider T, Roschitzki B, Schmid E, Eberl L, Riedel K, Zechmeister-Boltenstern S (2012) Soil metaproteomics - comparative evaluation of protein extraction protocols. Soil Biol Biochem 54(15–10):14–24. https://doi.org/10.1016/j.soilbio.2012.05.014
69. Wada OZ, Rashid N, Wijten P, Thornalley P, Mckay G, Mackey HR (2024) Evaluation of cell disruption methods for protein and coenzyme Q10 quantification in purple non-sulfur bacteria. Front Microbiol 15:1324099. https://doi.org/10.3389/fmicb.2024.1324099
70. Fowler CB, Waybright TJ, Veenstra TD, O'Leary TJ, Mason JT (2012) Pressure-assisted protein extraction: a novel method for recovering proteins from archival tissue for proteomic analysis. J Proteome Res 11(4):2602–2608. https://doi.org/10.1021/pr201005t
71. Zhang Y, Wang J, Chen D, Ge S, Fu J, Wu X, Li H, Wang RI, Han R, Tian X, Ma H (2025) A smart plugging system of temperature-sensitive gels based on cellulose ether. ACS Omega. https://doi.org/10.1021/acsomega.5c01828
72. Nozari B, Kander R (2025) Supercritical CO2 technology for biomass extraction: review. Ind Crop Prod 233:121348. https://doi.org/10.1016/j.indcrop.2025.121348
73. Uwineza PA, Waśkiewicz A (2020) Recent advances in supercritical fluid extraction of natural bioactive compounds from natural plant materials. Molecules 25(17):3847. https://doi.org/10.3390/molecules25173847

Sample Preparation Strategies for Microbial Cell Surface Proteomics: Integrating Shaving and Shotgun Approaches

Dorota Satala, Katarzyna Kowalik, and Justyna Karkowska-Kuleta

Abstract A range of mass spectrometry–based proteomic approaches is now employed to characterize the surface proteome—surfaceome—of pathogenic microorganisms, including fungi capable of causing infections in humans. Nevertheless, the preparation of surface-enriched samples remains technically challenging because of the persistent risk of intracellular protein contamination. Cell surface shaving represents a focused sample treatment approach in which proteolytic enzymes act on intact cells to release surface-exposed peptides, thereby enriching this fraction for high-resolution liquid chromatography coupled tandem mass spectrometry (LC-MS/MS) analysis. Optimizing sample preparation for fungal surface proteomics is critical, as reducing sample complexity, enriching surface-exposed components, and ensuring LC-MS/MS compatibility must be achieved while avoiding matrix interferences, partial lysis, and biases toward only trypsin-accessible epitopes. These methodological challenges, as well as strategies for validating sample-treatment selectivity with appropriate controls, are discussed in this chapter.

Keywords Surfaceome · Fungi · Shaving · Trypsin · Surface · Sample preparation

D. Satala · J. Karkowska-Kuleta (✉)
Department of Comparative Biochemistry and Bioanalytics, Faculty of Biochemistry, Biophysics and Biotechnology, Jagiellonian University, Kraków, Poland
e-mail: justyna.karkowska@uj.edu.pl

K. Kowalik
Department of Comparative Biochemistry and Bioanalytics, Faculty of Biochemistry, Biophysics and Biotechnology, Jagiellonian University, Kraków, Poland

Doctoral School of Exact and Natural Sciences, Faculty of Biochemistry, Biophysics and Biotechnology, Jagiellonian University, Kraków, Poland

J.-L. Capelo-Martínez (ed.), *Emerging Sample Treatments in Proteomics*, Advances in Experimental Medicine and Biology 1510,
https://doi.org/10.1007/978-3-032-21638-0_7

1 Introduction

In the analysis of the virulence traits of pathogenic microorganisms, a critical component is the identification of surface-exposed proteins, which play central roles in host cell adhesion, tissue invasion, modulation of immune responses, and the initiation of infection. One of the methods currently employed for characterizing molecules displayed exclusively on the surface of bacterial or fungal cells is the cell surface shaving technique, in which specific proteolytic enzymes are used to catalyze the hydrolysis of peptide bonds within accessible protein moieties embedded in the polysaccharide matrix of microbial cell walls [1, 2]. Cell surface shaving serves as a specialized sample treatment strategy that enables the selective enrichment of surface-exposed peptides, and this approach is predicated on the principle that, under conditions preserving cellular integrity, enzymatic digestion of surface-exposed proteins results in the release of peptides, which are subsequently collected and subjected to analysis via shotgun proteomic approach and consequent identification by liquid chromatography coupled tandem mass spectrometry (LC-MS/MS), enabling detailed profiling of the surface proteome—the so-called surfaceome [3, 4]. Enzymatic digestion applied to intact, non-permeabilized cells constitutes a highly selective sample treatment step, and by restricting proteolytic activity to the cell surface, this approach enables the targeted enrichment of extracellular peptides while preserving membrane integrity, thereby minimizing cytoplasmic leakage. This controlled processing enhances both the specificity and interpretability of downstream peptide profiling.

The significance of this methodological approach for sample preparation and analysis is well supported by evidence presented in the scientific literature. A term co-occurrence network was constructed using VOSviewer [5], based on bibliographic data retrieved from the Web of Science Core Collection database up to the year 2025. The search strategy included the topic query "shaving" combined with terms such as "surface shaving", "trypsin shaving", "surfaceome", and "cell surface shaving" as well as research areas including microbiology, cell biology, physiology, and biochemistry and molecular biology. The resulting visualization revealed a notable pattern of interconnections among terms related to surface shaving, surfaceome, microorganisms, both bacteria and fungi, and their virulence factors (Fig. 1).

These associations demonstrate the increasing number of literature reports emphasizing the role of cell surface profiling techniques in elucidating microbial pathogenicity and host–pathogen interactions, including the identification of antigenic determinants suitable for future vaccine design, alongside the indication of emerging molecular targets for the development of novel and effective antibacterial and antifungal treatment [4].

It is particularly important in the case of the identification of proteins exposed at the cell surface of pathogenic fungi from the genus *Candida*, including species *C. albicans*, *C. auris*, *C. parapsilosis*, *C. tropicalis*, and *C. glabrata* (currently reclassified as *Nakaseomyces glabratus*), since the recent report published by World Health Organization in October 2022 identifies the first two species as critical

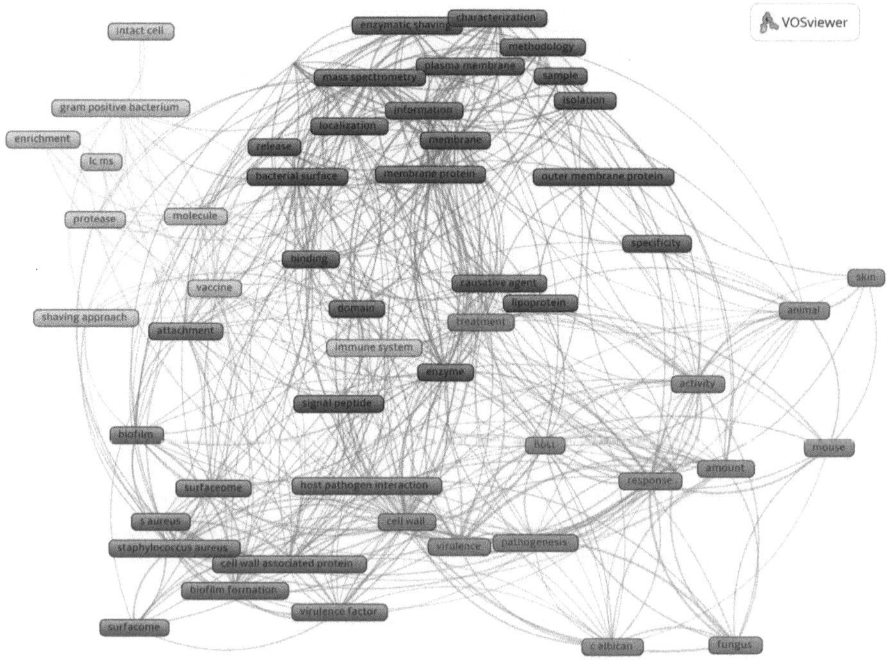

Fig. 1 Term co-occurrence network illustrating conceptual linkages between surface shaving, surfaceome, and microorganisms and their virulence factors, derived from scientific publications indexed in the Web of Science database till 2025. The visualization was created using VOSviewer (version 1.6.20, Centre for Science and Technology Studies, Leiden University, The Netherlands). Frequently co-occurring terms appear in close proximity, and thematically related terms are grouped into four thematic clusters

priority pathogens and the next three as high-risk pathogens [6], and this classification underscores the necessity for comprehensive and rigorous research on these organisms [7]. Fungi from genus *Candida* are responsible for both superficial and invasive, disseminated fungal infections, and the cell wall surface proteins they produce may play key roles in host interactions and overall pathogenic mechanisms [1, 8].

2 Conceptual Background of Surface Shaving

One of the first studies of microbial cell surface involved the technique of cell surface shaving with proteolytic enzymes focused on the identification of surface-exposed proteins of bacterial cells from group A *Streptococcus* including *S. pyogenes* strains M1_SF370 and M23_DSM2071 and was performed with the use of trypsin and proteinase K treatment of intact bacteria to indicate proteinaceous candidates for vaccine development [9]. Preceding studies that aimed at the identification of surface-associated molecules have typically involved workflows based on the controlled cell disruption, followed by the isolation of membrane or cell wall proteins and their subsequent identification using gel-based or gel-free proteomic methods [10].

However, these approaches revealed numerous limitations, including challenges in separating surface proteins from cytoplasmic contaminants, poor solubility of membrane-associated proteins, and interference from membrane or cell wall non-proteinaceous fragments, as well as other intracellular molecules and proteins. Critically, it is essential to ensure unequivocal identification of exclusively surface-localized proteins, free from intracellular contamination [11]. Moreover, in the case of membrane or transmembrane proteins, as well as proteins anchored or penetrating the cell wall, post-isolation analysis often fails to determine which regions of the molecule are exposed extracellularly and which are oriented toward the intracellular space. This limitation can be addressed by selectively labeling surface-exposed molecules prior to isolation, for example with biotin moiety; nevertheless, such approach may also affect the subsequent identification of labeled peptides through the introduction of steric hindrance for proteolytic enzymes, blocking or chemical inactivation of cleavage sites, and incorporation of structural modifications or mass-altering moieties [10].

Therefore, shaving-based proteomics offers a unique advantage in ensuring the selective identification of surface-exposed proteins, overcoming limitations of other strategies. Nonetheless, in the cell surface shaving approach, it is equally essential to maintain strict control over cell viability throughout the entire procedure, as any disruption of the cell, including compromise of the cell wall or membrane integrity, or complete cellular lysis will result in the release of cytoplasmic contents and contamination of the samples with intracellular proteins. Consequently, the validation of sample treatment effectiveness is of key importance together with the use of controls applied to validate the specificity of the shaving, including trypsin-treated lysed fungal cells, the use of membrane-impermeable labeling, as well as indication and verification of the presence of cytoplasmic markers.

3 Sample Preparation

Cell surface shaving is widely used as a comparative proteomic strategy to investigate condition-dependent remodeling of the microbial surfaceome [2, 12–16]. The approach enables the characterization of surface-exposed proteins across different morphological states (e.g., yeast versus hyphae) and allows the analysis of how environmental and physiological cues—such as nutrient availability, pH, oxygen levels or temperature—as well as lifestyle transitions, including biofilm versus planktonic growth, shape the extracellular protein landscape. A schematic overview of the shaving workflow is presented in Fig. 2.

The diagram summarizes the major stages of the procedure—including cell harvesting, controlled proteolysis with trypsin, peptide recovery, and sample clean-up

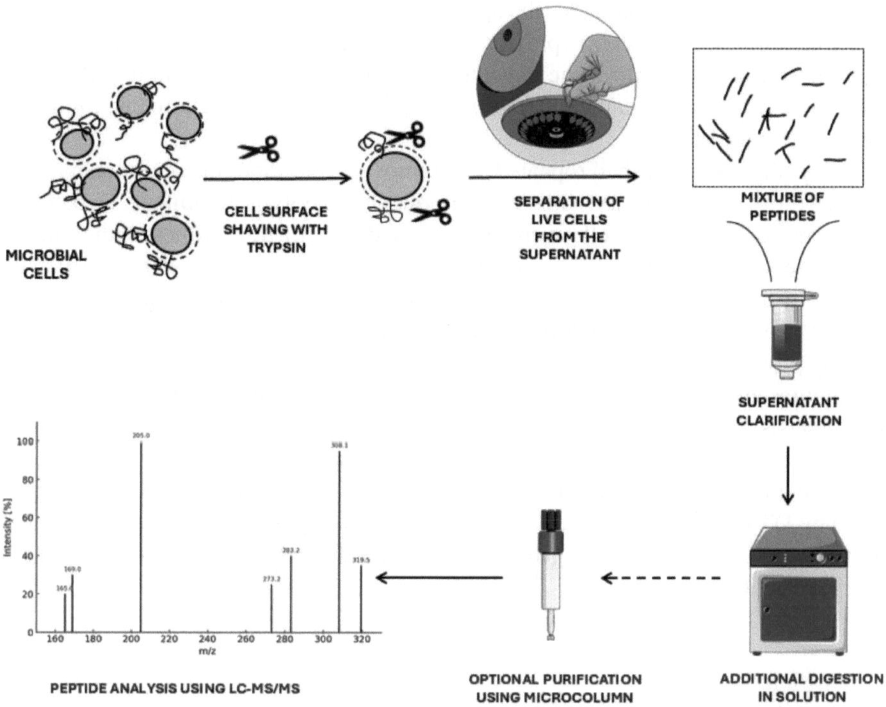

Fig. 2 Schematic overview of the surface shaving workflow. Microbial cells are incubated with proteolytic enzymes (e.g., trypsin) under conditions preserving cell integrity, resulting in the release of surface-accessible peptides. The peptide mixture is then collected by centrifugation and, if required, subjected to additional in-solution digestion to improve peptide recovery and coverage. The resulting peptide pool is analyzed by liquid chromatography–coupled tandem mass spectrometry (LC-MS/MS), enabling the identification of proteins exposed at the microbial surface. The figure was partly generated using Servier Medical Art, provided by Servier, licensed under a Creative Commons Attribution 3.0 unported license

prior to LC-MS/MS analysis—and highlights the steps that are shared across classical shaving protocols.

The sample-preparation workflow described below is based on methodologies reported for *Candida* species in several independent studies and reflects procedural elements that are consistently shared across classical shaving protocols [2, 12, 14, 15].

3.1 Cell Harvesting and Preparation

Preparation of samples for surface proteomics was carried out using controlled proteolysis of intact *Candida* cells with sequencing-grade trypsin. Cells of *C. albicans*, *C. glabrata*, *C. parapsilosis*, or *C. tropicalis* (typically 5×10^8 per condition) were harvested after cultivation under the appropriate growth conditions and collected by centrifugation at 3000–6000 rpm for 3–5 min. Microscopic inspection was routinely performed to confirm the expected morphological state, including the presence of yeast-like or filamentous forms depending on the culture medium and environmental conditions.

Cell pellets were washed repeatedly to remove culture medium components and to ensure full compatibility with the subsequent proteolytic treatment. Depending on the protocol variant, cells were first washed two to three times with 1 ml of phosphate-buffered saline (PBS, pH 7.4) to remove extracellular debris, followed by two to three additional washes with 1 ml of 25 mM ammonium bicarbonate (NH_4HCO_3). This stepwise washing strategy, reported consistently across published shaving protocols, ensures that the final cell suspension is free of interfering salts or metabolites and is fully compatible with trypsin digestion and downstream LC-MS/MS analysis. After the final wash, cell pellets were resuspended in a small volume (typically 100–800 μl) of 25 mM NH_4HCO_3 supplemented with dithiothreitol (DTT) at concentrations ranging from 0.1 to 5 mM. The reducing environment provided by DTT facilitated gentle loosening of the outer proteinaceous layers without compromising membrane integrity and prepared the cells for controlled, surface-restricted proteolysis.

3.2 Surface Shaving with Trypsin

Proteolytic shaving was initiated by adding 1–5 μg of sequencing-grade trypsin to the resuspended cells, followed by incubation at 37 °C for 5–10 min under gentle agitation (typically 600 rpm). These conditions were selected to specifically hydrolyze surface-exposed peptide regions without compromising cell integrity. Subsequently, microbial cells were separated from the supernatant containing protein fragments, peptides, and trypsin by centrifugation at 5000 rpm for 5 min, and cell pellet was discarded. The reaction was then stopped by adding trifluoroacetic acid (TFA) to the obtained supernatant to a final concentration of 0.1% and incubating the suspensions on ice for 10–15 min to ensure complete inactivation of enzyme activity. Samples were then centrifuged at 10000–12000 rpm, and the peptide-rich supernatants were clarified by filtration through 0.22 μm membranes. The clarified supernatants were dried directly in a vacuum concentrator (Speed-Vac) and stored at −20 °C until LC-MS/MS analysis.

3.3 Extended Digestion of Released Peptides (Optional Optimization)

To increase peptide yield and enhance the analytical depth of the surface-derived proteome, an optional extended digestion step was applied. After the removal of intact cells, and before adding TFA, the clarified supernatants were incubated with trypsin for an additional 5 h or up to 24 h at 37 °C. This step enabled more extensive proteolysis of partially cleaved extracellular fragments released during the initial shaving phase. Importantly, extended digestion was carried out only on cell-free supernatants to avoid damaging membrane integrity during the primary enzymatic shaving step.

3.4 Sample Clean-Up and Preparation for LC-MS/MS

Before LC-MS/MS analysis, dried peptide extracts were reconstituted in 10% acetonitrile (ACN) containing 0.1% formic acid (FA), clarified by centrifugation, and transferred to fresh vials to remove particulate material. When additional purification was required, an optional clean-up step was introduced prior to drying, during which peptide supernatants were passed through Poros 50 R2 microcolumns to remove salts and low-molecular-weight contaminants known to suppress electrospray ionization efficiency. Peptides were subsequently eluted with an organic solvent containing 0.1% TFA, dried under vacuum, and either stored at −20 °C or resuspended in 0.1% FA immediately before LC-MS/MS acquisition. This clean-up produced peptide mixtures fully compatible with reversed-phase LC separation and high-sensitivity MS/MS acquisition.

3.5 Assessment of Cell Integrity

A critical component of the surface-shaving procedure was rigorous monitoring of cell viability and membrane integrity. Because the aim of shaving is to release exclusively extracellular peptide fragments, any indication of membrane permeabilization resulted in exclusion of the sample. Integrity assays included the use of impermeant fluorescent dyes such as SYTOX Green, Trypan Blue, and propidium iodide (PI) or a combination of methods and dyes to strengthen visualization. These markers were evaluated by flow cytometry for yeast-form cells or by fluorescence microscopy for filamentous morphotypes as described below in detail. Only preparations lacking intracellular staining signals were accepted for downstream proteomic analysis, ensuring that peptide collections originated solely from the cell surface.

3.6 Protein Identification and Database Searching

Raw LC-MS/MS spectra were processed using instrument-specific preprocessing software (DataAnalysis 4.0, Bruker) and exported in Mascot Generic Format (.mgf). Peptide and protein identification was performed using Mascot (v.2.3.0, Matrix

Science), either directly through the standalone search engine or via Proteome Discoverer (v.1.4.1.14, Thermo Scientific). Searches were conducted against the non-redundant NCBI database and the manually curated SwissProt database, both restricted to fungal taxonomy, or alternatively against the Candida Genome Database (CGD) (as custom database). For selected experiments, searches were additionally performed against the SwissProt (Homo sapiens) database to verify the presence of potential host-derived contaminants.

Search parameters included:

- enzyme specificity set to trypsin (up to two missed cleavages),
- carbamidomethylation of cysteine as a fixed modification,
- methionine oxidation as a variable modification,
- precursor mass tolerance 0.3–10 ppm (depending on instrument),
- fragment mass tolerance 0.3–0.8 Da,
- monoisotopic masses and charge states +1 to +3.

Protein identifications were filtered using a strict false discovery rate (FDR <1%) calculated by Mascot Percolator ($q = 0.01$). Only proteins exceeding the Mascot significance threshold (Mascot Score >67, $p < 0.05$) were included. Redundant protein groups were consolidated using parsimony rules to avoid inflation of spectral counts before subsequent semi-quantitative analysis.

3.7 Semi-quantitative Analysis of Surface-Derived Proteins

Semi-quantitative estimation of relative protein abundances was performed using the normalized spectral abundance factor (NSAF), a well-established label-free metric based on spectral counting [17]. After peptide identification, only proteins surpassing the Mascot significance threshold (Mascot Score >67, $p < 0.05$) were included in the quantitative dataset. For each protein, the spectral count (SpC) was divided by its amino-acid length (L) to correct for protein size, and this ratio was normalized to the sum of all SpC/L values within the sample:

$$\mathrm{NSAF} = \frac{\left(\frac{\mathrm{SpC}}{L}\right)}{\sum_{i=1}^{N}\left(\frac{\mathrm{SpC}}{L}\right)}$$

This metric provides a semi-quantitative approximation of protein abundance, suitable for relative comparisons between samples or experimental conditions. Importantly, NSAF reflects the number of MS/MS spectra assigned to peptides and thus depends on peptide detectability, tryptic cleavage site availability, and MS/MS sampling bias.

Because NSAF does not measure absolute quantities, its interpretive power is limited by several intrinsic constraints:

- proteins with few accessible tryptic sites or with heavily glycosylated/exposed regions may generate fewer detectable peptides,
- structural features of the fungal cell wall influence peptide release during shaving,
- shaving-based data disproportionately represent proteins with surface-exposed Lys/Arg residues,
- spectral counting saturates for highly abundant proteins and is less sensitive for low-abundance species.

Therefore, NSAF values should be interpreted as relative indicators of peptide-level detectability, rather than absolute protein concentrations, and conclusions must consider enzyme-specific, structural, and analytical sources of bias.

4 Additional Applications of Cell Surface Shaving in Host–Pathogen and Interkingdom Interactions

As demonstrated in several studies, cell surface shaving has also been applied to characterize how external biological factors—including immune components, host-derived proteins, and microbial partners—associate with or remodel the fungal cell surface. This broader use of shaving highlights its value as an experimental strategy for interrogating biologically relevant interactions at the fungal cell wall [18–21].

One line of research has shown that shaving can be used to identify immune effector molecules that bind to pathogenic fungi. In studies examining interactions between *Candida* species and neutrophil extracellular traps (NETs), NET-derived granular proteins—including myeloperoxidase, neutrophil elastase, cathepsin G, proteinase-3, lactotransferrin, and the antimicrobial peptide LL-37—were demonstrated to bind stably to the fungal surface. Surface shaving enabled their selective recovery, providing direct evidence of surface association and offering mechanistic insight into antifungal immune responses [22].

As further demonstrated in serum-interaction studies, shaving has been used to investigate how components of human serum remodel the fungal cell surface during bloodstream infection. Using gel-free shaving strategies, researchers showed that serum proteins readily adsorb to *Candida* cells, while the fungal surfaceome undergoes profound, condition-dependent reorganization, including the induction of immune-evasion proteins and serum-responsive adhesins [19]. These findings illustrate how host physiological fluids dynamically modulate the extracellular protein landscape.

Shaving-based approaches have also shown that exposure to extracellular matrix proteins induces selective remodeling of the fungal surface. When *C. glabrata* cells were incubated with human fibronectin, the host protein was detected on the fungal surface, accompanied by the selective appearance of adhesins such as Epa6, indicating that direct interaction with matrix components can reshape the repertoire of exposed cell-wall proteins [21].

Interkingdom interactions have likewise been elucidated using shaving-based proteomics. In dual-species biofilms containing *C. albicans* and *Porphyromonas gingivalis*, surface shaving revealed a protein profile distinct from that of single-species biofilms, suggesting that bacterial cohabitation induces structural and compositional reprogramming of the fungal cell wall [20]. These observations support the view that polymicrobial environments are potent modulators of fungal surface architecture.

Collectively, these examples demonstrate that surface shaving is not only a method for characterizing the endogenous fungal surface proteome but also a versatile strategy for elucidating how host factors, immune molecules, and microbial partners dynamically modulate the protein composition of the fungal cell surface. This broad applicability underscores the utility of shaving-based proteomics in studying ecological, pathogenic, and immunological contexts in which surface-exposed proteins play critical functional roles.

5 Technical Challenges in Shaving-Based Proteomics and Complementary Validation Approaches

To monitor and confirm microbial cell viability and avoid the risk of partial lysis, a range of selective staining techniques are employed complemented by subsequent microscopic visualization or flow cytometric analysis [23]. These methods should be carefully chosen taking into consideration the characteristic of investigated microorganism, growth conditions, and experimental setup, and the most commonly used techniques for microorganisms include staining with PI, which penetrates only cells with compromised membranes, and SYTO 9, which stains all cells but is used in conjunction with PI to distinguish live from dead populations, as well as for fungal cells, particularly yeasts, a permeant nucleic acid–specific staining agent FUN-1 ([2-chloro-4-(2,3-dihydro-3-methyl-(benzo-1,3-thiazol-2-yl)-methylidene)-1-phenylquinolinium iodide]) is used to assess metabolic activity, while Trypan Blue staining remains a classic method for identifying non-viable cells [24–26]. Additionally, calcein-AM (calcein acetoxymethyl) and CFDA-SE (CFSE, carboxyfluorescein diacetate *N*-succinimidyl ester) provide fluorescent readouts of esterase activity and membrane integrity in live cells, suitable for both microscopy and flow cytometry. These staining approaches are often complemented by assays based on colony-forming units (CFUs) counting, which offer a functional readout of viability and are especially reliable for microorganisms cultured under controlled and well-characterized growth conditions [26].

Together, these methods provide robust validation of cell viability, which is essential for accurate surface proteome profiling and downstream analytical fidelity. The identification of cytoplasmic proteins in the investigated sample containing peptides shaved from the cell surface of microbial cells, however, does not necessarily indicate cytoplasmic contamination, although their extracellular presence is still a matter of debate. However, the increasing number of reports indicating the presence of the so-called moonlighting proteins—those that perform a completely different function from their primary one in a different location—within the surface-identified proteome suggests functional heterogeneity and reveals additional levels of complexity in the protein landscape, beyond canonical localization and activity [27–29].

In the bottom-up proteomics, particularly in surface shaving approaches, the most commonly employed enzyme is trypsin, a serine protease that preferentially cleaves peptide bonds at the carboxyl side of lysine (K) and arginine (R) residues within polypeptides and proteins, and this specificity facilitates the generation of predictable peptide fragments suitable for mass spectrometric identification and quantification [30–32]. In addition, proteinase K was frequently used in the surface shaving approach, which exhibits relatively broad substrate specificity, with a preference for cleaving peptide bonds adjacent to aliphatic or aromatic amino acid residues at the P position, and the presence of alanine (A) at the P2 position has been shown to enhance proteolytic efficiency [33]. However, under the conditions of treating living cells with particular protease, the observed cleavage pattern may differ from that established under standard in vitro conditions, and this discrepancy arises from the requirement to maintain cellular viability and structural integrity, which imposes constraints on conditions that differ significantly from those optimized for isolated proteinaceous substrates, including parameters such as temperature, pH, ionic strength, reaction time, and enzyme concentration [3]. Beyond these operational constraints, additional limitations arise from the architecture and composition of the cell surface itself. All proteases used in surface shaving are inherently limited by their accessibility to solvent-exposed regions of surface proteins, meaning that deeply embedded, shielded or heavily glycosylated domains may remain uncleaved. Moreover, protease-specific cleavage preferences may introduce representation bias: proteins containing few accessible cleavage sites tend to be underrepresented, whereas those enriched in compatible sites may appear disproportionately abundant. Such accessibility and specificity-driven biases are particularly relevant for spectral-counting-based quantification and should be carefully considered when interpreting comparative surfaceome datasets [17].

Due to the limited duration of proteolytic treatment required to preserve cell viability and membrane integrity, which typically ranges from 5 to 10 min, some protocols introduce an additional digestion step targeting microbial proteins previously released from the cell surface in the absence of intact cells. This additional digestion enhances the extent of surface proteome coverage while minimizing cellular stress caused by hydrolase. It is therefore important to optimize the treatment parameters, including enzymatic digestion conditions, such as time and trypsin concentration in order to maximize surface peptide recovery while preserving cell integrity.

Other important issue is also differential accessibility of proteins in different fungal morphotypes, such as yeast-like unicellular forms and elongated hyphal forms [34]. Fungal morphological forms exhibit notable differences in cell wall composition [16, 35]. Extensive N- and O-glycosylation, the dense β-glucan–chitin scaffold and outer-layer mannan fibrils can sterically restrict protease accessibility, leading to systematic underrepresentation of rigid, deeply embedded or heavily glycosylated proteins [16, 35]. Furthermore, environmental cues—including nutrient availability, pH, oxygen levels, temperature and host-derived factors—are known to remodel the fungal cell wall and modulate the exposure of surface proteins [2, 12]. Such condition-dependent remodeling directly affects protease accessibility during shaving and may account for variation in shaving efficiency that reflects biological differences rather than technical inconsistencies.

Another important improvement to the analysis is the introduction of a false-positive control, which involves incubating microbial cells under the same conditions as the test sample, but without the presence of a hydrolytic enzyme [36, 37]. Proteins released into the solution after cell centrifugation are then hydrolyzed with the same enzyme and analyzed using mass spectrometry. These proteins should then be excluded from the analyzed dataset, as they may result from cell lysis and the release of cytoplasmic contents into the solution during incubation; however, still this control does not account for potential lysis associated with proteinase activity on the cells. The strategy of applying the false-positive control has been successfully employed in profiling the surface proteome of *Staphylococcus aureus*, resulting in a marked enhancement of surface-exposed peptide recovery within the trypsin-generated dataset; specifically, approximately 80% of the identified peptides (124 out of 155) corresponded to surface-localized sequences, underscoring the effectiveness of the approach in improving proteomic resolution [36].

Additionally, combining cell surface shaving techniques with labeling, such as biotinylation, can strengthen the conclusion that proteins are unequivocally surface-exposed [10, 38, 39]. Surface biotinylation employs membrane-impermeable modified biotin N-hydroxysuccinimide (NHS) esters, which covalently label primary amine groups on proteins accessible at the cell surface, whereas membrane-permeable NHS biotin esters can be used as controls to evaluate labeling specificity and monitor potential intracellular access. If the biotinylation reagent used for surface molecules labeling contains a sulfonate group, it is highly hydrophilic and carries a negative charge, which prevents it from diffusing across the plasma membrane. As a result, the compound remains strictly extracellular during the labeling step. After labeling, cells are lysed and biotinylated proteins are enriched using streptavidin-coated matrices, ensuring that only proteins exposed at the time of labeling are recovered. By cross-referencing datasets obtained from the two approaches, one based on proteolytic accessibility and the other on chemical tagging, the cell surface localization with noticeably higher probability might be assigned [10, 38, 39]. Proteins identified by both shaving and non-permeable biotinylation are highly likely to be genuinely surface-exposed, whereas discrepancies between methods may indicate intracellular contaminants released due to mechanical or enzymatic stress or may highlight potential moonlighting proteins with dual intra- and extracellular functions [28, 40]. The combined strategy therefore increases analytical confidence, reduces false positives, and provides a more comprehensive and robust characterization of the fungal surface proteome. However, for *Candida* species, it is important to consider that avidin exhibits some binding affinity to the fungal cell surface, which can interfere with subsequent analytical procedures and potentially confound downstream results [41].

These abovementioned limitations underline the necessity of rigorous protocol validation and the use of complementary approaches—such as membrane-impermeant labeling, morphological controls, and extended supernatant digestion—to ensure high-confidence identification of truly surface-exposed proteins.

6 Conclusions

The cell surface shaving technique represents a powerful and highly selective strategy for probing the extracellular proteome of microbial cells under near-native conditions. By restricting proteolysis to intact, non-permeabilized cells, the method enables the targeted enrichment of surface-exposed peptides while minimizing contamination from intracellular components. This approach has provided important insights into fungal biology, including the identification of adhesion factors, stress-responsive cell-wall proteins, and dynamically regulated components of the *Candida* surfaceome [2, 12–15]. Beyond its classical use in characterizing endogenous surface proteins, shaving has proven valuable for investigating host–pathogen and interkingdom interactions, enabling the identification of host-derived molecules bound to fungal cells and revealing how environmental cues reshape surface architecture [18, 20, 21]. These applications underscore the versatility of the technique and its relevance to infection biology, immunology, and microbial ecology.

Nevertheless, surface shaving has inherent methodological constraints, including limited access to deeply embedded or heavily glycosylated proteins, protease-specific biases, and the requirement for rigorous control of cell integrity. Semi-quantitative interpretation of spectral data must therefore be performed with caution, acknowledging structural, enzymatic, and analytical sources of bias.

Taken together, when implemented under optimized and carefully validated conditions, cell surface shaving provides a robust and informative framework for dissecting surface-associated proteomes and for exploring the molecular determinants of microbial interactions in complex biological environments. The continued integration of shaving with advanced LC-MS/MS platforms, quantitative proteomics, and complementary labeling strategies will further expand its analytical potential and enhance our understanding of microbial surface biology.

References

1. Karkowska-Kuleta J, Rapala-Kozik M, Kozik A (2009) Fungi pathogenic to humans: molecular bases of virulence of Candida albicans, Cryptococcus neoformans and Aspergillus fumigatus. Acta Biochim Pol 56:211–224
2. Karkowska-Kuleta J, Zajac D, Bochenska O, Kozik A (2015) Surfaceome of pathogenic yeasts, Candida parapsilosis and Candida tropicalis, revealed with the use of cell surface shaving method and shotgun proteomic approach. Acta Biochim Pol 62:807–819. https://doi.org/10.18388/abp.2015_1140
3. Olaya-Abril A, Jiménez-Munguía I, Gómez-Gascón L, Rodríguez-Ortega MJ (2014) Surfomics: shaving live organisms for a fast proteomic identification of surface proteins. J Proteome 97:164–176. https://doi.org/10.1016/j.jprot.2013.03.035
4. Pauwels J, Fijałkowska D, Eyckerman S, Gevaert K (2022) Mass spectrometry and the cellular surfaceome. Mass Spectrom Rev 41:804–841. https://doi.org/10.1002/mas.21690
5. van Eck NJ, Waltman L (2010) Software survey: VOSviewer, a computer program for bibliometric mapping. Scientometrics 84:523–538. https://doi.org/10.1007/s11192-009-0146-3
6. WHO fungal priority pathogens list to guide research, development and public health action. https://www.who.int/publications/i/item/9789240060241. Accessed 25 Sept 2025.
7. Casalini G, Giacomelli A, Antinori S (2024) The WHO fungal priority pathogens list: a crucial reappraisal to review the prioritisation. Lancet Microbe 5:717–724. https://doi.org/10.1016/S2666-5247(24)00042-9
8. Chaffin WL (2008) Candida albicans cell wall proteins. Microbiol Mol Biol Rev 72:495–544. https://doi.org/10.1128/MMBR.00032-07
9. Rodríguez-Ortega MJ, Norais N, Bensi G et al (2006) Characterization and identification of vaccine candidate proteins through analysis of the group A Streptococcus surface proteome. Nat Biotechnol 24:191–197. https://doi.org/10.1038/nbt1179
10. Li Y, Qin H, Ye M (2020) An overview on enrichment methods for cell surface proteome profiling. J Sep Sci 43:292–312. https://doi.org/10.1002/jssc.201900700
11. Rodríguez-Ortega MJ (2018) "Shaving" live bacterial cells with proteases for proteomic analysis of surface proteins. In: Boheler KR, Gundry RL (eds) The surfaceome: methods and protocols. Springer, New York, pp 21–29
12. Karkowska-Kuleta J, Satala D, Bochenska O et al (2019) Moonlighting proteins are variably exposed at the cell surfaces of Candida glabrata, Candida parapsilosis and Candida tropicalis under certain growth conditions. BMC Microbiol 19:149. https://doi.org/10.1186/s12866-019-1524-5
13. Hernáez ML, Ximénez-Embún P, Martínez-Gomariz M et al (2010) Identification of Candida albicans exposed surface proteins in vivo by a rapid proteomic approach. J Proteome 73:1404–1409. https://doi.org/10.1016/j.jprot.2010.02.008
14. Vialás V, Perumal P, Gutierrez D et al (2012) Cell surface shaving of Candida albicans biofilms, hyphae, and yeast form cells. Proteomics 12:2331–2339. https://doi.org/10.1002/pmic.201100588
15. Gil-Bona A, Parra-Giraldo CM, Hernáez ML et al (2015) Candida albicans cell shaving uncovers new proteins involved in cell wall integrity, yeast to hypha transition, stress response and host-pathogen interaction. J Proteome 127:340–351. https://doi.org/10.1016/j.jprot.2015.06.006
16. Gil-Bona A, Amador-García A, Gil C, Monteoliva L (2018) The external face of Candida albicans: a proteomic view of the cell surface and the extracellular environment. J Proteome 180:70–79. https://doi.org/10.1016/j.jprot.2017.12.002
17. Zybailov B, Mosley AL, Sardiu ME et al (2006) Statistical analysis of membrane proteome expression changes in Saccharomyces cerevisiae. J Proteome Res 5:2339–2347. https://doi.org/10.1021/pr060161n

18. Karkowska-Kuleta J, Wronowska E, Satala D et al (2021) Als3-mediated attachment of enolase on the surface of Candida albicans cells regulates their interactions with host proteins. Cell Microbiol 23:e13297. https://doi.org/10.1111/cmi.13297
19. Marín E, Parra-Giraldo CM, Hernández-Haro C et al (2015) Candida albicans shaving to profile human serum proteins on hyphal surface. Front Microbiol 6. https://doi.org/10.3389/fmicb.2015.01343
20. Bartnicka D, Karkowska-Kuleta J, Zawrotniak M et al (2019) Adhesive protein-mediated cross-talk between Candida albicans and Porphyromonas gingivalis in dual species biofilm protects the anaerobic bacterium in unfavorable oxic environment. Sci Rep 9:4376. https://doi.org/10.1038/s41598-019-40771-8
21. Zajac D, Karkowska-Kuleta J, Bochenska O et al (2016) Interaction of human fibronectin with Candida glabrata epithelial adhesin 6 (Epa6). Acta Biochim Pol 63:417–426. https://doi.org/10.18388/abp.2016_1328
22. Karkowska-Kuleta J, Smolarz M, Seweryn-Ozog K et al (2021) Proteinous components of neutrophil extracellular traps are arrested by the cell wall proteins of Candida albicans during fungal infection, and can be used in the host invasion. Cells 10:2736. https://doi.org/10.3390/cells10102736
23. Shailaja A, Bruce TF, Gerard P et al (2022) Comparison of cell viability assessment and visualization of Aspergillus niger biofilm with two fluorescent probe staining methods. Biofilm 4:100090. https://doi.org/10.1016/j.bioflm.2022.100090
24. Millard PJ, Roth BL, Thi HP et al (1997) Development of the FUN-1 family of fluorescent probes for vacuole labeling and viability testing of yeasts. Appl Environ Microbiol 63:2897–2905. https://doi.org/10.1128/aem.63.7.2897-2905.1997
25. Netuschil L, Auschill TM, Sculean A, Arweiler NB (2014) Confusion over live/dead stainings for the detection of vital microorganisms in oral biofilms – which stain is suitable? BMC Oral Health 14:2. https://doi.org/10.1186/1472-6831-14-2
26. Kwolek-Mirek M, Zadrag-Tecza R (2014) Comparison of methods used for assessing the viability and vitality of yeast cells. FEMS Yeast Res 14:1068–1079. https://doi.org/10.1111/1567-1364.12202
27. Jeffery CJ (1999) Moonlighting proteins. Trends Biochem Sci 24:8–11. https://doi.org/10.1016/s0968-0004(98)01335-8
28. Karkowska-Kuleta J, Kozik A (2014) Moonlighting proteins as virulence factors of pathogenic fungi, parasitic protozoa and multicellular parasites. Mol Oral Microbiol 29:270–283. https://doi.org/10.1111/omi.12078
29. Satala D, Karkowska-Kuleta J, Zelazna A et al (2020) Moonlighting proteins at the candidal cell surface. Microorganisms 8:1046. https://doi.org/10.3390/microorganisms8071046
30. Olsen JV, Ong S-E, Mann M (2004) Trypsin cleaves exclusively C-terminal to arginine and lysine residues. Mol Cell Proteomics 3:608–614. https://doi.org/10.1074/mcp.T400003-MCP200
31. Woessmann J, Petrosius V, Üresin N et al (2023) Assessing the role of trypsin in quantitative plasma and single-cell proteomics toward clinical application. Anal Chem 95:13649–13658. https://doi.org/10.1021/acs.analchem.3c02543
32. Sun J, McAlister J, Geddes-McAlister J (2025) Temporal and thermal optimization of trypsin digestion for the cryptococcal proteome. Microbiol Resour Announc 14:e0113724. https://doi.org/10.1128/mra.01137-24
33. Keil B (2012) Specificity of proteolysis. Springer
34. Casanova M, Lopez-Ribot JL, Martinez JP, Sentandreu R (1992) Characterization of cell wall proteins from yeast and mycelial cells of Candida albicans by labelling with biotin: comparison with other techniques. Infect Immun 60:4898–4906. https://doi.org/10.1128/iai.60.11.4898-4906.1992
35. Gow NAR, Latge J-P, Munro CA (2017) The fungal cell wall: structure, biosynthesis, and function. Microbiol Spectrum 5. https://doi.org/10.1128/microbiolspec.FUNK-0035-2016

36. Solis N, Larsen MR, Cordwell SJ (2010) Improved accuracy of cell surface shaving proteomics in Staphylococcus aureus using a false-positive control. Proteomics 10:2037–2049. https://doi.org/10.1002/pmic.200900564
37. Solis N, Cordwell SJ (2016) Cell shaving and false-positive control strategies coupled to novel statistical tools to profile gram-positive bacterial surface proteomes. Methods Mol Biol 1440:47–55. https://doi.org/10.1007/978-1-4939-3676-2_4
38. Elia G (2008) Biotinylation reagents for the study of cell surface proteins. Proteomics 8:4012–4024. https://doi.org/10.1002/pmic.200800097
39. Roh JW, Choi HW, Gee HY (2025) Guidelines for plasma membrane protein detection by surface biotinylation. Mol Cells 48:100174. https://doi.org/10.1016/j.mocell.2024.100174
40. Karkowska-Kuleta J, Kozik A (2015) Cell wall proteome of pathogenic fungi. Acta Biochim Pol 62:339–351. https://doi.org/10.18388/abp.2015_1032
41. Masuoka J, Guthrie LN, Hazen KC (2002) Complications in cell-surface labelling by biotinylation of Candida albicans due to avidin conjugate binding to cell-wall proteins. Microbiology (Reading) 148:1073–1079. https://doi.org/10.1099/00221287-148-4-1073

Methods for the Investigation of Protein–Ligands Interactions

Ilaria Iacobucci, Flora Cozzolino, Irene Cipollone, and Maria Monti

Abstract Biomolecular interactions involving proteins, nucleic acids, and small molecules constitute the molecular foundation of cellular regulation, signaling, and therapeutic intervention. Advances in mass spectrometry–based proteomics have enabled the systematic characterization of these interactions at unprecedented depth, sensitivity, and structural resolution. This chapter provides a comprehensive overview of state-of-the-art proteomics methodologies developed to investigate protein–protein, protein–nucleic acid, and protein–drug interactions, with particular emphasis on experimental design, sample preparation, and data quality control. Targeted and untargeted strategies are discussed, including affinity purification–mass spectrometry, proximity-dependent labeling, cross-linking mass spectrometry, blue native electrophoresis, and size-exclusion chromatography–mass spectrometry for protein–protein interactions; affinity capture, EMSA-MS, chromatin immunoprecipitation–mass spectrometry, CRISPR-based locus-specific enrichment, and CLIP-based approaches for protein–nucleic acid complexes; and chemoproteomics, thermal proteome profiling, and label-free structural proteomics for protein–drug interaction analysis. The chapter further highlights recent technological innovations, computational tools, and integrative multi-omics strategies that enhance interaction mapping across biological scales. By critically evaluating the strengths, limitations, and appropriate applications of each methodology, this work aims to provide practical guidance for researchers seeking to design robust interactomics experiments and to interpret complex molecular networks in both basic and translational research contexts.

Keywords Protein–protein interactions · Protein–nucleic acid interactions · Protein–ligand interactions · Interactomics · Mass spectrometry · Affinity purification · Cross-linking ms · Proximity labeling · Chemoproteomics · Thermal proteome profiling · Structural proteomics · Systems biology

I. Iacobucci · F. Cozzolino · I. Cipollone · M. Monti (✉)
Department of Chemical Sciences, University of Naples Federico II, Naples, Italy
e-mail: ilaria.iacobucci@unina.it; flora.cozzolino@unina.it; irene.cipollone@unina.it; montimar@unina.it

J.-L. Capelo-Martínez (ed.), *Emerging Sample Treatments in Proteomics*,
Advances in Experimental Medicine and Biology 1510,
https://doi.org/10.1007/978-3-032-21638-0_8

1 A General Overview

This chapter explores the proteomics-based methodologies developed to study different biomolecular interactions, with attention to the challenges researchers face and the choice of the most suitable techniques.

Techniques such as affinity purification, cross-linking mass spectrometry, and immunoprecipitation coupled to mass spectrometry have been widely adopted to map protein–protein interaction (PPI) networks. A key distinction among these methods lies in the differentiation between targeted and untargeted approaches [49]. Targeted techniques focus on identifying the physical interactions of a specific protein of interest, often referred to as the "bait," to understand its molecular partners. In contrast, untargeted methods aim to uncover the composition of protein complexes within a given biological condition, providing a broader view of interactomes in specific cellular states. However, both approaches face challenges, such as the transient or weak nature of many PPIs and the difficulty in ensuring physiological relevance when such interactions are explored in vitro. To overcome these limitations, numerous complementary techniques have been developed over time.

As examples, Proximity-Dependent Labeling (PDL) methods have become a powerful complement to AP-MS and Co-IP in living cells, thereby stabilizing transient contacts prior to cell lysis and preserving physiological context (BioID: [123]; TurboID: [5]; APEX: [96]; and TransitID: [111, 178]). In parallel, in-cell cross-linking mass spectrometry has substantially contributed to PPI investigation stabilizing weak interactions allowing spatial restraints directly from complex samples [47, 55, 74].

Similarly, protein–nucleic acid interactions are at the heart of almost every fundamental process in cell biology. These include DNA replication and repair, transcriptional regulation, RNA splicing and editing, translation, chromatin remodeling, RNA transport, and degradation. They can be classified in several meaningful ways. One way is by the type of nucleic acid involved, distinguishing between DNA, which may include genomic DNA, promoter regions, or enhancers, and RNA, which may encompass pre-mRNA or various noncoding RNAs. Another important distinction lies in the dynamics of the interaction itself; some are highly stable, such as histones binding DNA, while others are transient or conditionally inducible, as often observed with RNA-binding proteins during splicing.

Similar to what is observed in protein–protein interaction studies, the investigation of protein–nucleic acid interactions, as -omic approach, mainly relies on affinity-based, cross-linking, and mass spectrometry approaches. Techniques like Ch-IP and CLIP are functionally analogous to co-immunoprecipitation or BioID strategies, as they aim to isolate and identify interacting molecules in a specific biological context [53, 125]. Moreover, many proteins involved in nucleic acid binding, such as transcription factors and RNA-binding proteins, also participate in protein–protein complexes or act as drug targets, highlighting the molecular overlap between these interaction networks.

As research increasingly adopts multi-omics strategies, integrative methodologies that combine protein–protein and protein–nucleic acid interaction data are emerging as crucial tools for understanding regulatory networks. For example, proximity labeling methods originally developed for PPI studies, such as APEX or TurboID, have been successfully adapted to profile RNA–protein interactions in living cells with high spatial and temporal resolution [5, 59]. This convergence underscores the need for a unified analytical framework that accommodates the dynamic interplay between different classes of biomolecular interactions.

The study of protein–drug interactions is another critical area, as the binding of therapeutic molecules to proteins underpins drug efficacy and specificity. In recent years, numerous proteomic approaches coupled with mass spectrometry have emerged as reliable methods to characterize drug–protein interactions and elucidate their mechanisms of action. In this field, a range of strategies can be employed, including chemoproteomics [22] and thermal proteome profiling [174]. Nevertheless, such approaches can be challenging due to non-specific interactions and the quantification of binding affinities under physiological and native conditions. Furthermore, understanding the biological significance of these interactions and validating them in vivo remain essential for translational relevance.

Regardless of the application and purification methods, all approaches converge in a final step of identification and, in some cases, also of relative quantification between sample and control. Each of the methods discussed in this chapter is commonly applied in combination with bottom-up proteomics procedures such as shotgun or in situ hydrolysis, sometimes including intermediate electrophoretic fractionation steps. The resulting peptide mixtures are then analyzed by liquid chromatography–tandem mass spectrometry (LC-MS/MS), typically using high-resolution instruments. Bioinformatics tools like MaxQuant [150] or Proteome Discoverer [102] are then used to analyze the mass spectrometry data, filtering the results with stringent false discovery rates (FDR <1%) to confidently identify protein interactors.

In addition to the above-mentioned platform, a broad spectrum of proteomics software is routinely used for protein identification and/or quantification. Open-source pipelines such as FragPipe [173] and OpenMS [120] provide fast database searching, rigorous target–decoy FDR estimation, and label-free quantification. For DIA mode, freely available engines including DIA-NN [17], EncyclopeDIA [129], MaxDIA [135], and OpenSWATH [119] enable spectral-library or library-free identification and robust protein-level quantification. Widely used commercial search/quantification solutions include Mascot Server [109], PEAKS Studio [80], and Spectronaut [6].

By delving into these methodologies and addressing the inherent challenges, this chapter aims to provide a comprehensive overview for researchers investigating biomolecular interactions, with a section dedicated to the advantages and limitations of each method. Many methods are the result of the integration of classical biochemical techniques with modern high-throughput and -omics approaches. The success of each approach not only depends on the availability of state-of-the-art instruments ensuring the highest sensitivity and reliability but also on the experience and care of the operators in sample handling, as well as on using the best reference/control to distinguish false-positive interactors from the right ones. Finally, the presence of an internal reference can be employed as a good QC element, helping in the evaluation of method performance and purification yield.

Before detailing the individual analytical strategies, it is essential to consider the principles of sample treatment and data quality control, which provide the foundation for reproducible and physiologically meaningful proteomic analyses across all experimental contexts.

1.1 Sample Treatment and Data Quality

Sample preparation is a critical methodological phase that fundamentally determines the sensitivity, specificity, and reproducibility of proteomics-based analyses. Despite being often implicit in methodological descriptions, rigorous sample treatment represents the cornerstone for obtaining reliable and physiologically meaningful interaction data [20, 117]. Within affinity purification, cross-linking mass spectrometry, chromatin immunoprecipitation, and label-free proteome profiling workflows, experimental success depends on careful optimization of lysis conditions, buffer composition, and handling procedures that preserve the native state of biomolecules while maintaining compatibility with downstream LC-MS/MS analysis. Critical parameters include the use of volatile, MS-compatible buffers (e.g., ammonium acetate or bicarbonate), avoidance of harsh detergents such as SDS prior to digestion, and incorporation of protease and phosphatase inhibitors to prevent post-lysis degradation or dephosphorylation [20, 58, 168]. Mechanical or enzymatic disruption such as mild sonication or nuclease trimming should be empirically optimized to ensure complete extraction without disrupting protein–protein or protein–nucleic acid assemblies [49]. Temperature control (typically 4 °C throughout the workflow) and pre-clearing of extracts to remove nonspecific binders further enhance specificity and reproducibility. Quality control (QC) assessment prior to MS analysis is equally essential and should include: (i) evaluation of protein integrity and purity by SDS-PAGE or CE-SDS; (ii) protein concentration quantification by Bradford or BCA assays; (iii) assessment of sample homogeneity or aggregation by DLS or SEC; and (iv) verification of labeling efficiency or tag removal for proximity-labeling or activity-based assays [20, 117, 158]. Establishing such standardized QC checkpoints ensures that observed molecular interactions reflect biological reality rather than artifacts of sample manipulation. These principles are universally applicable across the techniques discussed in the following sections from affinity-based and cross-linking approaches to chromatin- and RNA-centered assays and label-free drug-binding workflows providing a conceptual framework for reproducible interactome analysis. Given the central role of sample handling in determining the accuracy, reproducibility, and biological relevance of proteomics-based interaction studies, the principal experimental parameters and recommended best practices for sample preparation and quality control are summarized in Table 1.

Table 1 Key experimental parameters and recommended best practices for sample treatment and quality control in proteomics workflows investigating biomolecular interactions. Parameters include lysis conditions, buffer composition, temperature control, and quality assurance checks prior to LC-MS/MS analysis. These standardized principles provide a methodological foundation for the experimental approaches discussed in the following sections, ensuring consistency and reproducibility across diverse proteomic platforms

Parameter	Recommended practice	Rationale/notes	References
Lysis conditions	Perform extraction at 4 °C using mild, non-denaturing buffers (e.g., 20 mM HEPES, 150 mM NaCl)	Preserves native protein–protein and protein–nucleic acid interactions	Rogers and Bomgarden [117]; Duong and Lee [20]
Detergent use	Avoid SDS and other harsh surfactants; use MS-compatible detergents (e.g., octyl-β-D-glucopyranoside $\leq 1\%$, NP-40 $\leq 0.5\%$)	Maintains weak interactors and ensures LC-MS/MS compatibility	Yang et al. [168]; Kachuk et al. [58]
Protease and phosphatase inhibitors	Add inhibitor cocktails during lysis and washing	Prevents post-lysis degradation and dephosphorylation	Duong and Lee [20]
Mechanical disruption	Apply controlled sonication or nuclease trimming; avoid overheating	Ensures extraction efficiency without damaging complexes	Iacobucci et al. [49]
Pre-clearing step	Incubate extracts with blank beads before immunoprecipitation	Removes nonspecific binders, improving specificity	Kaboord and Perr [57]
Temperature control	Maintain all steps at 4 °C (except digestion)	Limits proteolysis and complex dissociation	Rogers and Bomgarden [117]
QC by electrophoresis	Check protein integrity by SDS-PAGE or CE-SDS	Detects degradation or contamination before MS	Duong and Lee [20]
Concentration assessment	Quantify proteins by Bradford, or BCA assays	Ensures reproducible loading and normalization	Duong and Lee [20]
Homogeneity/aggregation	Verify by dynamic light scattering (DLS) or size-exclusion chromatography (SEC)	Confirms monodispersity of complexes	Rogers and Bomgarden [117]
Labeling verification	Assess labeling or tag removal efficiency; eliminate free label	Prevents artifacts in fluorescence or proximity-based assays	Wang et al. [158]

2 Methods for the Investigation of Protein–Protein Interactions

Protein interactions are fundamental to cellular functions, underlying both physiological and pathological processes. Proteomics techniques, empowered by mass spectrometry (MS) and bioinformatics, have revolutionized the study of protein–protein interactions (PPIs). This section explores the main methodologies employed to investigate protein interactions, highlighting recent innovations. Despite significant advancements, researchers face several challenges in studying protein–protein interactions (PPIs), and the selection of the most appropriate method often depends on the specific research question and the nature of the biological sample.

2.1 *Targeted Methods*

Affinity Purification Mass Spectrometry (AP-MS) Approaches

AP-MS is a widely used approach for studying PPIs in a near-native state, offering a robust approach to identify both stable and dynamic protein complexes in biological systems. AP-MS relies on isolating protein complexes using a bait protein immobilized on a solid support, followed by mass spectrometry analysis to identify co-purified interactors [49]. Within the AP-MS framework, two key experimental approaches are widely used: pull-down [77] and co-immunoprecipitation assays (Co-IP) [99].

The pull-down assay is an in vitro approach that exploits the affinity of a bait protein, often tagged with a fusion tag (e.g., GST, His-tag, or biotin) to facilitate purification. The tagged bait protein is immobilized on beads functionalized with a complementary ligand, such as glutathione for GST-tagged proteins or nickel ions for His-tagged proteins. Once immobilized, the bait is incubated with a cellular or tissue lysate, allowing the formation of non-covalent interactions with potential partners under quasi-physiological conditions. In Table 2, we summarized the mail pull-down systems. After a series of stringent washing steps to remove non-specific binders, the bait and its associated proteins are eluted and identified by mass spectrometry–based approaches. In particular, in a typical pull-down assay, the tagged bait protein is either pre-purified or overexpressed in cell lines and immobilized onto beads pre-functionalized with bait ligands such as glutathione-Sepharose for GST-tagged proteins or Ni-NTA agarose for His-tagged proteins.

To start the experimental procedure, cellular lysates are prepared under conditions that preserve protein–protein interactions, utilizing mild detergents like NP-40 or digitonin to maintain native conformations. The lysate is incubated with the immobilized bait protein under optimized conditions, such as controlled temperature (typically 4 °C to prevent degradation) and gentle agitation to facilitate binding while minimizing nonspecific interactions. Protease and phosphatase inhibitors are commonly added to preserve proteins and post-translational modifications.

Following incubation, non-specific proteins are removed through multiple washing steps using progressively stringent buffers. Initial washes typically contain a low concentration of non-ionic detergent (e.g., 0.1–0.5% NP-40, Triton X-100, or Tween20) to remove weakly associated proteins while preserving native interactions [85]. Increasing salt concentrations (e.g., 150–500 mM NaCl or KCl) can further reduce background binding [9], and additional washes with mild chaotropic agents (e.g., urea at 1–2 M) may help disrupt transient interactions if needed [156].

Table 2 Summary of the main pull-down systems. The table reports the corresponding immobilized ligands, representative protein counterparts, typical elution strategies, and practical notes highlighting critical considerations for each approach

Tag/ Approach	Immobilized ligand	Example of protein counterpart	Elution method	Practical notes
GST-tag	Glutathione	GST-fused proteins	Competitive with reduced glutathione	Glutathione must be absent from binding and wash buffers to avoid competition with the resin.
His-tag (6xHis)	Ni^{2+} or Co^{2+} ions	Recombinant his-tagged proteins	Competitive with imidazole	Co^{2+} reduces background but may have lower binding capacity than Ni^{2+}.
Biotin	Streptavidin or avidin (magnetic/ agarose beads)	Biotinylated proteins or small molecules	Denaturing conditions (SDS, urea) or free biotin (competitive)	Extremely strong binding (Kd ≈ 10^{-15} M), often requiring harsh conditions for elution
FLAG-tag	Anti-FLAG M2 antibody beads	FLAG-epitope fused proteins (DYKDDDDK)	Free 3xFLAG peptide (competitive)	Allows mild/ nondenaturing elution
HA-tag / Myc-tag/ V5-tag	Anti-HA, anti-Myc, or anti-V5 antibodies immobilized beads	HA (YPYDVPDYA), Myc (EQKLISEEDL), or V5 (GKPIPNPLLGLDST) tagged proteins	Free peptide competitor or low pH elution	Peptide competition enables gentle elution

The bound protein complexes are then eluted using methods tailored to the nature of bait and experimental goals. Competitive elution can be achieved by adding excess ligand or affinity competitor, such as free glutathione for GST-tagged proteins or imidazole for His-tagged proteins. Alternatively, more stringent elution methods include changing pH (e.g., glycine-HCl pH 2.5–3.0) or using denaturing conditions (e.g., SDS buffer or 8 M urea), which are particularly useful for recovering tightly bound complexes.

The eluted proteins are then concentrated and buffer-exchanged, if necessary, by protein precipitation, followed by enzymatic digestion. Trypsin is the most commonly used protease, typically added at a 1:50 to 1:100 enzyme-to-substrate ratio, but alternative proteases (e.g., Lys-C or Glu-C) can be used to generate complementary peptide fragments for enhanced coverage. Digestion is performed at 37 °C for 4–16 h, sometimes preceded by reduction (e.g., 5 mM DTT) of disulfide bridges and alkylation (e.g., 10–20 mM iodoacetamide) of thiol-free cysteine residues to improve peptide recovery.

Pros and Cons The main advantages of this approach consist of the fact that it is straightforward, classically quite standardized, cheap, and in the commercial availability of many tagged proteins at highest purity levels. The standardization of each step is largely due to the widespread use of well-characterized tags such as GST, His, or FLAG. This flexibility allows researchers to shift protocols efficiently from one bait protein to another, reducing the need for extensive re-optimization and enabling rapid experimental setup across diverse biological contexts. However, the technique is not without limitations. It could retrieve that not all PPIs detected by AP occur in vivo because of bait/interactor different cell localization; moreover, also the unbalanced molecular ratio of the protein of interest and its binders, being the first in large excess, might force the occurrence of non-native interactions. These conditions can basically favor the presence of false positives, additionally enriched by non-specific binding proteins because of their interaction with the beads, tag, or other assay components, obscuring genuine interactions. To address this, the inclusion of appropriate control is fundamental to distinguish bona fide interactors from background contaminants. For instance, control experiments using beads alone or lysates from cells expressing only the affinity tag can reveal proteins that bind non-specifically to the matrix, the tag, or other assay components. Similarly, the meticulous optimization of washing conditions is essential to strike a balance between removing contaminants and retaining true interactors. Bioinformatics resources, such as CRAPome interface, can be also used to discriminate between bona fide and false-positive interactors [71]. Quantitative proteomics can further distinguish true interactors from non-specific binders by comparing protein abundances in the bait-enriched interactome and in appropriate controls, incorporating replicates to apply statistical filters. The enrichment of a potential interactor in the bait-containing setup is assessed by calculating the fold change (protein intensity in the pull-down vs. control) [50]. Finally, the pull-down assay is less effective for capturing transient or weak interactions, as these are often disrupted during the washing and elution steps.

Co-immunoprecipitation (Co-IP) is a widely used ex vivo technique that relies on antigen–antibody recognition to purify protein complexes containing the protein of interest directly from cellular lysates. An antibody specific to the bait protein is immobilized on beads (e.g., protein A/G), which are then incubated with lysates to capture the bait and its interactors. The immune complex is subsequently eluted and analyzed by mass spectrometry, as previously described for the pull-down assay.

In a typical Co-IP experiment, the selection of high-affinity, specific antibodies for the bait protein is critical for successful interaction studies. When such antibodies are unavailable, epitope-tagged versions of the bait protein, such as FLAG, HA, or Myc tags, can be used with corresponding commercially available antibodies. The chosen antibody is immobilized on beads through its interaction with protein A or G, depending on the antibody subclass. The binding, washing, and elution steps are similar to those described for pull-down assays.

However, in the Co-IP case, when denaturing buffers are used to elute the bait and its interactors, antibody chains may contaminate the eluate, complicating protein identification. To minimize contamination from antibody heavy and light chains during elution, cross-linking reagents like DSS (disuccinimidyl suberate) can be used to covalently bind the antibody to the beads [16].

Pros and Cons Differently from pull-down, Co-IP has the advantage of capturing protein complexes as they naturally occur in vivo, preserving their physiological relevance. This makes it a powerful tool for studying interactions in their native cellular context. Similarly to the pull-down assay, Co-IP can detect not only direct interactions between the bait protein and its partners but also indirect interactions mediated through larger protein complexes, providing a broader view of the interactome.

However, the technique has some limitations. The quality and specificity of the antibody used are critical to the success of the experiment; low-quality or non-specific antibodies can lead to unreliable results. Moreover, each target protein and sample type require optimization of lysis and washing conditions to ensure reproducible results. In fact, similarly to the pull-down approach, false-positive binders can compromise the robustness of Co-IP data.

Equally critical is the inclusion of proper negative controls, such as immunoprecipitation with an unrelated antibody or with beads lacking antibody, which help discriminate genuine interactors from proteins binding non-specifically to the beads, antibody Fc regions, or the tag itself. These controls, together with replicate experiments, provide a baseline to improve the reliability of Co-IP-derived interactomes.

Unlike AP-MS, where the bait tag provides maximal surface exposure, Co-IP may mask the native epitope recognized by the antibody, potentially affecting complex recovery. This limitation is overcome when the bait is expressed with a tag such as FLAG, HA, or Myc, since the non-native region (the tag) is bound by the specific antibody. In this context, an additional advantage is the possibility of performing competitive elution using an excess of free peptide tag, thereby reducing non-specific interactors.

In both pull-down and Co-IP approaches, the purification methods lead to dissociation of protein complexes, thus losing information about the native complex composition. Other methods, such as BN-PAGE or SEC purification (see below), when coupled with AP or Co-IP before protein identification, allow retention of the supramolecular arrangement of complexes.

2.2 Methods to Investigate Transient Interactions

Proximity-Dependent Labeling (PDL) Techniques

Proximity-Dependent Labeling (PDL) methods, such as BioID [123] and APEX [96], have revolutionized the ability to capture transient and weak PPIs, particularly in their native cellular context. PDL methods such as BioID, TurboID [5], and APEX allow in vivo labeling of proteins in close proximity to a bait protein. These techniques involve fusing an enzyme, such as biotin ligase (BioID) or ascorbate peroxidase (APEX), to a bait protein. The enzyme catalyzes the labeling of nearby proteins through biotinylation, marking those within a specific radius (e.g., 10–20 nm). Biotinylated proteins are enriched by pulling them down with streptavidin-derivatized beads. The advantage of PDL lies in its ability to snapshot interactions within living cells, capturing both stable and fleeting interactions that would otherwise be lost during sample preparation.

The main PDL variants include:

1. BioID, which utilizes a mutated biotin ligase (BirA*) that promiscuously biotinylates lysine residues on proteins within a 10-nm radius. The biotinylated proteins are subsequently purified using streptavidin-coated beads and identified by mass spectrometry. While BioID is highly effective, it requires long labeling times (16–24 h) due to slow BirA* kinetics.
2. *TurboID and miniTurboID* are optimized versions of BioID. TurboID dramatically reduces labeling time to 10–30 min by increasing the catalytic efficiency of the biotin ligase. MiniTurboID employs a truncated BirA* variant that minimizes potential steric hindrance while maintaining fast labeling.
3. *APEX* labels proteins through catalysis of biotin-phenol in the presence of hydrogen peroxide. APEX operates on a different principle than BioID, generating short-lived radicals that label electron-rich residues such as tyrosines. This reaction is extremely fast (within 1 min), making APEX ideal for capturing highly dynamic interactions.
4. *TurboID-APEX,* a TurboID–APEX hybrid approach [111, 178] combines features of both systems, allowing researchers to tailor labeling kinetics and subcellular compatibility to experimental needs.

A typical PDL experiment follows a well-defined workflow designed to capture and analyze protein–protein interactions in a native cellular context. The process begins with the design of a construct in which the gene encoding the protein of interest is fused to the coding sequence of the labeling enzyme (APEX or biotin ligase). This fusion protein is then expressed in the cell line under investigation, ensuring that the bait protein retains its natural ability to interact with partners in its native environment. The correct localization and biological activity of the fusion protein must be validated to exclude interference from the tag.

Once validated, the next step is labeling the interacting proteins. For BioID or TurboID-based experiments, cells are treated with biotin, which serves as the substrate for the biotin ligase. The ligase biotinylates nearby lysine residues on proteins in close proximity to the bait. In APEX-based experiments, a combination of biotin-phenol and hydrogen peroxide is used to generate reactive radicals that label neighboring proteins. Optimization of labeling time and reagent concentration is essential to minimize non-specific background.

After labeling, cells are lysed, and proteins are extracted under denaturing conditions to ensure complete solubilization and to disrupt non-specific interactions that may form during preparation. Biotinylated proteins are then enriched using streptavidin affinity columns, which specifically bind biotin with high affinity. This purification step is critical for isolating the labeled interactors from the complex cellular lysate.

The final step involves mass spectrometry (MS) analysis to identify the purified proteins. This workflow enables comprehensive analysis of protein interaction networks, especially transient or weak ones often missed by other methods.

Pros and Cons BioID and APEX are both powerful proximity-labeling techniques, each with distinct advantages and limitations. *BioID* is well-suited for mapping stable PPIs and subcellular environments, as it uses a mutant BirA* biotin ligase to label proteins within ~10 nm. Its main advantages include high specificity, compatibility with live cells, and a non-toxic labeling process using biotin. However, BioID requires long labeling times (6–24 h), making it unsuitable for capturing transient interactions. Additionally, its relatively large enzyme size (~35 kDa) may cause steric hindrance.

APEX, in contrast, offers high temporal resolution, labeling nearby proteins within seconds using biotin-phenol and hydrogen peroxide (H_2O_2). This makes it ideal for studying fast and transient interactions, such as those in signal transduction [103] or synaptic activity [147]. APEX is also effective in live and fixed cells, allowing precise subcellular mapping. However, it presents some drawbacks: H_2O_2 toxicity must be tightly controlled, its labeling radius (~20 nm) is less defined, and oxidative stress can induce unwanted protein modifications that complicate MS analysis.

Table 3 summarizes the main features of BioID and APEX. Briefly, BioID is best suited for long-term interactions and stable compartmentalization studies, whereas APEX excels in capturing rapid and dynamic processes. Ultimately, the choice between them depends on the biological question: BioID provides stability and specificity, while APEX offers speed and sensitivity but requires careful optimization to avoid cellular toxicity.

Cross-Linking Mass Spectrometry (XL-MS)

Cross-Linking Mass Spectrometry (XL-MS) is a versatile technique that combines the power of covalent chemistry and mass spectrometry (MS) to study protein–protein interactions (PPIs) at a molecular level [110]. By covalently linking amino acid residues that are spatially close within or between proteins, XL-MS not only identifies interactors but also provides structural insights into the spatial arrangement of proteins within a complex. This makes it particularly suited for probing transient or weak interactions and mapping interaction interfaces within multiprotein assemblies.

The experimental setup begins with the selection of appropriate cross-linking reagents, typically bifunctional molecules containing reactive groups targeting specific amino acid side chains, such as lysines, cysteines, or carboxylic groups. Table 3 summarizes the most common reagents and their properties.

Proteins or protein complexes, either in solution or in intact cells, are incubated with the cross-linking reagent under controlled conditions. The reactive groups

Table 3 Main cross-linkers commonly used in cross-linking mass spectrometry (XL-MS). The table summarizes their chemical reactivity, spacer arm length, MS-cleavability, solubility, and key features, highlighting the advantages and limitations that influence their applicability in structural proteomics workflows

Cross-linker	Reactive groups/ primary targets	Spacer arm (Å)	MS-cleavable	Aqueous solubility	Key features/notes
DSSO [51, 60]	NHS esters (primarily lysines)	10–30 Å range	Yes	Standard (non-sulfo)	Cleavable cross-linker that generates diagnostic fragment ions, simplifying XL-MS identification
BS3 [132]	NHS esters (primarily lysines)	10–30 Å range	No	Water-soluble (sulfo-NHS)	Widely used non-cleavable reagent; favorable in aqueous buffers due to sulfonate groups.
DSSBU [51, 143]	NHS esters (primarily lysines)	10–30 Å range	Yes	Standard (non-sulfo)	Newer MS-cleavable analogue of DSS/DSBU-type linkers; produces characteristic cleavage patterns that aid identification
DSBSO [55]	NHS esters (primarily lysines)	10–30 Å range	Yes	Standard (non-sulfo)	Clickable (alkyne) handle enables streptavidin-based enrichment of cross-linked peptides for improved sensitivity in complex samples.

form covalent bonds between residues that are spatially close (typically within 10–30 Å), thereby stabilizing transient or weak interactions and capturing a "snapshot" of the interaction network.

After cross-linking, the proteins are digested into peptides with trypsin, producing a complex mixture of linear, intra-, and inter-protein cross-linked peptides. Because cross-linked peptides are scarce compared to linear ones, enrichment strategies such as SCX or SEC are often employed.

Recently, the conjugation of biotin to cross-linked peptides via a clickable cross-linker (alkyne-DSBSO) has been used to study the interactome in patient-derived xenograft breast cancer models [55]. In this approach, biotin-labeled peptides are enriched using streptavidin-based pull-down and analyzed by high-resolution LC-MS/MS.

MS-cleavable cross-linkers like DSSO and DSBSO have gained popularity because they generate diagnostic fragment ions that simplify peptide identification. Specialized software such as XlinkX [74], MeroX [47], or pLink [24] enables identification of cross-linked peptides and mapping of linkage sites for structural modeling.

Quantitative XL-MS approaches allow dynamic comparison of interaction networks under different conditions. Photoactivatable cross-linkers now permit in vivo capture of PPIs, revealing that interactions in XL-MS have undergone major evolution through automation, high-throughput workflows, and computational innovation, enabling proteome-wide analysis of protein interactions and structures with unprecedented speed and resolution. Key advances include integration of automated pipelines, new reagents, and improved search algorithms.

Liao et al. [72] demonstrated how novel reagents and automated workflows have extended XL-MS to complex biological systems. Jiang et al. [54] showed that streamlined in vivo cross-linking can reduce acquisition time to ~15 h without compromising data quality. Enhancements in separation techniques, such as two-dimensional fractionation [46, 56], have further expanded the structural proteomic landscape.

In parallel, software developments have been crucial. Tools like eXL-MS [97, 115] and the 4D-diaXLMS workflow [39] now allow high-accuracy data processing even for low-abundance interactions, combining intuitive interfaces with advanced analysis features.

Furthermore, integration of structural modeling with XL-MS exemplified by pipelines like Sidewinder [141] enhances structural interpretation even from minimal samples. Building upon these advances, AlphaCross-XL [131] represents a major step forward, automating the mapping of cross-linked peptides onto 3D protein structures. This integration improves mapping accuracy and streamlines large-scale XL-MS data interpretation.

Pros and Cons One of XL-MS's greatest strengths is its ability to provide spatial information about protein complexes, revealing how proteins interact at the molecular level. This makes it invaluable in structural biology, complementing cryo-EM and X-ray crystallography by providing distance constraints for modeling large assemblies.

Additionally, XL-MS captures transient or weak interactions often missed by conventional affinity purification methods. By covalently linking proteins in close proximity, it stabilizes dynamic complexes that exist only briefly in cells. Another major advantage is its applicability to intact cells and tissues, preserving the native environment of protein complexes and enabling study of endogenous PPIs.

Despite these advantages, XL-MS faces challenges that have limited its widespread adoption. Sample preparation remains complex, as cross-linking reactions generate heterogeneous peptide mixtures that complicate purification and enrichment.

Data analysis is also demanding, requiring sophisticated bioinformatics tools like XlinkX, MeroX, and pLink for accurate identification and interpretation. Specificity of cross-linkers is another limitation, as most reagents (e.g., DSSO) primarily target lysines, potentially missing low-lysine regions. Furthermore, high reagent costs and the need for advanced mass spectrometers make XL-MS technically demanding.

To overcome these limitations, recent advances including MS-cleavable cross-linkers such as DSSO and DSBU have simplified data analysis by generating characteristic fragment ions that aid identification and increase throughput.

2.3 *Untargeted Methods to Study PPIs*

Blue Native Polyacrylamide Gel Electrophoresis (BN-PAGE)

Blue Native Polyacrylamide Gel Electrophoresis (BN-PAGE) is a powerful technique for studying PPIs under native conditions, preserving protein complexes and allowing their separation based on size and shape rather than just molecular weight. Unlike denaturing electrophoresis methods such as SDS-PAGE, BN-PAGE maintains the integrity of protein complexes, making it particularly useful for analyzing membrane proteins, large protein assemblies, and dynamic protein interactions [94]. The primary reagent in this technique, Coomassie Brilliant Blue G-250 dye, interacts with proteins to confer a negative charge without fully unfolding them, allowing their separation according to native mass-to-charge ratios [64]. Higher resolution can be achieved through a second denaturing electrophoresis step (native 2D electrophoresis) that separates individual protein components by molecular weight. This method has become a key tool in proteomics, often coupled with mass spectrometry (MS) to identify protein complex components and understand their functional roles.

The extraction of protein complexes, especially from membrane sources, presents a critical challenge due to the need to preserve native interactions. Mild nonionic detergents such as digitonin or dodecyl maltoside are typically used to solubilize membrane proteins while keeping them functional [142]. Cytosolic complexes, instead, are extracted with mild ionic buffers that sustain interactions and prevent aggregation [163].

BN-PAGE relies on Coomassie Brilliant Blue G-250 dye, which binds to proteins and imparts a uniform negative charge. Unlike SDS, which completely unfolds proteins, Coomassie dye stabilizes protein complexes through hydrophobic interactions, preventing aggregation while allowing migration through the gel based on native mass-to-charge ratio. Additionally, the dye enables direct visualization of protein bands. Proteins are loaded onto a gradient polyacrylamide gel and migrate under an electric field; larger complexes move more slowly, while smaller ones migrate further. The choice of gradient concentration (typically 3–16% acrylamide) is crucial for resolving complexes of different sizes [64].

Coupling BN-PAGE with MS enhances its proteomic applications, particularly for identifying individual components of protein complexes. For instance, bands can be excised, digested with trypsin, and analyzed via LC-MS/MS to determine each component's identity [92].

Pros and Cons BN-PAGE is especially suitable for studying membrane-bound protein complexes, which are often challenging to analyze using traditional proteomics methods. Unlike affinity purification, BN-PAGE does not require prior knowledge of complex composition, enabling unbiased interactome analysis. Additionally, it provides information on the stoichiometry of protein complexes, allowing distinction between different oligomeric states. Despite its advantages, BN-PAGE has several limitations. Resolution can be problematic for complexes of similar sizes that migrate closely together, making gradient optimization (3–16%) essential [19]. Another drawback is that Coomassie dye, while stabilizing complexes, can partially denature some proteins and disrupt weak or transient interactions [104]. To mitigate this, BN-PAGE can be combined with cross-linking mass spectrometry (XL-MS) to stabilize weak interactions and achieve a more accurate representation of native complexes [94].

Several advancements have improved BN-PAGE resolution, sensitivity, and applicability in proteomics. Recent optimization of gel composition and buffer systems has led to high-resolution BN-PAGE (hrBN-PAGE), offering better separation of closely migrating complexes. A major breakthrough was the development of cryo-slicing BN-MS (csBN-MS), which freezes BN-PAGE gels and slices them into ultra-thin sections for MS analysis [91]. This approach provides unprecedented resolution in mapping protein complex distributions and allows integration of quantitative proteomics for comparative studies of protein complexes under different conditions enhancing the investigation of disease-related interactome changes.

Size Exclusion Chromatography–Mass Spectrometry (SEC-MS)

Size Exclusion Chromatography coupled with Mass Spectrometry (SEC-MS) has become an essential tool for studying protein–protein interactions (PPIs), particularly in complex and dynamic biological systems. This technique separates proteins and protein complexes according to their hydrodynamic size before subjecting them to mass spectrometry analysis, while maintaining their native conformation during separation [10]. The integration of SEC with mass spectrometry enhances analytical sensitivity and preserves the structural integrity of protein complexes, which is critical for accurately identifying interaction partners without prior knowledge of their identities [149]. As such, SEC-MS stands out as a non-biased approach, allowing for the discovery of novel protein complexes that might otherwise remain unnoticed in traditional affinity-based methods [155].

The sample preparation phase is critical for successful SEC-MS applications. Proteins must be extracted under native conditions to ensure that their interactions are preserved [149]. The selection of a buffer compatible with mass spectrometry is essential; ammonium acetate or other volatile salts are typically preferred to avoid ion suppression caused by non-volatile phosphate or Tris buffers [23]. Mild detergents or stabilizing agents may be used when working with membrane proteins.

During SEC, larger complexes elute first, as they cannot penetrate the porous structure of the stationary phase, while smaller molecules diffuse through the pores and elute later. This separation enables resolution of different oligomeric states, supporting studies of complex assembly and disassembly [106]. This is especially useful in drug discovery, where SEC-MS can monitor how small molecules influence protein complex stability. The SEC-separated fractions are directly analyzed by MS, either in native MS mode (preserving protein complexes) or after enzymatic digestion to identify the composition of individual subunits. High-resolution instruments such as Orbitrap and Q-TOF allow for precise mass determination, revealing stoichiometry and heterogeneity among protein complexes. The elution profiles can be quantitatively assessed to map co-eluting species, which suggest stable interactions and thus indicate biological relevance [14]. Advanced data processing tools, such as PCprophet [27] and CCprofiler [4], enable comprehensive reconstruction of protein complex architecture through algorithmic deconvolution and computational modeling. Recent years have seen significant improvements in SEC-MS technology, addressing many of its limitations and expanding its scope for PPI studies. Advances in column design and ultrahigh-performance liquid chromatography (UHPLC) have substantially improved SEC resolution, allowing separation of complexes with similar hydrodynamic radii [172]. Additionally, SEC-MS is well-suited for high-throughput interactome analyses, particularly when combined with Data-Independent Acquisition (DIA)-MS techniques such as SWATH-MS [43]. This integration enables quantitative monitoring of protein complex dynamics across different conditions, such as healthy versus diseased or treated versus untreated samples.

Pros and Cons SEC-MS has emerged as a powerful and versatile platform for investigating protein–protein interactions under near-native conditions. Its ability to preserve structural integrity, combined with high analytical sensitivity and resolution, makes it particularly effective for characterizing the composition, stoichiometry, and dynamics of both stable and transient interactions. Despite its advantages, SEC-MS entails several challenges that complicate data interpretation. Co-elution of similarly sized complexes can impair the identification of individual species. Additionally, dilution effects during SEC may occur. As protein complexes pass through the column, they are distributed across multiple fractions, reducing concentration and making low-abundance interactions difficult to detect. This may lead to the loss of weak or transient interactors due to insufficient signal intensity. Buffer composition and chromatography conditions must therefore be carefully optimized for each sample type, since incompatible conditions can destabilize protein interactions. Moreover, transient interactions that characterize dynamic PPIs may dissociate during SEC separation, necessitating complementary techniques such as cross-linking mass spectrometry (XL-MS) to capture these fleeting events.

Overall, AP-MS, BioID, and XL-MS address complementary aspects of PPI biology: affinity capture is optimal for stable complexes, proximity labeling improves recovery of weak/transient contacts in vivo, and XL-MS provides structural constraints by “freezing” interactions in space (Fig. 1). Their combined use strengthens confidence in interaction calls and supports more robust network reconstruction from LC-MS/MS data.

To provide a comparative view, Table 4 summarizes the main methodologies employed to study protein–protein interactions (PPIs), highlighting their resolution, experimental context, major strengths, and inherent limitations. This overview complements the detailed sections above and supports the selection of the most suitable method based on experimental goals and system characteristics.

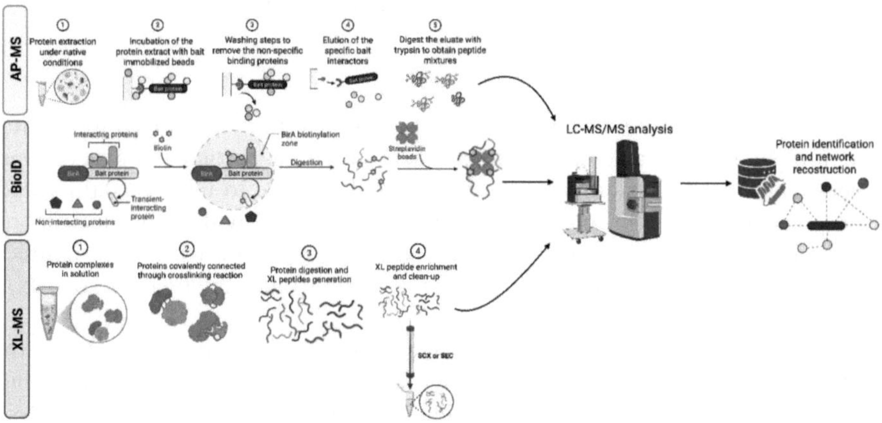

Fig. 1 Proteomics workflows to map protein–protein interactions (PPIs) by mass spectrometry

Table 4 Comparative overview of the main techniques used to investigate protein–protein interactions. The table highlights resolution, experimental context, strengths, and limitations of each method

Technique	Resolution	Context	Strengths	Limitations
AP-MS (affinity purification–mass spectrometry) [49]	Moderate	Ex vivo/in vitro	Robust, standardized, suitable for stable complexes	Limited for transient/ weak interactions; false positives if controls are lacking
Co-IP (co-immunoprecipitation) [99]	Moderate	Ex vivo	Captures interactions in near-native complexes, preserves cellular context	Dependent on antibody quality; antibody contamination; weaker for transient interactions
Proximity-dependent labeling (PDL: BioID, APEX, TurboID) [5, 96, 123]	High	In vivo	Captures transient/ weak interactions; high temporal/ spatial resolution	Requires bait fusion; possible artifacts; labeling efficiency varies
XL-MS (cross-linking mass spectrometry) [110]	High (structural)	In vitro/in vivo	Provides spatial constraints; stabilizes weak/ transient PPIs	Data analysis complex; depends on cross-linker chemistry
BN-PAGE (Blue native PAGE) [94]	Moderate	In vitro/ex vivo	Preserves protein complexes in native state; useful for large assemblies	Limited resolution for similar-sized complexes; may disrupt weak interactions
SEC-MS (size exclusion chromatography–MS) [149]	Moderate	In vitro	Separates complexes by size under native conditions; unbiased discovery	Dilution effect; transient interactions often lost; resolution issues for co-eluting complexes

3 Methods for the Investigation of Protein–Nucleic Acid Interactions

Over the years, a variety of powerful techniques have been developed to investigate protein–nucleic acid interactions, each suited to particular contexts and experimental needs. Among these, several stand out due to their versatility and widespread adoption. One of the most commonly used methods is the electrophoretic mobility shift assay (EMSA), employed to detect and measure in vitro direct binding between a protein and a nucleic acid molecule, either DNA or RNA [41, 83].

For in vivo studies of protein–DNA interactions within chromatin, chromatin immunoprecipitation (ChIP), followed by either quantitative PCR or sequencing (ChIP-qPCR/seq), allows for genome-wide mapping of protein-binding sites [15]. To study protein–RNA interactions in living cells, techniques like RNA immunoprecipitation (RIP) or enhanced CLIP (eCLIP) offer transcriptome-wide resolution.

More targeted approaches have emerged with the advent of CRISPR-based technologies, including methods such as CAPTURE and dCas9 fusion systems, which enable locus- or transcript-specific recruitment of protein complexes in vivo [76]. Finally, to better understand how RNA structure influences binding, methods like SHAPE-MaP can be applied to map RNA secondary structures either in vitro or in living cells.

Subsequent sections will explore these techniques in detail, describing the principles behind each method, the requirements for their implementation, and criteria for selecting the most appropriate approach for a given biological question. In this context, in vitro affinity-based strategies have proven particularly powerful for capturing and characterizing nucleic acid–binding proteins under controlled experimental conditions [145].

3.1 In Vitro Affinity Approaches for the Investigation of DNA/RNA–Protein Interactions

Affinity Purification–Based (AP) Methods Coupled with MS

Affinity purification–based (AP) methods are widely applied to study protein–nucleic acid (NA) interactions in proteomics. Here, DNA or RNA oligonucleotides are employed as molecular baits to selectively capture their interacting proteins from cell or nuclear extracts. Analogously to other affinity-based technologies, the key lies in recreating biologically meaningful molecular contexts that reflect the native physicochemical features of the interaction surfaces.

A critical step in any nucleic acid–affinity approach is the design of the bait sequence. The probe must be long enough to accommodate the necessary structural and sequence elements for the selection of specific interaction but not so extended as to increase non-specific background [112]. Some designs include multimerized versions of a known motif to favor the capture of low-abundance or transiently bound proteins, such as transcription factors or RNA-binding proteins [8]. In studies involving structured RNAs, it is often necessary to account for folding predictions, as secondary and tertiary structures can either facilitate or obstruct protein recognition [122].

Once the probe has been designed, it must be immobilized in a way that maintains accessibility and binding capacity. Biotinylated oligonucleotides coupled to streptavidin-coated magnetic beads are among the most commonly used systems because of their robustness and ease of use [15]. Alternatively, covalent immobilization on derivatized resins offers higher stability, particularly when harsher washing steps are required [145]. In both cases, linkers are often used to distance the functional sequence from the bead surface, reducing steric hindrance and promoting more native-like binding.

The success of the capture experiment is largely dictated by buffer composition, which must be optimized to favor specific interactions without compromising protein or nucleic acid integrity [112]. Buffers are typically formulated to reflect physiological pH, often incorporating monovalent salts to moderate nonspecific ionic interactions. Divalent cations such as Mg^{2+} or Ca^{2+} may be added when the structure of the nucleic acid bait depends on their presence, as is the case for certain folded RNAs or DNA quadruplexes [15, 145].

Protease and RNase inhibitors are usually included to prevent degradation of either bait or prey molecules, especially during prolonged incubations [79]. Additionally, non-specific competitors such as poly(dI-dC) or tRNA may be introduced to reduce background binding, while pre-clearing the extract with uncoupled beads can further enhance specificity [145]. The incubation itself is generally performed at 4 °C to preserve protein conformation and complex integrity, with duration adjusted based on expected binding kinetics.

After binding and washing steps, retained proteins are eluted using conditions that disrupt the protein–nucleic acid interaction or the probe–bead linkage. Elution strategies may involve changes in ionic strength or pH or the introduction of competitor oligonucleotides [8]. The recovered proteins are then digested, typically with trypsin, and the resulting peptides are analyzed via LC-MS/MS [1]. Quantitative proteomics methods such as SILAC, label-free quantification, or tandem mass tags are especially valuable for comparative studies, allowing researchers to assess interaction dynamics across experimental conditions [101]. Recent methodological improvements have led to more refined workflows, including on-bead digestion protocols that minimize sample loss and enrichment strategies to concentrate low-abundance interactors. Computational pipelines now assist in filtering out common contaminants using databases like CRAPome, while functional annotation tools help interpret the biological significance of identified interactors [15].

When integrated with cross-linking (e.g., UV or formaldehyde) or high-resolution genomic platforms (e.g., CRISPR-based recruitment or eCLIP), affinity-based nucleic acid pull-downs provide orthogonal confirmation and deeper insight into the spatial and temporal dynamics of nucleic acid–protein complexes [112]. Emerging CRISPR-based strategies, such as Mega-CRISPR, further expand the functional profiling of protein–nucleic acid complexes by enabling modular, programmable recruitment of regulatory complexes in living cells [37].

Pros and Cons The main limitation of this in vitro approach is the loss of three-dimensional genome organization and absence of competing cofactors. Artificial or misfolded RNA secondary structures may obscure binding sites or promote non-native interactions, while overly stringent washes can remove weak but biologically relevant partners [41]. Careful control experiments are therefore essential. Scrambled sequences, mock beads, and RNase or DNase treatments help assess specificity. Validation through complementary assays such as EMSA for affinity, RIP or CLIP for in vivo validation, or CRISPR-based techniques for locus-specific recruitment strengthens conclusions from affinity capture experiments [15, 37, 112].

Despite these limitations, the combination of rational oligonucleotide design, rigorous optimization, and modern proteomic analysis renders affinity purification a uniquely in vitro informative approach, although it often lacks the resolution to dissect individual binding events or quantify interaction strengths. For these purposes, alternative biochemical assays such as the Electrophoretic Mobility Shift Assay (EMSA) offer complementary insights, enabling detection and analysis of direct binding interactions at the single-molecule level under near-native conditions [41].

Electrophoretic Mobility Shift Assay–Mass Spectrometry (EMSA-MS)

The Electrophoretic Mobility Shift Assay (EMSA), also known as a gel retardation or gel shift assay, is a classical technique used to detect direct interactions between proteins and nucleic acids. The central idea is that when a protein binds to a nucleic acid molecule, whether DNA or RNA, the resulting complex migrates more slowly during non-denaturing gel electrophoresis than the free nucleic acid. This difference in mobility allows researchers to distinguish bound from unbound molecules (Fig. 2) [29, 35, 41].

This assay provides qualitative and semi-quantitative insights into several aspects of protein–nucleic acid interactions. It can be used to determine whether a binding interaction occurs, to assess relative affinity and specificity, to examine how mutations in the nucleic acid or protein affect binding, and to explore the sequence dependence of the interaction [98]. EMSA is particularly well suited for experiments that aim to confirm direct interactions under near-physiological conditions without the need for precise mapping of binding sites. Moreover, EMSA can be integrated into proteomic workflows to identify probe–interacting proteins directly from in-gel digestion (EMSA-MS).

An EMSA-MS protocol typically combines sample preparation for native electrophoretic analysis of protein–nucleic acid complexes, followed by an in situ hydrolysis step for the identification of protein partners. The nucleic acid probe is usually short double-stranded for DNA or single-stranded for RNA and includes the sequence motif of interest, flanked by additional nucleotides to support proper folding and stability. These probes are labeled to enable detection, commonly using fluorescent dyes (e.g., Cy5) or biotin. Once the probe is ready, it is incubated with protein or nuclear extracts (typically 10–20 μL reaction volume) in the presence of non-specific competitors like poly(dI-dC). For RNA-based reactions, RNase

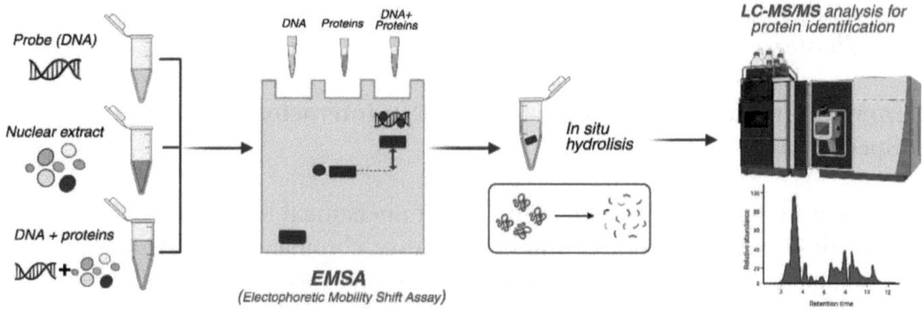

Fig. 2 Schematic representation of the Electrophoretic Mobility Shift Assay (EMSA) for DNA–protein interaction analysis. DNA (probe), nuclear extract, and DNA probes incubated with nuclear extracts are separated by electrophoresis, followed by in situ hydrolysis and LC-MS/MS analysis for protein identification

inhibitors are added to prevent degradation. After incubation, samples are loaded onto a non-denaturing polyacrylamide gel (usually 4–8% depending on probe size) and run under native conditions in TBE buffer. Fluorescent or biotin-labeled samples are visualized by chemiluminescence or fluorescence scanning [41]. Protein–nucleic acid complexes resolved by EMSA can also be excised directly from native gels and subjected to in-gel digestion followed by liquid chromatography–mass spectrometry (LC-MS/MS) to identify bound proteins in an unbiased manner. This EMSA-MS workflow enables the discovery of sequence-specific DNA- or RNA-binding proteins directly from cell or tissue extracts and is particularly useful when the binding partners are unknown or complex [32].

Validation of proteomics results can be achieved through a "supershift assay," where the protein–nucleic acid mixture is incubated with an antibody specific to a known interactor. Successful formation of a larger complex (nucleic acid–protein–antibody) causes a further mobility reduction on the gel, indicating specific binding [41].

Pros and Cons EMSA-MS is a relatively low-cost, technically straightforward, and high-throughput compatible method. It does not require cross-linking and can detect even low-affinity interactions under carefully optimized conditions [29, 41]. Despite its versatility, EMSA-MS shares the same limitations as other in vitro approaches: (i) it does not account for chromatin context or cellular cofactors); (ii) at high protein concentrations, non-specific interactions or aggregation can occur, leading to ambiguous results.

Moreover, working with RNA probes introduces additional complications. RNA is inherently less stable than DNA, so stringent RNase-free conditions and the use of RNase inhibitors are essential. Furthermore, undesired RNA secondary structures may obscure binding motifs, reducing sensitivity and complicating data interpretation.

3.2 *In Vivo Approaches for the Investigation of DNA/RNA–Protein Interactions*

Chromatin Immunoprecipitation (ChIP-MS): Interactome Mapping at Specific *Loci*

Chromatin Immunoprecipitation (ChIP) is a foundational technique for investigating protein–DNA interactions within their native chromatin context. It is particularly valuable for mapping the genomic binding sites of transcription factors, chromatin-modifying enzymes, and histone marks. When coupled with high-throughput sequencing (ChIP-seq), this approach enables genome-wide identification of protein occupancy with high spatial resolution [31].

An important advancement of classical ChIP has been its integration with mass spectrometry, giving rise to ChIP-MS, a hybrid approach that enables the identification of protein complexes localized at defined genomic loci. Rather than isolating the associated DNA for sequencing, ChIP-MS focuses on eluting and identifying proteins cross-linked to chromatin, which are subsequently processed according to adapted proteomics protocols for protein identification [89].

Typically, ChIP-MS employs a dual cross-linking strategy using formaldehyde in combination with longer bifunctional reagents such as DSG (disuccinimidyl glutarate) or EGS (ethylene glycol bis(succinimidyl succinate)), which extend the cross-linking range (7–16 Å) and enable capture of proteins that interact indirectly with DNA as part of larger chromatin assemblies [67, 100].

Ultraviolet cross-linking (typically 254–365 nm) can also be applied in hybrid protocols, particularly for studying transient RNA–protein interactions. Critical for success is the selection of elution buffers compatible with mass spectrometry and free from harsh detergents such as SDS, which can interfere with proteomic analysis [89]. Upon cross-linking treatment, nuclei are purified via fractionated lysis [26], and chromatin is fragmented by controlled sonication to produce ~200–500 bp fragments. Sonication power and duty cycle must be empirically optimized to avoid over- or under-shearing and to minimize inadvertent cross-link reversal during shearing. Dual cross-linking (e.g., FA + DSG/EGS) also enhances resistance to sonication-induced de-cross-linking and preserves indirect interactors [67].

The fragmented chromatin is then subjected to immunoprecipitation with high-affinity, ChIP-validated antibodies (or affinity tags) immobilized on protein A/G magnetic beads, followed by stringent washes to remove nonspecific material. In ChIP-MS and related approaches such as RIME, the immunocaptured, cross-linked complexes provide the proteomic read-out of the interactome at the target locus or around the bait protein [89].

For elution and cross-link reversal (de-cross-linking) prior to proteomic analysis, widely used conditions include incubation of bead-bound complexes in high-salt buffer at 65 °C (often overnight); these MS-compatible workflows are adapted from robust ChIP protocols and validated across multiple systems. Alternatively, on-bead protease digestion can be performed without prior de-cross-linking, minimizing sample loss and simplifying handling [89].

Finally, the released proteins (or on-bead-generated peptides) are cleaned up in detergent-free buffers and analyzed by LC-MS/MS, yielding locus-resolved protein inventories and interaction networks, as originally demonstrated in early ChIP-MS studies and now widely adopted [89]. Optimization of cross-linking conditions is crucial. For example, formaldehyde is typically applied at 1% concentration for 5–15 min, whereas bifunctional cross-linkers require longer incubations (30–60 min).

Pros and Cons Chromatin immunoprecipitation and its modern extensions, such as ChIP-seq and ChIP-MS, remain indispensable tools for investigating protein–DNA interactions with both spatial precision and biological relevance. The reliability of ChIP-based experiments is tightly linked to key experimental parameters including antibody specificity, consistency in chromatin fragmentation, and appro priate signal normalization strategies, such as spike-in controls or input DNA comparison [2, 18, 62, 68]. The integration of ChIP with proteomics has further expanded the ability to characterize chromatin-associated protein complexes with high biochemical resolution [153, 162].

Over-crossl-inking can reduce chromatin accessibility, impair sonication efficiency, and hinder antibody binding, ultimately compromising signal quality [31].

CRISPR-Cas9-Based Approaches

The advent of CRISPR-Cas9 technologies has revolutionized genome engineering, and their applications now extend well beyond genome editing into chromatin biology and molecular interaction mapping. One of the most innovative adaptations involves the use of a catalytically dead Cas9 (dCas9), which retains DNA-targeting ability but lacks endonuclease activity. When dCas9 is fused to affinity tags or functional domains and guided to specific loci via single-guide RNAs (sgRNAs), it acts as a programmable scaffold for molecular recruitment, enabling locus-specific interrogation of chromatin-associated proteins [34].

A prominent example of this strategy is the CAPTURE method (CRISPR Affinity Purification in situ of Regulatory Elements), which uses dCas9 fused to biotin ligases or streptavidin-binding peptides to purify protein complexes bound at defined genomic regions [76]. Once recruited to a genomic locus of interest, the dCas9 fusion protein acts as a molecular anchor, enabling affinity purification of the associated chromatin fraction. Following in vivo recruitment and cross-linking, the associated chromatin fraction is isolated, and the co-purified proteins are then identified using quantitative mass spectrometry, allowing elucidation of protein networks assembled at regulatory elements such as promoters, enhancers, or insulators. In practical applications, dCas9-based enrichment strategies rely on optimized sgRNA design, efficient cross-linking conditions, and stringent purification steps to minimize background binding and ensure high locus specificity. This approach

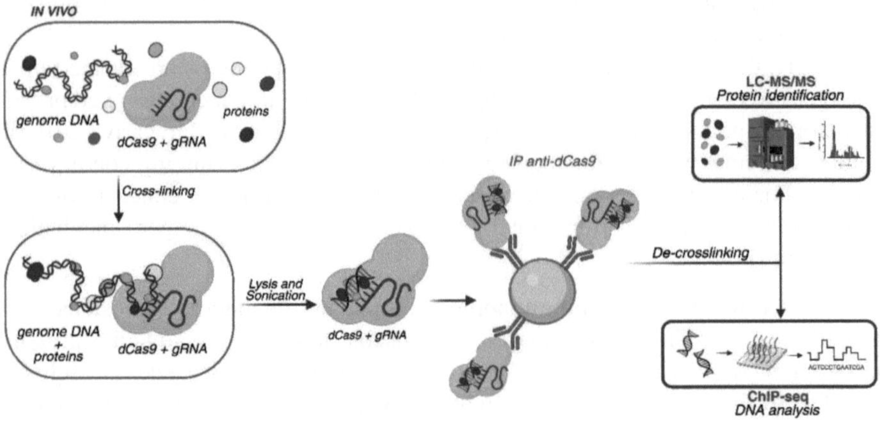

Fig. 3 Schematic workflow of dCas9-based chromatin immunoprecipitation. In vivo cross-linking preserves DNA–protein interactions, followed by cell lysis and sonication. dCas9-gRNA-bound chromatin is immunoprecipitated using anti-dCas9 antibodies. After de-cross-linking, associated proteins are identified by LC-MS/MS, while bound DNA is analyzed by ChIP-seq

enables the capture of native chromatin-associated assemblies under near-physiological conditions. CAPTURE thus complements classical chromatin immunoprecipitation (ChIP) by offering greater spatial precision, flexibility, and independence from antibody-based detection. The experimental workflow of the CAPTURE approach is schematically illustrated in Fig. 3.

Building on this versatility, RNA-centric adaptations further expand the functional scope of dCas9-based systems. CRISPR-display (CRISP-Disp) is one such strategy, in which sgRNAs are engineered to include RNA aptamer sequences such as MS2 or PP7 hairpins that bind RNA-binding proteins fused to fluorescent tags or affinity handles [130]. This modular design enables both visualization and biochemical isolation of RNA-associated complexes at defined loci.

This approach enables recruitment of specific RNA molecules or RNP complexes to targeted genomic regions, facilitating functional analyses of noncoding RNAs, enhancer RNAs, and lncRNAs. Moreover, by tethering structured RNA motifs to dCas9, researchers can explore how specific RNAs contribute to chromatin architecture or transcriptional regulation [30]. The resolution and sensitivity of these approaches are further enhanced by the integration of quantitative proteomics workflows, such as label-free or isobaric tagging strategies, enabling comparative analysis of chromatin-associated protein complexes across different cellular states. These RNA-centric methods are particularly useful for studying interactions involving epigenetic modifiers, scaffolding RNAs, or chromatin-modifying complexes, in a locus-specific, condition-dependent context [130].

Pros and Cons Compared to traditional ChIP-based methods, CRISPR-guided interaction mapping offers several compelling advantages. First, this method bypasses the need for antibodies and can operate under milder conditions, preserving transient or weak interactions [76]. It reduces variability introduced by immunoprecipitation and enables analysis of loci difficult to target with conventional ChIP, such as repetitive or low-occupancy noncoding regions [34].

Additionally, dCas9-based methods can be used in living cells and, when coupled with inducible expression systems, allow temporal control of protein recruitment and interaction mapping. The use of engineered systems such as SunTag, VPR, or scaffold RNAs further expands the modularity of CRISPR-based recruitment, making it a flexible platform for probing transcriptional regulation, chromatin remodeling, and epigenetic states [12, 44].

Despite these advantages, CRISPR-based mapping approaches are not without limitations. Off-target binding remains a concern, especially with sgRNAs of suboptimal specificity or targeting repetitive DNA regions [148]. Additionally, chromatin accessibility at the target site can significantly influence recruitment efficiency, and the presence of nucleosomes or DNA-binding proteins may sterically hinder dCas9 access [65].

There is also a need for careful experimental controls to distinguish bona fide interactors from those recruited artifactually or via overexpression. The integration of orthogonal validation strategies such as ChIP-MS, CUT&RUN, or proximity ligation assays is strongly recommended [136]. Furthermore, establishing stable cell lines expressing dCas9 fusions or delivering large CRISPR constructs can be technically and economically demanding, particularly in in vivo systems or primary cells [128]. Finally, while CRISPR-based affinity approaches provide insight into local chromatin or RNP environments, they may perturb the native architecture or transcriptional output of the targeted locus. As such, functional validation and minimal-perturbation design principles are essential components of a rigorous experimental pipeline.

RNA Immunoprecipitation (RIP) and Enhanced Cross-Linking (eCLIP)

RNA immunoprecipitation (RIP) is a foundational method used to investigate the interaction between RNA molecules and RNA-binding proteins (RBPs) in their native cellular context. Conceptually analogous to ChIP, RIP employs antibodies specific to an RBP of interest to isolate RNA–protein complexes from cell lysates. Following immunoprecipitation, both RNA and proteins are analyzed: RNAs are analyzed through reverse transcription quantitative PCR (RT-qPCR) or high-throughput sequencing (RIP-seq), while proteins can be identified by proteomic workflows [63]. To achieve higher resolution and specificity in mapping RNA–protein interactions, cross-linking-based techniques have emerged as the gold standard. Among these, enhanced cross-linking and immunoprecipitation (eCLIP) have become widely adopted approach. This method begins with ultraviolet (UV) irradiation of living cells, inducing covalent bonds between RBPs and the RNA molecules they contact. Following lysis, immunoprecipitation of the RBP is performed under highly stringent conditions, enabling the identification of direct and transient protein–RNA interactions [154].

Pros and Cons Conventional RIP typically captures both direct and indirect interactions, making it a valuable discovery tool for initial target identification and condition-specific interactome profiling, particularly when integrated with transcriptomic or functional datasets [63, 73]. However, integration of cross-linking significantly improves the detection of transient or low-affinity interactions [151]. Importantly, for in vivo applications such as RIP or eCLIP, the expression level, subcellular localization, and turnover rate of the target RNA must be considered, as these factors influence immunoprecipitation efficiency and the biological relevance of the detected interactions. Novel fluorescent RNA-selective probes have also been developed for real-time monitoring of RNA–protein condensates such as stress granules in living cells, improving resolution and specificity in live imaging [169]. In addition, immunoprecipitated RNA–protein complexes from RIP or cross-link-based assays such as eCLIP can be subjected to proteolytic digestion and analyzed by liquid chromatography–tandem mass spectrometry (LC-MS/MS) to identify bound proteins in an unbiased and high-throughput manner. This approach, often termed RIP-MS or CLIP-MS, enhances the molecular resolution of interactome studies and allows the simultaneous detection of both RNA targets and their associated protein partners [42].

Table 5 summarizes the main methodologies employed to study protein–nucleic acid interactions, highlighting their resolution, experimental context, major strengths, and inherent limitations. This overview complements the detailed sections above and supports the selection of the most suitable method based on experimental goals and system characteristics.

4 Methods for the Investigation of Protein–Drug Interactions

In the course of the drug discovery process, significant advances in technology have been made to characterize the interaction between small molecules and proteins. Many of these techniques rely on computational predictions [66, 166], while others are based on crystallographic [88, 161] or NMR [36, 69] approaches. A further group is predicated on disparate biophysical characteristics, including Surface Plasmon Resonance [157] or Isothermal Titration Calorimetry [124]. In this scenario, the development of focused proteomics protocols has given a paramount contribution by enabling detailed examination of protein–small molecule interactions through experimentally tailored strategies. Collectively, these methodologies constitute the domain of chemical proteomics, whose primary goal is to determine which proteins interact with a compound of interest (drug target identification) and, increasingly, to reveal the specific interacting regions [165].

The most common approaches can be categorized as either labeled or label-free probes [40], depending on the specific objective and workflows.

Labeled strategies rely on covalent binding of the drug to potential protein targets during purification, enabled by the addition of chemical or photochemical activating labels positioned away from the pharmacophore. These labels, typically electrophilic groups, react with nucleophilic sites on the protein targets via Michael

Table 5 Comprehensive overview of the main techniques used to investigate protein–nucleic acid interactions. The table highlights resolution, experimental context, strengths, and limitations of each method

Technique	Resolution	Context	Nucleic acid target	Strengths	Limitations
Affinity purification–based (AP) methods coupled with MS [15, 145]	Moderate	In vitro	DNA/ RNA	Detects RNA/ DNA–protein complexes using rational probe design; compatible with quantitative MS	Lacks chromatin context; RNA folding affects results; stringent washes may remove weak interactors
Pull-down [8]	Moderate	In vitro	RNA	Specific capture of binding partners	Loss of dynamic context, non-specific binding
EMSA [41]	Low	In vitro	DNA/ RNA	Simple, fast, direct binding	No chromatin context, no site mapping
EMSA-MS [41, 164]	Low to moderate	In vitro	DNA/ RNA	Direct interaction detection with proteomic identification	RNA instability, low-affinity interactions may be missed
ChIP/ChIP-MS [31, 89]	Moderate to high	In vivo	DNA	Maps protein–DNA interactions genome-wide	Epitope masking, chromatin shearing
CRISPR–CAPTURE [75]	High	In vivo	DNA	Locus-specific interactome isolation	Off-targets, technical complexity
CRISPR-display and RNA-centric targeting [130]	High	In vivo	RNA	Enables RNA-guided recruitment; useful for studying lncRNAs and enhancer RNAs	Engineering complexity; potential structural perturbation
RIP [63]	Moderate	In vivo	RNA	Captures RNA–protein complexes	No precise binding site; risk of reassortment
eCLIP [154]	High	In vivo	RNA	Nucleotide-level resolution	Complex protocol; requires strong antibodies and strict controls

addition. Labeled probes are employed in a variety of approaches, including—but not limited to—Compound-Centric Chemical Proteomics (CCCP), Activity-Based Protein Profiling (ABPP), click chemistry–based proteomics, protein chip technology, and Proteolysis Targeting Chimeras (PROTACs). The utility of these probes lies in their ability to track and identify specific protein targets. Often, these methodologies incorporate an affinity purification step prior to identification through conventional shotgun proteomics protocols [40].

A limitation of probe-based methods is that the chemical modifications introduced can alter binding properties or accessibility, potentially leading to artifactual interactions. To circumvent this issue, numerous label-free technologies have been developed, leveraging the ability to detect target proteins by observing binding-induced changes in their native conformation or thermal stability. The main approaches in this category include Drug Affinity Responsive Target Stability (DARTS), Thermal Proteome Profiling (TPP), and Limited Proteolysis–Mass Spectrometry [40].

4.1 Targeted Methods

In all targeted approaches, it is imperative to focus on the design and synthesis of the modified probes, as this is fundamental to constructing a Structure–Activity Relationship (SAR) model. The objective is to produce probes that are both effective and specific. SAR studies involve systematic modification of the probe structure to optimize its binding affinity and specificity for the target protein(s). These alterations may encompass the affinity/specificity unit, the linker, the photoreactive moiety, and the identification/reporter tag [33]. For instance, if the linker is too short, the probe may self-cross-link, while an excessively long linker may position the photoreactive group too far from the target to react efficiently [137]. A further issue that may emerge from these methodologies is the occurrence of elevated false-positive rates. To mitigate this, a competitive binding assay is often employed, in which the cell lysate is pre-incubated with the free small-molecule compound prior to its exposure to the immobilized drug on the beads. This strategy allows the saturation of selective binding sites, so that the resulting target profile reflects only non-selective binding partners [95]. It is also imperative to recognize the pivotal role those critical experimental parameters, including the selection of detergents [78], temperature control [38, 139], and the employment of bio-orthogonal chemistries, play in ensuring specificity, preserving protein integrity, and reducing background interactions [87]. As these technologies continue to evolve, they are becoming increasingly instrumental in target deconvolution and drug discovery efforts.

Compound-Centric Chemical Proteomics (CCCP)

In this affinity-based approach, also known as drug-pull-down [180], the target molecule, i.e., the pharmacophore, is labeled to introduce chemical groups that enable immobilization onto Sepharose or magnetic beads. Once the labeled drug is immobilized on a solid-phase matrix, it is incubated with cell lysates for the selective capture of target proteins. These proteins are then eluted, enriched, and identified via mass spectrometry (MS). This method is effective and low-cost for broad analysis and rapid enrichment of target proteins [116].

In many cases, the matrices are activated with reagents such as cyanogen bromide or carbodiimide to allow covalent attachment of CCCP probes [177]. Optimal probe selectivity is achieved through adjustment of factors including the matrix type, temperature, and reaction time [133]. Alternatively, probes can be conjugated to an affinity tag, such as biotin, for non-covalent immobilization on streptavidin-derivatized beads [152].

A range of approaches can be categorized within this domain. In the on-bead affinity matrix approach, a linker such as polyethylene glycol is utilized for covalent attachment of the small molecule to the solid support before incubation with the cell lysate. After several washes, potential protein targets bound to the derivatized matrix can be eluted competitively by adding an excess of free ligand and collected for direct mass spectrometry analysis using a shotgun approach [126]. Conversely, elution of target proteins can also be performed in loading buffer, enabling subsequent SDS-PAGE fractionation, in-gel digestion, and mass spectrometry analysis for protein identification [139].

An alternative approach involves the use of biotin, a compound that exhibits a high degree of binding affinity for streptavidin. In this method, a biotin molecule is attached to the small molecule of interest via chemical linkage, e.g., click chemistry [7]. The biotin-tagged small molecule is then incubated with cell lysates or living cells containing the target proteins, which are purified through streptavidin-bead affinity capture [45]. Mass spectrometry is then employed to identify target proteins isolated as drug interactors from the whole-cell extracts [118].

Pros and Cons This methodology offers several advantages, including cost-effectiveness and simplicity in target protein purification. The prevailing elution strategy involves subjecting the matrix to denaturing buffer, typically SDS-based, at elevated temperatures (95 °C–100 °C). However, such conditions can alter protein structure or activity. In some cases, milder elution conditions can release proteins in a quasi-native state, reducing contamination from matrix-associated proteins, albeit at lower yield compared to denaturing elution.

PhotoAffinity Labeling

Another method is the photoaffinity labeling (PAL) approach, where a chemical probe covalently binds to its target upon exposure to light. The probe design includes a photoreactive group, a linker connecting it to the small molecule, and an affinity tag for downstream purification [7].

The photoreactive moiety is activated by light irradiation, allowing the probe to form a permanent covalent bond with the target molecule at the site of interaction. The photoreactive group can be of different nature, such as diazo-[134], aryl azide-[93], diazirine [138], aryl ketones [7], and tetrazoles [61] groups (Table 6).

Upon activation, each of these groups produces a distinct reactive intermediate, which can be utilized in various ways [13, 175]. The selection of the photoreactive group for a PAL experiment depends on the specific objectives of the study and on the characteristics of the target protein or molecule of interest. Currently, aryldiazirines represent the most prevalent photoreactive group used in PAL experiments. This is attributable to their notable chemical stability and resistance to a broad range

Table 6 Principal photo groups used in PAL approach. For each group, the general structure, the reactive intermediate, the residue preferences, the activation wavelength, and the major advantages and limitations are reported

Photo group	General structure	Reactive intermediate	Residue preferences	Activation	Advantages	Limitations
Diazo compounds [134]	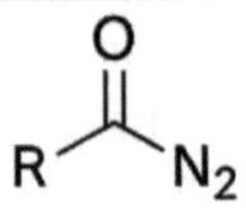	Carbene	Leu, Val, Ile, Met	~350 nm	Simple to synthetize	Highly reactive
Aryl azides [93]	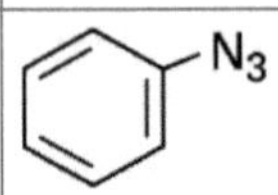	Nitrene (singlet/ triplet)	Trp, Tyr, Phe + Lys, His, Arg	~254–300 nm	Small, easy to incorporate	Needs high-energy UV; Nitrenes can rearrange
Diazirines [138]	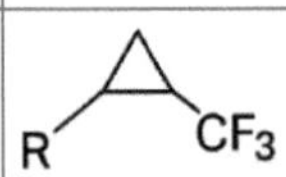	Carbene (singlet/ triplet equilibrium)	Asp/ Glu + Leu, Val, Ile, Met	~330–370 nm	Small. Stable until irradiation	Highly reactive
Aryl ketones [7]	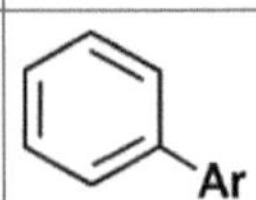	Triplet carbonyl diradical	Met, Leu, Ile, Val, Ala); Trp, Tyr, Phe	~350–365 nm	Less phototoxic than azides	Longer irradiation
Tetrazoles [61]	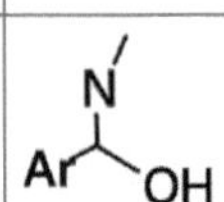	Nitrile imine	Lys, His, Tyr	~300–365 nm	Selective	More complex synthesis

of environmental factors, including temperature, nucleophiles, acidic or basic conditions, and oxidizing or reducing agents [144]. This approach is characterized by its high degree of specificity, enabling precise labeling of small molecules of interest while effectively eliminating false positives [167]. Furthermore, it has been demonstrated to exhibit high sensitivity, allowing detection of low-abundance protein–ligand interactions [179].

Pros and Cons PAL is a powerful technique for the covalent capture of protein–ligand interactions with high specificity and sensitivity. Incorporation of photoreactive groups such as diazirines enables permanent tagging of target proteins upon light activation, thereby facilitating downstream identification. The versatility and robustness of this method make it ideal for probing transient or low-abundance interactions, although careful probe design is essential to account for steric hindrance and to ensure selectivity and functional compatibility with the protein target. Conversely, photoreactive groups employed for labeling may interfere with binding to the target protein, leading to false-negative results [70]. Furthermore, the use of ultraviolet light to initiate covalent bond formation between the small molecule and the target protein can compromise cell integrity, potentially resulting in artifactual outcomes [81].

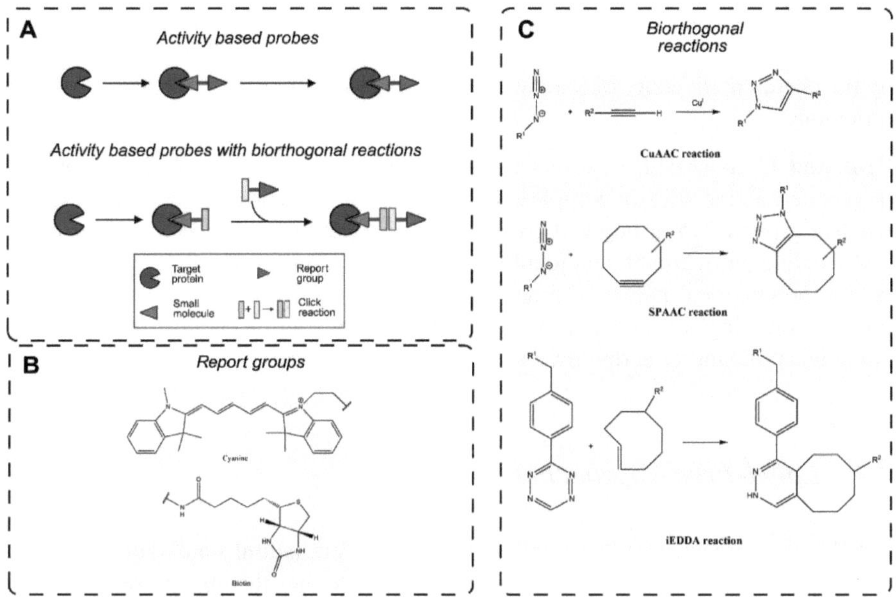

Fig. 4 Schematic representation of ABPP methodology. (**a**) Diagram of the activity-based probes and the activity-based probes with biorthogonal reactions (click reactions). (**b**) Structures of the two major report groups, cyanine and biotin. (**c**) General representations of most frequently employed bio-orthogonal reactions (CuAAC, SPAAC, and iEDDA reactions)

Affinity-Based Protein Profiling

Affinity-Based Protein Profiling (ABPP) has emerged as a powerful approach for identifying protein targets directly in complex biological systems. This approach integrates principles from both chemical proteomics and activity-based labeling, as Activity-Based Probes (ABPs) are designed to contain a reactive group, a reporter group (usually an alkyne) attached to the compound of interest, and a linker [105] (Fig. 4a).

The reactive group is often derived from a known covalent inhibitor of the enzyme. Reporter groups such as fluorescent dyes (e.g., cyanine) or biotin (Fig. 4, Panel B) enable visualization and subsequent enrichment of labeled proteins for downstream analysis. The linker provides adequate spacing between the reactive and reporter groups and can be composed of hydrophilic or lipophilic chains or short peptide sequences.

The probes used in this methodology are incubated with proteins extracted from cell lysates or living cells, where the inhibitor moiety forms covalent adducts with its interactor(s) [25]. To enhance probe selectivity and cell permeability, recent advances have incorporated bio-orthogonal chemistry, employing small and non-perturbing functional groups.

The most frequently employed bio-orthogonal reactions include copper(I)-catalyzed azide–alkyne cycloaddition (CuAAC) [121], copper-free strain-promoted azide–alkyne cycloaddition (SPAAC) [52], and inverse electron-demand Diels–Alder (iEDDA) [3] (Fig. 4, Panel C). However, ABPP applicability remains limited by the chemical diversity of reactive groups, which restricts its coverage across the proteome.

Pros and Cons ABPP is a versatile and selective strategy for mapping protein targets and active sites in complex biological systems. Through the integration of reactive groups, bio-orthogonal chemistry, and reporter tags, ABPP enables covalent labeling, enrichment, and identification of target proteins with high specificity. While the chemical nature of reactive groups can limit the breadth of detectable proteins, ongoing advancements in probe design and reaction chemistries are broadening its applicability across diverse proteomic landscapes.

4.2 Label-Free Approaches

Label-free approaches utilize small molecules in their natural state, without chemical modification for immobilization on inert matrices, thereby preserving their native activity. These methodologies provide complementary strategies for identifying and characterizing protein–ligand interactions, each with distinct strengths in terms of specificity, sensitivity, and structural insight. It is evident that, from live-cell analysis to in vitro techniques, these methods empower researchers to probe conformational dynamics, stability, and functional changes in proteins upon ligand binding. This, in turn, offers insights into drug–target interactions, protein stability shifts, and functional perturbations. While many methods can be applied to purified proteins, their extension to proteome-wide studies or complex biological systems remains challenging. Despite these limitations, label-free methods represent a robust and diverse toolkit for probing the molecular mechanisms underlying drug action and protein dynamics. Integrating these approaches with orthogonal methods enhances the depth of biological insight and accelerates advancements in proteomics-driven drug discovery.

Methods Based on Thermal Denaturation Monitoring Experiments

Thermal Proteome Profiling

Thermal Proteome Profiling (TPP) measures the extent of drug–target interaction by monitoring the effects of pharmacological treatment on protein denaturation and solubility as a function of temperature [84]. After cell lysis under native conditions, the protein extract is incubated in parallel with both vehicle and drug. Aliquots of each sample are then heated at increasing temperatures. Soluble fractions are collected by ultracentrifugation or multi-well filter plates, digested with trypsin, and labeled with TMT isotopic tags. The samples are analyzed by LC-MS/MS for protein identification and quantification using bioinformatic tools [48]. Sigmoidal melting curves are fitted, and melting temperatures are calculated for all proteins in both control and treated samples. Binding of a protein to a small molecule can alter its thermal stability; therefore, detecting a shift in denaturation temperature enables identification of drug targets [87]. Cell lysis is typically performed by mechanical disruption (e.g., freeze–thaw cycles), as detergents can alter protein melting points and affect the accuracy of thermal stability measurements [170]. When performing the heat cycle, samples are aliquoted into a thermocycler to ensure uniform temperature transfer. The temperature range is chosen between the point where the proteome remains largely soluble and the point where most proteins become insoluble—typically 10–12 points spaced by 3 °C–5 °C, within a 30 °C–50 °C range [113]. Several adaptations of the standard workflow have been developed, including Temperature-Range TPP (TPP-TR) [28, 181], TPP compound concentration range (TPP-CCR) [28] and a combination of both within two-dimensional TPP (2D-TPP) [146].

Recently, a simplified version known as STPP (Simplified Thermal Proteome Profiling), which relies on dimethyl labeling rather than TMT tagging, has been proposed as a more cost-effective and less technically demanding alternative for protein and peptide quantification [176].

Pros and Cons TPP is a powerful technique for identifying drug–protein interactions by exploiting thermal stability shifts that occur upon ligand binding. The generation of proteome-wide thermal denaturation profiles provides valuable insight into target engagement and off-target effects under native conditions. However, the workflow is time- and resource-intensive due to the need for multiple temperature points and labeling steps. Some proteins require extreme temperature conditions [86] while others exhibit minimal stability changes upon ligand binding. In addition, TPP shows limited sensitivity for membrane proteins or low-abundance targets [127]. To overcome these challenges, high-throughput and simplified variants have been developed, providing adaptable and scalable solutions that optimize cost, throughput, and resolution.

Cellular Thermal Shift Assay

The Cellular Thermal Shift Assay (CETSA) is based on the principle of ligand-induced thermodynamic stabilization of protein targets. Unlike TPP-based workflows, the CETSA approach is performed in living cells. Cells are exposed to either a small molecule or vehicle control and subsequently heated across a temperature gradient. The cells are then lysed, and in accordance with the classical workflow, Western blotting is used to determine whether potential target proteins are denatured in a temperature-dependent manner and whether their melting curves shift upon ligand binding. By comparing the thermal stability of proteins with and without the compound, it is possible to assess target engagement and estimate the affinity of the interaction [90]. It is important to note that several modified versions of this approach are currently under development, such as Thermal Proteome Profiling–Cellular Context Thermal Shift Assay (TTP or MS-CETSA) [127]; IsoThermal

Dose-Response–Mass Spectrometry–Cellular Context Thermal Shift Assay (ITDR-MS-CETSA [21]; Thermal stability shift–based fluorescence difference in two-dimensional gel electrophoresis (2DE-CETSA) [107].

Pros and Cons CETSA represents a major advance in target engagement studies, as it allows the assessment of protein–ligand interactions directly in living cells, thereby preserving the physiological context. Although classical CETSA relies on Western blotting to monitor individual target stability (targeted format), newer adaptations now enable proteome-wide and quantitative applications. These innovations improve sensitivity, allow dose–response profiling, and enhance the precision of target identification. However, CETSA and its derivatives share common challenges with thermal profiling methods, including the interpretation of non-binding-related thermal shifts and the limited detectability of membrane or low-abundance proteins. Despite these constraints, CETSA remains a robust and versatile tool for validating drug–target interactions in a native cellular environment.

Limited Proteolysis-Coupled to Mass Spectrometry

Limited Proteolysis coupled to Mass Spectrometry (LiP-MS) is a combined strategy of limited proteolysis and mass spectrometry for the label-free quantification of peptides released under controlled hydrolysis conditions [82] (Fig. 5), allowing the detection of structural changes resulting from ligand interaction, unfolding, aggregation, or small-molecule allostery. LiP-MS is a highly versatile method, applicable both to in vitro conformational studies of isolated protein–ligand complexes and to complex

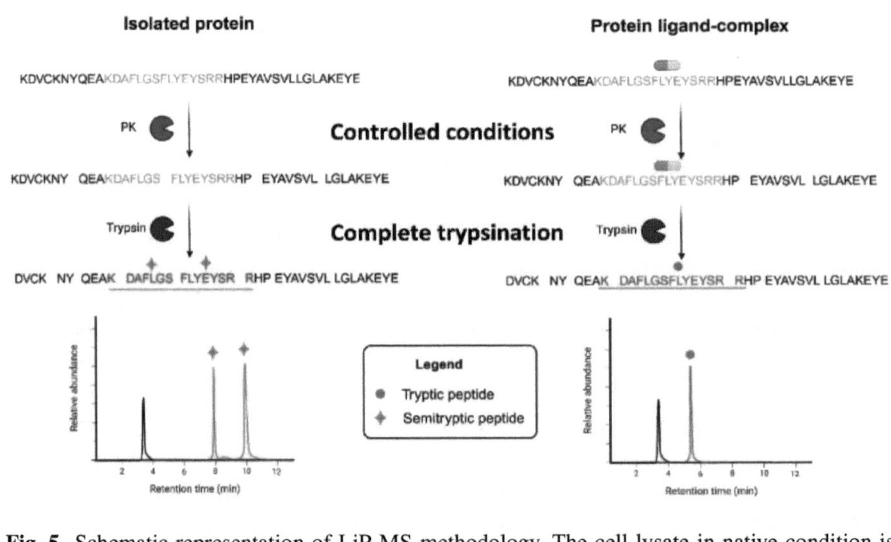

Fig. 5 Schematic representation of LiP-MS methodology. The cell lysate in native condition is incubated with the compound of interest and then an unspecific digestion is performed with Proteinase K, followed by a classic trypsinization

proteomes, where it enables identification of protein targets by quantitatively comparing peptide release patterns in the presence or absence of a ligand. In drug–protein interaction studies, isolated proteins or whole proteomes are subjected to limited digestion with non-specific proteases—most commonly proteinase K (PK)—under carefully controlled, short incubation times. This first step is followed by extensive digestion, typically with trypsin, as in a conventional bottom-up proteomics workflow. The resulting peptides are then identified and quantified by LC–MS/MS, followed by bioinformatic and statistical analysis in a label-free manner. By comparing proteolytic patterns between drug- and vehicle-treated samples using a peptide-centric approach, potential target proteins can be identified. When applied to complex proteomes, parallel trypsin-only digestions are performed to quantify total protein abundance and serve as normalization references [11]. The analysis of LiP-MS data relies on the observation that protein regions containing LiP cleavage sites are detected either by an increase in the abundance of fully tryptic peptides or by a decrease in half-tryptic peptides (i.e., peptides with one non-tryptic terminus) in drug-treated samples relative to controls. Changes in peptide intensities are expressed as fold differences, reflecting altered accessibility or flexibility of the corresponding protein regions under native conditions. This approach achieves peptide-level resolution, allowing detection of structural alterations with an approximate spatial precision of 12 amino acids—sufficient to infer potential small-molecule-binding sites. The method can discriminate among diverse classes of protein–small molecule interactions, including allosteric, enzyme–substrate, enzyme–product, and drug–target associations.

Pros and Cons LiP-MS is a powerful label-free method for detecting small molecule–protein interactions through drug-induced conformational changes at peptide-level resolution. Its ability to monitor both global and local structural rearrangements in complex proteomes makes it a highly versatile tool for structural proteomics. However, the sensitivity of LiP-MS depends on two key parameters: sequence coverage and protein abundance. Identification of binding sites requires detection of peptides mapping to the binding interface; consequently, the method favors medium- to high-abundance proteins. Extensive conformational rearrangements can obscure the precise binding site, as LiP-MS cannot easily distinguish between local and long-range structural effects. Moreover, target identification depends on the assumption that ligand binding alters protease accessibility if no accessibility change occurs, target detection becomes unfeasible [108]. Despite these limitations, LiP MS stands out for its structural insight, high resolution, and broad applicability across systems biology and drug discovery.

Method Based on Induced Protein Modification: Monitoring Methionine Oxidation Rates

The hypothesis underpinning the SPROX (Stability of Proteins from Rates of Oxidation) experiment is that protein stability increases upon binding to a small molecule, thereby rendering the protein less susceptible to oxidation than its denatured counterpart. In this assay, the small molecule is first incubated with a cell lysate, after which the proteins are chemically denatured [159]. Following denaturation, all samples are treated with identical amounts of hydrogen peroxide to oxidize accessible methionine side chains. The reaction proceeds for a defined period before being quenched [160]. Ligand binding alters the protein's resistance to unfolding, thereby changing the concentration of denaturant required to expose methionine residues to oxidation. Quantification of oxidized and non-oxidized peptides can be achieved through several analytical strategies, including chemical labeling [140]. Typically, oxidation reactions are carried out for 1–2 min using 1–2 mM H_2O_2, although the optimal duration depends on the specific folding–unfolding kinetics of each protein [159]. Cysteine side chains can also react with H_2O_2 to form oxidized products; however, their oxidation rate is generally slower than that of methionine residues [171].

Pros and Cons SPROX is a chemoproteomic approach designed to evaluate protein stability changes induced by ligand binding by monitoring methionine oxidation under denaturing conditions in the presence or absence of a compound. The strength of this method lies in its capacity to provide quantitative information on protein–small molecule interactions within complex mixtures. Nevertheless, its applicability is limited by dependence on methionine content since this residue is not widespread across all proteins and by the influence of protein unfolding kinetics. Therefore, careful optimization of oxidation parameters is required to ensure accurate and reproducible detection.

Table 7 provides a comparative overview of the principal techniques used to identify and characterize protein–small molecule interactions, summarizing their approach type, advantages, limitations, and a specific note highlighting critical considerations for each methodology.

5 Integrative Design Strategies Across Interaction Modalities

Across protein–protein, protein–nucleic acid, and protein–drug investigations, rigorous experimental design begins by aligning the capture strategy with the biological question and the kinetic or structural properties of the interaction. Native workflows such as affinity capture or co-fractionation preserve activity and binding thermodynamics, whereas in vivo stabilization with formaldehyde, UV, or bifunctional cross-linkers can "freeze" weak or transient contacts provided that cross-linker type, concentration, and exposure time are carefully calibrated to avoid epitope masking, impaired shearing, or buffer incompatibilities with proteomics readouts.

When transient or compartmentalized interactions are of interest, proximity-labeling methods (BioID, TurboID, APEX, and TransitID) and in-cell cross-linking MS extend coverage by tagging neighboring molecules before lysis or by generating MS-interpretable spatial restraints. Combining these with targeted immunocapture increases depth without losing physiological context. Locus- or sequence-resolved

Table 7 Summary of each technique presented for the characterization of the binding between protein and small molecules. The table contrasts resolution, biological context, key strengths, and limitations of commonly used affinity-based and label-free approaches

Technique	Resolution	Context	Strengths	Limitations
Photo Affinity Labeling (PAL) [7]	Moderate	In vivo/in vitro	High specificity and sensitivity; diazirines widely used for stability and low phototoxicity	Requires UV activation (potential cell damage); probe design is critical
Affinity-Based Protein Profiling (ABPP) [105]	Moderate	In vitro/ex vivo	Selective mapping of active sites; click chemistry increases probe selectivity	Limited by reactive group choice; requires careful probe optimization
Thermal Proteome Profiling (TPP) [84]	Moderate to high	*In cell/ proteome-wide*	Broad, unbiased proteome coverage; adaptable to high-throughput (2D-TPP, STPP)	Time- and resource-intensive; low sensitivity for membrane or low-abundance proteins
Cellular Thermal Shift Assay (CETSA) [127]	Moderate	In vivo (cells, tissues)	Measures protein–ligand interactions in living cells; adaptable to dose–response profiling (ITDR-MS-CETSA)	Limited sensitivity for low-abundance and membrane proteins
Drug Affinity Responsive Target Stability (DARTS) [114]	Low to moderate	In vitro	Simple, cost-effective; no chemical modification of ligands needed	Low sensitivity; relies on protease susceptibility; validation often needed (e.g., Western blot)
Limited Proteolysis–MS (LiP-MS) [82]	High (local, ~12 aa)	In vitro/ex vivo	Detects local conformational changes; peptide-level structural insights	Requires high protein abundance and sequence coverage
Stability of Proteins from Rates of Oxidation (SPROX) [159]	Moderate	In vitro/ *proteome-wide*	Quantitative stability profiling; sensitive to folding/unfolding	Limited to methionine-containing peptides; oxidation conditions require fine-tuning

questions may instead employ CRISPR/dCas9-guided enrichment approaches such as CAPTURE or RNA-centric display, which offer programmable specificity but require careful guide design, stringent washing, and low-expression systems to minimize artifacts. Complementary native separations BN-PAGE (including hrBN- and csBN-MS) and SEC-MS add architectural information and preserve supramolecular organization under MS-compatible, volatile buffers. These separations can also be coupled with cross-linking to stabilize labile assemblies.

Across all modalities, lysis and wash buffers must balance stringency with retention of biologically meaningful, low-affinity partners while remaining LC-MS/MS compatible (e.g., volatile salts, cleavable or low-residue surfactants, and avoidance of harsh detergents). Physical processing steps such as sonication or nuclease trimming should be tuned to achieve optimal fragment sizes without disrupting complexes. Downstream, quantitative mass spectrometry supported by stringent identification thresholds and error control ensures data reliability. Modern pipelines such as MaxQuant/Perseus, FragPipe/OpenMS, Skyline, and DIA-NN/EncyclopeDIA/OpenSWATH, combined with interactome-scoring tools (SAINTexpress, MiST, and CompPASS) and contaminant-filtering resources (CRAPome), help distinguish bona fide interactors from background noise. Finally, robust inference relies on careful study design (biological replicates, matched negatives or competition controls, pre-clearing) and orthogonal validation through reciprocal pulldown, site-directed mutagenesis, or complementary assays (e.g., ChIP-MS, eCLIP, or biophysical binding). Iterative pilot runs allow fine-tuning of cross-linking, shearing, and wash parameters. Framed this way by selecting a fit-for-purpose capture method, engineering MS-compatible chemistries, enforcing quantitative and statistical rigor, and corroborating findings through independent validation, the resulting interaction maps are both physiologically meaningful and reproducible across PPI, protein–nucleic acid, and protein–drug systems.

References

1. Aebersold R, Mann M (2016) Mass-spectrometric exploration of proteome structure and function. Nature 537:347–355. https://doi.org/10.1038/nature19949
2. Bailey T, Krajewski P, Ladunga I, Lefebvre C, Li Q, Liu T, Madrigal P, Taslim C, Zhang J (2013) Practical guidelines for the comprehensive analysis of ChIP-seq data. PLoS Comput Biol 9:e1003326. https://doi.org/10.1371/journal.pcbi.1003326
3. Blackman ML, Royzen M, Fox JM (2008) Tetrazine ligation: fast bioconjugation based on inverse-electron-demand Diels-Alder reactivity. J Am Chem Soc 130:13518–13519. https://doi.org/10.1021/ja8053805
4. Bludau I, Heusel M, Frank M, Rosenberger G, Hafen R, Banaei-Esfahani A, Van Drogen A, Collins BC, Gstaiger M, Aebersold R (2020) Complex-centric proteome profiling by SEC-SWATH-MS for the parallel detection of hundreds of protein complexes. Nat Protoc 15:2341–2386. https://doi.org/10.1038/s41596-020-0332-6
5. Branon TC, Bosch JA, Sanchez AD, Udeshi ND, Svinkina T, Carr SA, Feldman JL, Perrimon N, Ting AY (2018) Efficient proximity labeling in living cells and organisms with TurboID. Nat Biotechnol 36:880–887. https://doi.org/10.1038/nbt.4201
6. Bruderer R, Bernhardt OM, Gandhi T, Miladinović SM, Cheng L-Y, Messner S, Ehrenberger T, Zanotelli V, Butscheid Y, Escher C, Vitek O, Rinner O, Reiter L (2015) Extending the limits of quantitative proteome profiling with data-independent acquisition and application to acetaminophen-treated three-dimensional liver microtissues. Mol Cell Proteomics 14:1400–1410. https://doi.org/10.1074/mcp.M114.044305
7. Burton NR, Kim P, Backus KM (2021) Photoaffinity labelling strategies for mapping the small molecule-protein interactome. Org Biomol Chem 19:7792–7809. https://doi.org/10.1039/d1ob01353j
8. Butter F, Scheibe M, Mörl M, Mann M (2009) Unbiased RNA–protein interaction screen by quantitative proteomics. Proc Natl Acad Sci USA 106:10626–10631. https://doi.org/10.1073/pnas.0812099106

9. Cadoni E, Moerman H, Madder A (2025) Development of a His-Tag-mediated pull-down and quantification assay for G-quadruplex containing DNA sequences. RSC Chem Biol 6:56–64. https://doi.org/10.1039/d4cb00185k
10. Cai W, Tucholski T, Chen B, Alpert AJ, McIlwain S, Kohmoto T, Jin S, Ge Y (2017) Top-down proteomics of large proteins up to 223 kDa enabled by serial size exclusion chromatography strategy. Anal Chem 89:5467–5475. https://doi.org/10.1021/acs.analchem.7b00380
11. Cappelletti V, Hauser T, Piazza I, Pepelnjak M, Malinovska L, Fuhrer T, Li Y, Dörig C, Boersema P, Gillet L, Grossbach J, Dugourd A, Saez-Rodriguez J, Beyer A, Zamboni N, Caflisch A, de Souza N, Picotti P (2021) Dynamic 3D proteomes reveal protein functional alterations at high resolution in situ. Cell 184:545–559.e22. https://doi.org/10.1016/j.cell.2020.12.021
12. Chavez A, Scheiman J, Vora S, Pruitt BW, Tuttle M, Iyer E PR, Lin S, Kiani S, Guzman CD, Wiegand DJ, Ter-Ovanesyan D, Braff JL, Davidsohn N, Housden BE, Perrimon N, Weiss R, Aach J, Collins JJ, Church GM (2015) Highly efficient Cas9-mediated transcriptional programming. Nat Methods 12:326–328. https://doi.org/10.1038/nmeth.3312
13. Chen S, Do JT, Zhang Q, Yao S, Yan F, Peters EC, Schöler HR, Schultz PG, Ding S (2006) Self-renewal of embryonic stem cells by a small molecule. Proc Natl Acad Sci USA 103:17266–17271. https://doi.org/10.1073/pnas.0608156103
14. Cline EN, Alvarez C, Duan J, Patrie SM (2021) Online μSEC^2 -nRPLC-MS for improved sensitivity of intact protein detection of IEF-separated nonhuman primate cerebrospinal fluid proteins. Anal Chem 93:16741–16750. https://doi.org/10.1021/acs.analchem.1c00396
15. Cozzolino F, Iacobucci I, Monaco V, Monti M (2021) Protein–DNA/RNA interactions: an overview of investigation methods in the -omics era. J Proteome Res 20:3018–3030. https://doi.org/10.1021/acs.jproteome.1c00074
16. DeCaprio J, Kohl TO (2019) Cross-linking antibodies to beads with disuccinimidyl suberate (DSS). Cold Spring Harb Protoc 2019:pdb.prot098632. https://doi.org/10.1101/pdb.prot098632
17. Demichev V, Messner CB, Vernardis SI, Lilley KS, Ralser M (2020) DIA-NN: neural networks and interference correction enable deep proteome coverage in high throughput. Nat Methods 17:41–44. https://doi.org/10.1038/s41592-019-0638-x
18. Derveaux S, Vandesompele J, Hellemans J (2010) How to do successful gene expression analysis using real-time PCR. Methods 50:227–230. https://doi.org/10.1016/j.ymeth.2009.11.001
19. Dewson G (2015) Blue native PAGE and antibody gel shift to assess Bak and Bax conformation change and oligomerization. Cold Spring Harb Protoc 2015:pdb.prot086488. https://doi.org/10.1101/pdb.prot086488
20. Duong V-A, Lee H (2023) Bottom-up proteomics: advancements in sample preparation. IJMS 24:5350. https://doi.org/10.3390/ijms24065350
21. Dziekan JM, Wirjanata G, Dai L, Go KD, Yu H, Lim YT, Chen L, Wang LC, Puspita B, Prabhu N, Sobota RM, Nordlund P, Bozdech Z (2020) Cellular thermal shift assay for the identification of drug-target interactions in the Plasmodium falciparum proteome. Nat Protoc 15:1881–1921. https://doi.org/10.1038/s41596-020-0310-z
22. Edwards AN, Hsu K-L (2025) Emerging opportunities for intact and native protein analysis using chemical proteomics. Anal Chim Acta 1338:343551. https://doi.org/10.1016/j.aca.2024.343551
23. Ehkirch A, D'Atri V, Rouviere F, Hernandez-Alba O, Goyon A, Colas O, Sarrut M, Beck A, Guillarme D, Heinisch S, Cianferani S (2018) An online four-dimensional HIC×SEC-IM×MS methodology for proof-of-concept characterization of antibody drug conjugates. Anal Chem 90:1578–1586. https://doi.org/10.1021/acs.analchem.7b02110
24. Fan S-B, Meng J-M, Lu S, Zhang K, Yang H, Chi H, Sun R-X, Dong M-Q, He S-M (2015) Using pLink to analyze cross-linked peptides. Curr Protoc Bioinformatics 49:8.21.1–8.21.19. https://doi.org/10.1002/0471250953.bi0821s49
25. Fang H, Peng B, Ong SY, Wu Q, Li L, Yao SQ (2021) Recent advances in activity-based probes (ABPs) and affinity-based probes (AfBPs) for profiling of enzymes. Chem Sci 12:8288–8310. https://doi.org/10.1039/d1sc01359a

26. Federico A, Sepe R, Cozzolino F, Piccolo C, Iannone C, Iacobucci I, Pucci P, Monti M, Fusco A (2019) The complex CBX7-PRMT1 has a critical role in regulating E-cadherin gene expression and cell migration. Biochim Biophys Acta Gene Regul Mech 1862:509–521. https://doi.org/10.1016/j.bbagrm.2019.02.006
27. Fossati A, Li C, Sykacek P, Heusel M, Frommelt F, Uliana F, Hallal M, Bludau I, Klemens CT, Xue P, Purcell AW, Gstaiger M, Aebersold R (2020) Systematic protein complex profiling and differential analysis from co-fractionation mass spectrometry data
28. Franken H, Mathieson T, Childs D, Sweetman GMA, Werner T, Tögel I, Doce C, Gade S, Bantscheff M, Drewes G, Reinhard FBM, Huber W, Savitski MM (2015) Thermal proteome profiling for unbiased identification of direct and indirect drug targets using multiplexed quantitative mass spectrometry. Nat Protoc 10:1567–1593. https://doi.org/10.1038/nprot.2015.101
29. Fried M, Crothers DM (1981) Equilibria and kinetics of lac repressor-operator interactions by polyacrylamide gel electrophoresis. Nucl Acids Res 9:6505–6525. https://doi.org/10.1093/nar/9.23.6505
30. Fu Y, Rocha PP, Luo VM, Raviram R, Deng Y, Mazzoni EO, Skok JA (2016) CRISPR-dCas9 and sgRNA scaffolds enable dual-colour live imaging of satellite sequences and repeat-enriched individual loci. Nat Commun 7:11707. https://doi.org/10.1038/ncomms11707
31. Furey TS (2012) ChIP–seq and beyond: new and improved methodologies to detect and characterize protein–DNA interactions. Nat Rev Genet 13:840–852. https://doi.org/10.1038/nrg3306
32. Fusco S, Aulitto M, Iacobucci I, Crocamo G, Pucci P, Bartolucci S, Monti M, Contursi P (2020) The interaction between the F55 virus-encoded transcription regulator and the RadA host recombinase reveals a common strategy in archaea and bacteria to sense the UV-induced damage to the host DNA. Biochim Biophys Acta Gene Regul Mech 1863:194493. https://doi.org/10.1016/j.bbagrm.2020.194493
33. Gao M, Nettles RE, Belema M, Snyder LB, Nguyen VN, Fridell RA, Serrano-Wu MH, Langley DR, Sun J-H, O'Boyle DR, Lemm JA, Wang C, Knipe JO, Chien C, Colonno RJ, Grasela DM, Meanwell NA, Hamann LG (2010) Chemical genetics strategy identifies an HCV NS5A inhibitor with a potent clinical effect. Nature 465:96–100. https://doi.org/10.1038/nature08960
34. Gao XD, Tu L-C, Mir A, Rodriguez T, Ding Y, Leszyk J, Dekker J, Shaffer SA, Zhu LJ, Wolfe SA, Sontheimer EJ (2018) C-BERST: defining subnuclear proteomic landscapes at genomic elements with dCas9–APEX2. Nat Methods 15:433–436. https://doi.org/10.1038/s41592-018-0006-2
35. Garner MM, Revzin A (1981) A gel electrophoresis method for quantifying the binding of proteins to specific DNA regions: application to components of the Escherichia coli lactose operon regulatory system. Nucl Acids Res 9:3047–3060. https://doi.org/10.1093/nar/9.13.3047
36. Gee CT, Arntson KE, Koleski EJ, Staebell RL, Pomerantz WCK (2018) Dual labeling of the CBP/p300 KIX domain for 19 F NMR leads to identification of a new small-molecule binding site. Chembiochem 19:963–969. https://doi.org/10.1002/cbic.201700686
37. Gilbert LA, Larson MH, Morsut L, Liu Z, Brar GA, Torres SE, Stern-Ginossar N, Brandman O, Whitehead EH, Doudna JA, Lim WA, Weissman JS, Qi LS (2013) CRISPR-mediated modular RNA-guided regulation of transcription in eukaryotes. Cell 154:442–451. https://doi.org/10.1016/j.cell.2013.06.044
38. Ha J, Park H, Park J, Park SB (2021) Recent advances in identifying protein targets in drug discovery. Cell Chem Biol 28:394–423. https://doi.org/10.1016/j.chembiol.2020.12.001
39. Hao Y, Chen M, Huang X, Xu H, Wu P, Chen S (2023) 4D-diaXLMS: proteome-wide four-dimensional data-independent acquisition workflow for cross-linking mass spectrometry. Anal Chem 95:14077–14085. https://doi.org/10.1021/acs.analchem.3c02824
40. He S-J, Li J, Zhou J-C, Yang Z-Y, Liu X, Ge Y-W (2024) Chemical proteomics accelerates the target discovery of natural products. Biochem Pharmacol 230:116609. https://doi.org/10.1016/j.bcp.2024.116609

41. Hellman LM, Fried MG (2007) Electrophoretic mobility shift assay (EMSA) for detecting protein–nucleic acid interactions. Nat Protoc 2:1849–1861. https://doi.org/10.1038/nprot.2007.249
42. Hentze MW, Castello A, Schwarzl T, Preiss T (2018) A brave new world of RNA-binding proteins. Nat Rev Mol Cell Biol 19:327–341. https://doi.org/10.1038/nrm.2017.130
43. Heusel M, Frank M, Köhler M, Amon S, Frommelt F, Rosenberger G, Bludau I, Aulakh S, Linder MI, Liu Y, Collins BC, Gstaiger M, Kutay U, Aebersold R (2020) A global screen for assembly state changes of the mitotic proteome by SEC-SWATH-MS. Cell Systems 10:133–155.e6. https://doi.org/10.1016/j.cels.2020.01.001
44. Hilton IB, D'Ippolito AM, Vockley CM, Thakore PI, Crawford GE, Reddy TE, Gersbach CA (2015) Epigenome editing by a CRISPR-Cas9-based acetyltransferase activates genes from promoters and enhancers. Nat Biotechnol 33:510–517. https://doi.org/10.1038/nbt.3199
45. Hirota T, Lee JW, St John PC, Sawa M, Iwaisako K, Noguchi T, Pongsawakul PY, Sonntag T, Welsh DK, Brenner DA, Doyle FJ, Schultz PG, Kay SA (2012) Identification of small molecule activators of cryptochrome. Science 337:1094–1097. https://doi.org/10.1126/science.1223710
46. Hoenger Ramazanova RD, Roumeliotis TI, Wright JC, Choudhary JS (2024) PhoXplex: combining phospho-enrichable cross-linking with isobaric labeling for quantitative proteome-wide mapping of protein interfaces. J Proteome Res 23:5209–5220. https://doi.org/10.1021/acs.jproteome.4c00567
47. Iacobucci C, Götze M, Ihling CH, Piotrowski C, Arlt C, Schäfer M, Hage C, Schmidt R, Sinz A (2018) A cross-linking/mass spectrometry workflow based on MS-cleavable cross-linkers and the MeroX software for studying protein structures and protein-protein interactions. Nat Protoc 13:2864–2889. https://doi.org/10.1038/s41596-018-0068-8
48. Iacobucci I, La Manna S, Cipollone I, Monaco V, Canè L, Cozzolino F (2023) From the discovery of targets to delivery systems: how to decipher and improve the Metallodrugs' actions at a molecular level. Pharmaceutics 15:1997. https://doi.org/10.3390/pharmaceutics15071997
49. Iacobucci I, Monaco V, Cozzolino F, Monti M (2021) From classical to new generation approaches: an excursus of -omics methods for investigation of protein-protein interaction networks. J Proteome 230:103990. https://doi.org/10.1016/j.jprot.2020.103990
50. Iacobucci I, Monaco V, Hovasse A, Dupouy B, Keumoe R, Cichocki B, Elhabiri M, Meunier B, Strub J, Monti M, Cianférani S, Blandin SA, Schaeffer-Reiss C, Davioud-Charvet E (2024) Proteomic profiling of antimalarial plasmodione using 3-Benz(o)ylmenadione affinity-based probes. Chembiochem 25:e202400187. https://doi.org/10.1002/cbic.202400187
51. Ihling CH, Springorum P, Iacobucci C, Hage C, Götze M, Schäfer M, Sinz A (2020) The isotope-labeled, MS-cleavable cross-linker Disuccinimidyl Dibutyric urea for improved cross-linking/mass spectrometry studies. J Am Soc Mass Spectrom 31:183–189. https://doi.org/10.1021/jasms.9b00008
52. Jewett JC, Sletten EM, Bertozzi CR (2010) Rapid cu-free click chemistry with readily synthesized Biarylazacyclooctynones. J Am Chem Soc 132:3688–3690. https://doi.org/10.1021/ja100014q
53. Jiang N, Wen Z, Tao H, Liao H (2025) Improved ChIP sequencing for H3K27ac profiling and super-enhancer analysis assisted by fluorescence-activated sorting of formalin-fixed paraffin-embedded tissues. Biol Proced Online 27:1. https://doi.org/10.1186/s12575-025-00262-9
54. Jiang P, Wang C, Diehl A, Viner R, Etienne C, Nandhikonda P, Foster L, Bomgarden RD, Liu F (2022) A membrane-permeable and immobilized metal affinity chromatography (IMAC) enrichable cross-linking reagent to advance in vivo cross-linking mass spectrometry. Angew Chem Int Ed 61:e202113937. https://doi.org/10.1002/anie.202113937
55. Jiao F, Yu C, Wheat A, Chen L, Lih T-SM, Zhang H, Huang L (2024) DSBSO-based XL-MS analysis of breast cancer PDX tissues to delineate protein interaction network in clinical samples. J Proteome Res 23:3269–3279. https://doi.org/10.1021/acs.jproteome.3c00832
56. Jiao F, Yu C, Wheat A, Wang X, Rychnovsky SD, Huang L (2022) Two-dimensional fractionation method for proteome-wide cross-linking mass spectrometry analysis. Anal Chem 94:4236–4242. https://doi.org/10.1021/acs.analchem.1c04485

57. Kaboord B, Perr M (2008) Isolation of proteins and protein complexes by immunoprecipitation. Methods Mol Biol 424:349–364. https://doi.org/10.1007/978-1-60327-064-9_27
58. Kachuk C, Stephen K, Doucette A (2015) Comparison of sodium dodecyl sulfate depletion techniques for proteome analysis by mass spectrometry. J Chromatogr A 1418:158–166. https://doi.org/10.1016/j.chroma.2015.09.042
59. Kaewsapsak P, Shechner DM, Mallard W, Rinn JL, Ting AY (2017) Live-cell mapping of organelle-associated RNAs via proximity biotinylation combined with protein-RNA crosslinking. eLife 6:e29224. https://doi.org/10.7554/eLife.29224
60. Kao A, Chiu C, Vellucci D, Yang Y, Patel VR, Guan S, Randall A, Baldi P, Rychnovsky SD, Huang L (2011) Development of a novel cross-linking strategy for fast and accurate identification of cross-linked peptides of protein complexes. Mol Cell Proteomics 10:M110.002212. https://doi.org/10.1074/mcp.M110.002212
61. Karaj E, Sindi SH, Kuganesan N, Perera L, Taylor W, Tillekeratne LMV (2022) Tunable cysteine-targeting electrophilic heteroaromatic warheads induce Ferroptosis. J Med Chem 65:11788–11817. https://doi.org/10.1021/acs.jmedchem.2c00909
62. Kasinathan S, Orsi GA, Zentner GE, Ahmad K, Henikoff S (2014) High-resolution mapping of transcription factor binding sites on native chromatin. Nat Methods 11:203–209. https://doi.org/10.1038/nmeth.2766
63. Keene JD, Komisarow JM, Friedersdorf MB (2006) RIP-Chip: the isolation and identification of mRNAs, microRNAs and protein components of ribonucleoprotein complexes from cell extracts. Nat Protoc 1:302–307. https://doi.org/10.1038/nprot.2006.47
64. Killinger BA, Moszczynska A (2016) Characterization of α-Synuclein Multimer stoichiometry in complex biological samples by electrophoresis. Anal Chem 88:4071–4084. https://doi.org/10.1021/acs.analchem.6b00419
65. Knight SC, Xie L, Deng W, Guglielmi B, Witkowsky LB, Bosanac L, Zhang ET, El Beheiry M, Masson J-B, Dahan M, Liu Z, Doudna JA, Tjian R (2015) Dynamics of CRISPR-Cas9 genome interrogation in living cells. Science 350:823–826. https://doi.org/10.1126/science.aac6572
66. Kontoyianni M (2017) Docking and virtual screening in drug discovery. Methods Mol Biol 1647:255–266. https://doi.org/10.1007/978-1-4939-7201-2_18
67. Lambert J-P, Mitchell L, Rudner A, Baetz K, Figeys D (2009) A novel proteomics approach for the discovery of chromatin-associated protein networks. Mol Cell Proteomics 8:870–882. https://doi.org/10.1074/mcp.M800447-MCP200
68. Landt SG, Marinov GK, Kundaje A, Kheradpour P, Pauli F, Batzoglou S, Bernstein BE, Bickel P, Brown JB, Cayting P, Chen Y, DeSalvo G, Epstein C, Fisher-Aylor KI, Euskirchen G, Gerstein M, Gertz J, Hartemink AJ, Hoffman MM, Iyer VR, Jung YL, Karmakar S, Kellis M, Kharchenko PV, Li Q, Liu T, Liu XS, Ma L, Milosavljevic A, Myers RM, Park PJ, Pazin MJ, Perry MD, Raha D, Reddy TE, Rozowsky J, Shoresh N, Sidow A, Slattery M, Stamatoyannopoulos JA, Tolstorukov MY, White KP, Xi S, Farnham PJ, Lieb JD, Wold BJ, Snyder M (2012) ChIP-seq guidelines and practices of the ENCODE and modENCODE consortia. Genome Res 22:1813–1831. https://doi.org/10.1101/gr.136184.111
69. Laveglia V, Giachetti A, Cerofolini L, Haubrich K, Fragai M, Ciulli A, Rosato A (2021) Automated determination of nuclear magnetic resonance chemical shift perturbations in ligand screening experiments: the PICASSO web server. J Chem Inf Model 61:5726–5733. https://doi.org/10.1021/acs.jcim.1c00871
70. Lee H, Lee JW (2016) Target identification for biologically active small molecules using chemical biology approaches. Arch Pharm Res 39:1193–1201. https://doi.org/10.1007/s12272-016-0791-z
71. Lee PY, Low TY (2023) Identification and quantification of affinity-purified proteins with MaxQuant, followed by the discrimination of nonspecific interactions with the CRAPome Interface. In: Mukhtar S (ed) Protein-protein interactions. Springer US, New York, pp 299–310

72. Liao Q, Shu X, Sun W, Mandapaka H, Xie F, Zhang Z, Dai T, Wang S, Zhao J, Jiang H, Zhang L, Lin J, Li S, Coin I, Yang F, Peng J, Li K, Wu H, Zhou F, Yang B (2024) Capturing protein–protein interactions with acidic amino acids reactive cross-linkers. Small 20:2308383. https://doi.org/10.1002/smll.202308383
73. Licatalosi DD, Mele A, Fak JJ, Ule J, Kayikci M, Chi SW, Clark TA, Schweitzer AC, Blume JE, Wang X, Darnell JC, Darnell RB (2008) HITS-CLIP yields genome-wide insights into brain alternative RNA processing. Nature 456:464–469. https://doi.org/10.1038/nature07488
74. Liu F, Lössl P, Scheltema R, Viner R, Heck AJR (2017a) Optimized fragmentation schemes and data analysis strategies for proteome-wide cross-link identification. Nat Commun 8:15473. https://doi.org/10.1038/ncomms15473
75. Liu SJ, Horlbeck MA, Cho SW, Birk HS, Malatesta M, He D, Attenello FJ, Villalta JE, Cho MY, Chen Y, Mandegar MA, Olvera MP, Gilbert LA, Conklin BR, Chang HY, Weissman JS, Lim DA (2017b) CRISPRi-based genome-scale identification of functional long noncoding RNA loci in human cells. Science 355:eaah7111. https://doi.org/10.1126/science.aah7111
76. Liu X, Zhang Y, Chen Y, Li M, Zhou F, Li K, Cao H, Ni M, Liu Y, Gu Z, Dickerson KE, Xie S, Hon GC, Xuan Z, Zhang MQ, Shao Z, Xu J (2017c) In situ capture of chromatin interactions by biotinylated dCas9. Cell 170:1028–1043.e19. https://doi.org/10.1016/j.cell.2017.08.003
77. Louche A, Salcedo SP, Bigot S (2017) Protein-protein interactions: pull-down assays. Methods Mol Biol 1615:247–255. https://doi.org/10.1007/978-1-4939-7033-9_20
78. Lu X, Zhu H (2005) Tube-gel digestion: a novel proteomic approach for high throughput analysis of membrane proteins. Mol Cell Proteomics 4:1948–1958. https://doi.org/10.1074/mcp.M500138-MCP200
79. Lu Z, Jiang P, Xing Y (2012) Genetic variation of pre-mRNA alternative splicing in human populations. WIREs RNA 3:581–592. https://doi.org/10.1002/wrna.120
80. Ma B, Zhang K, Hendrie C, Liang C, Li M, Doherty-Kirby A, Lajoie G (2003) PEAKS: powerful software for peptide *de novo* sequencing by tandem mass spectrometry. Rapid Comm Mass Spectrometry 17:2337–2342. https://doi.org/10.1002/rcm.1196
81. MacKinnon AL, Garrison JL, Hegde RS, Taunton J (2007) Photo-leucine incorporation reveals the target of a cyclodepsipeptide inhibitor of cotranslational translocation. J Am Chem Soc 129:14560–14561. https://doi.org/10.1021/ja076250y
82. Malinovska L, Cappelletti V, Kohler D, Piazza I, Tsai T-H, Pepelnjak M, Stalder P, Dörig C, Sesterhenn F, Elsässer F, Kralickova L, Beaton N, Reiter L, de Souza N, Vitek O, Picotti P (2023) Proteome-wide structural changes measured with limited proteolysis-mass spectrometry: an advanced protocol for high-throughput applications. Nat Protoc 18:659–682. https://doi.org/10.1038/s41596-022-00771-x
83. Mansouri-Noori F, Bayfield MA (2024) Electrophoretic mobility shift assays (EMSAs) for in vitro detection of protein-nucleic acid interactions. STAR Protocols 5:103128. https://doi.org/10.1016/j.xpro.2024.103128
84. Martinez Molina D, Jafari R, Ignatushchenko M, Seki T, Larsson EA, Dan C, Sreekumar L, Cao Y, Nordlund P (2013) Monitoring drug target engagement in cells and tissues using the cellular thermal shift assay. Science 341:84–87. https://doi.org/10.1126/science.1233606
85. Maryáš J, Faktor J, Čápková L, Müller P, Skládal P, Bouchal P (2018) Pull-down assay on streptavidin beads and surface Plasmon resonance chips for SWATH-MS-based Interactomics. Cancer Genomics Proteomics 15:395–404. https://doi.org/10.21873/cgp.20098
86. Mateus A, Bobonis J, Kurzawa N, Stein F, Helm D, Hevler J, Typas A, Savitski MM (2018) Thermal proteome profiling in bacteria: probing protein state in vivo. Mol Syst Biol 14:e8242. https://doi.org/10.15252/msb.20188242
87. Mateus A, Kurzawa N, Becher I, Sridharan S, Helm D, Stein F, Typas A, Savitski MM (2020) Thermal proteome profiling for interrogating protein interactions. Mol Syst Biol 16:e9232. https://doi.org/10.15252/msb.20199232
88. Maveyraud L, Mourey L (2020) Protein X-ray crystallography and drug discovery. Molecules 25:1030. https://doi.org/10.3390/molecules25051030

89. Mohammed H, Taylor C, Brown GD, Papachristou EK, Carroll JS, D'Santos CS (2016) Rapid immunoprecipitation mass spectrometry of endogenous proteins (RIME) for analysis of chromatin complexes. Nat Protoc 11:316–326. https://doi.org/10.1038/nprot.2016.020
90. Morretta E, Esposito R, Festa C, Riccio R, Casapullo A, Monti M (2017) Discovering the biological target of 5-epi-Sinuleptolide using a combination of proteomic approaches. Mar Drugs 15:312. https://doi.org/10.3390/md15100312
91. Müller CS, Bildl W, Haupt A, Ellenrieder L, Becker T, Hunte C, Fakler B, Schulte U (2016) Cryo-slicing blue native-mass spectrometry (csBN-MS), a novel technology for high resolution complexome profiling. Mol Cell Proteomics 15:669–681. https://doi.org/10.1074/mcp.M115.054080
92. Müller CS, Bildl W, Klugbauer N, Haupt A, Fakler B, Schulte U (2019) High-resolution complexome profiling by Cryoslicing BN-MS analysis. JoVE:60096. https://doi.org/10.3791/60096
93. Murale DP, Hong SC, Haque MM, Lee J-S (2016) Photo-affinity labeling (PAL) in chemical proteomics: a handy tool to investigate protein-protein interactions (PPIs). Proteome Sci 15:14. https://doi.org/10.1186/s12953-017-0123-3
94. Negroni L, Zivy M, Lemaire C (2017) Mass spectrometry of mitochondrial membrane protein complexes. In: Lacapere J-J (ed) Membrane protein structure and function characterization. Springer New York, New York, pp 233–246
95. Neuditschko B, Legin AA, Baier D, Schintlmeister A, Reipert S, Wagner M, Keppler BK, Berger W, Meier-Menches SM, Gerner C (2021) Interaction with ribosomal proteins accompanies stress induction of the anticancer metallodrug BOLD-100/KP1339 in the endoplasmic reticulum. Angew Chem Int Ed Engl 60:5063–5068. https://doi.org/10.1002/anie.202015962
96. Nguyen TMT, Kim J, Doan TT, Lee M-W, Lee M (2020) APEX proximity labeling as a versatile tool for biological research. Biochemistry 59:260–269. https://doi.org/10.1021/acs.biochem.9b00791
97. Nie M, Li H (2023) Innovation in cross-linking mass spectrometry workflows: toward a comprehensive, flexible, and customizable data analysis platform. J Am Soc Mass Spectrom 34:1949–1956. https://doi.org/10.1021/jasms.3c00123
98. Nilsen TW (2014) Mapping RNA–protein interactions using hydroxyl-radical footprinting. Cold Spring Harb Protoc 2014:pdb.prot080952. https://doi.org/10.1101/pdb.prot080952
99. Noone J, Wallace RG, Rochfort KD (2023) Immunoprecipitation: variations, considerations, and applications. Methods Mol Biol 2699:271–303. https://doi.org/10.1007/978-1-0716-3362-5_15
100. Nowak DE, Tian B, Brasier AR (2005) Two-step cross-linking method for identification of NF-κB gene network by chromatin immunoprecipitation. BioTechniques 39:715–725. https://doi.org/10.2144/000112014
101. Ong S-E, Blagoev B, Kratchmarova I, Kristensen DB, Steen H, Pandey A, Mann M (2002) Stable isotope labeling by amino acids in cell culture, SILAC, as a simple and accurate approach to expression proteomics. Mol Cell Proteomics 1:376–386. https://doi.org/10.1074/mcp.M200025-MCP200
102. Orsburn BC (2021) Proteome discoverer-a community enhanced data processing suite for protein informatics. Proteomes 9:15. https://doi.org/10.3390/proteomes9010015
103. Paek J, Kalocsay M, Staus DP, Wingler L, Pascolutti R, Paulo JA, Gygi SP, Kruse AC (2017) Multidimensional tracking of GPCR signaling via peroxidase-catalyzed proximity labeling. Cell 169:338–349.e11. https://doi.org/10.1016/j.cell.2017.03.028
104. Pahlow S, Ostendorp A, Krüßel L, Kehr J (2018) Phloem sap sampling from Brassica napus for 3D-PAGE of protein and ribonucleoprotein complexes. JoVE:57097. https://doi.org/10.3791/57097
105. Pan S, Zhang H, Wang C, Yao SCL, Yao SQ (2016) Target identification of natural products and bioactive compounds using affinity-based probes. Nat Prod Rep 33:612–620. https://doi.org/10.1039/c5np00101c

106. Parikh RA, Miller LM, Draper BE, Kizekai L, Addepalli B, Chen M, Lauber MA, Jarrold MF (2025) Coupling of size exclusion chromatography to high throughput charge detection mass spectrometry for the analysis of large proteins and virus-like particles. Anal Chem 97:3036–3044. https://doi.org/10.1021/acs.analchem.4c06084
107. Park H, Ha J, Koo JY, Park J, Park SB (2017) Label-free target identification using in-gel fluorescence difference via thermal stability shift. Chem Sci 8:1127–1133. https://doi.org/10.1039/c6sc03238a
108. Pepelnjak M, de Souza N, Picotti P (2020) Detecting protein-small molecule interactions using limited proteolysis-mass spectrometry (LiP-MS). Trends Biochem Sci 45:919–920. https://doi.org/10.1016/j.tibs.2020.05.006
109. Perkins DN, Pappin DJC, Creasy DM, Cottrell JS (1999) Probability-based protein identification by searching sequence databases using mass spectrometry data. Electrophoresis 20:3551–3567. https://doi.org/10.1002/(SICI)1522-2683(19991201)20:18<3551::AID-ELPS3551>3.0.CO;2-2
110. Piersimoni L, Kastritis PL, Arlt C, Sinz A (2022) Cross-linking mass spectrometry for investigating protein conformations and protein–protein interactions—a method for all seasons. Chem Rev 122:7500–7531. https://doi.org/10.1021/acs.chemrev.1c00786
111. Qin W, Cheah JS, Xu C, Messing J, Freibaum BD, Boeynaems S, Taylor JP, Udeshi ND, Carr SA, Ting AY (2023) Dynamic mapping of proteome trafficking within and between living cells by TransitID. Cell 186:3307–3324.e30. https://doi.org/10.1016/j.cell.2023.05.044
112. Ramanathan M, Porter DF, Khavari PA (2019) Methods to study RNA–protein interactions. Nat Methods 16:225–234. https://doi.org/10.1038/s41592-019-0330-1
113. Reinhard FBM, Eberhard D, Werner T, Franken H, Childs D, Doce C, Savitski MF, Huber W, Bantscheff M, Savitski MM, Drewes G (2015) Thermal proteome profiling monitors ligand interactions with cellular membrane proteins. Nat Methods 12:1129–1131. https://doi.org/10.1038/nmeth.3652
114. Ren Y-S, Li H-L, Piao X-H, Yang Z-Y, Wang S-M, Ge Y-W (2021) Drug affinity responsive target stability (DARTS) accelerated small molecules target discovery: principles and application. Biochem Pharmacol 194:114798. https://doi.org/10.1016/j.bcp.2021.114798
115. Rey M, Dupré M, Lopez-Neira I, Duchateau M, Chamot-Rooke J (2018) eXL-MS: an enhanced cross-linking mass spectrometry workflow to study protein complexes. Anal Chem 90:10707–10714. https://doi.org/10.1021/acs.analchem.8b00737
116. Rodriguez EL, Poddar S, Iftekhar S, Suh K, Woolfork AG, Ovbude S, Pekarek A, Walters M, Lott S, Hage DS (2020) Affinity chromatography: a review of trends and developments over the past 50 years. J Chromatogr B Analyt Technol Biomed Life Sci 1157:122332. https://doi.org/10.1016/j.jchromb.2020.122332
117. Rogers JC, Bomgarden RD (2016) Sample preparation for mass spectrometry-based proteomics; from proteomes to peptides. In: Mirzaei H, Carrasco M (eds) Modern proteomics—sample preparation, analysis and practical applications. Springer International Publishing, Cham, pp 43–62
118. Rosania GR, Chang YT, Perez O, Sutherlin D, Dong H, Lockhart DJ, Schultz PG (2000) Myoseverin, a microtubule-binding molecule with novel cellular effects. Nat Biotechnol 18:304–308. https://doi.org/10.1038/73753
119. Röst HL, Rosenberger G, Navarro P, Gillet L, Miladinović SM, Schubert OT, Wolski W, Collins BC, Malmström J, Malmström L, Aebersold R (2014) OpenSWATH enables automated, targeted analysis of data-independent acquisition MS data. Nat Biotechnol 32:219–223. https://doi.org/10.1038/nbt.2841
120. Röst HL, Sachsenberg T, Aiche S, Bielow C, Weisser H, Aicheler F, Andreotti S, Ehrlich H-C, Gutenbrunner P, Kenar E, Liang X, Nahnsen S, Nilse L, Pfeuffer J, Rosenberger G, Rurik M, Schmitt U, Veit J, Walzer M, Wojnar D, Wolski WE, Schilling O, Choudhary JS, Malmström L, Aebersold R, Reinert K, Kohlbacher O (2016) OpenMS: a flexible open-source software platform for mass spectrometry data analysis. Nat Methods 13:741–748. https://doi.org/10.1038/nmeth.3959

121. Rostovtsev VV, Green LG, Fokin VV, Sharpless KB (2002) A stepwise huisgen cycloaddition process: copper(I)-catalyzed regioselective "ligation" of azides and terminal alkynes. Angew Chem Int Ed Engl 41:2596–2599. https://doi.org/10.1002/1521-3773(20020715)41:14<2596::AID-ANIE2596>3.0.CO;2-4
122. Rouskin S, Zubradt M, Washietl S, Kellis M, Weissman JS (2014) Genome-wide probing of RNA structure reveals active unfolding of mRNA structures in vivo. Nature 505:701–705. https://doi.org/10.1038/nature12894
123. Roux KJ, Kim DI, Burke B, May DG (2018) BioID: a screen for protein-protein interactions. Curr Protoc Protein Sci 91:19.23.1. https://doi.org/10.1002/cpps.51
124. Salim NN, Feig AL (2009) Isothermal titration calorimetry of RNA. Methods 47:198–205. https://doi.org/10.1016/j.ymeth.2008.09.003
125. Santamaría D, Barrière C, Cerqueira A, Hunt S, Tardy C, Newton K, Cáceres JF, Dubus P, Malumbres M, Barbacid M (2007) Cdk1 is sufficient to drive the mammalian cell cycle. Nature 448:811–815. https://doi.org/10.1038/nature06046
126. Sato S, Murata A, Shirakawa T, Uesugi M (2010) Biochemical target isolation for novices: affinity-based strategies. Chem Biol 17:616–623. https://doi.org/10.1016/j.chembiol.2010.05.015
127. Savitski MM, Reinhard FBM, Franken H, Werner T, Savitski MF, Eberhard D, Martinez Molina D, Jafari R, Dovega RB, Klaeger S, Kuster B, Nordlund P, Bantscheff M, Drewes G (2014) Tracking cancer drugs in living cells by thermal profiling of the proteome. Science 346:1255784. https://doi.org/10.1126/science.1255784
128. Schmid-Burgk JL, Höning K, Ebert TS, Hornung V (2016) CRISPaint allows modular base-specific gene tagging using a ligase-4-dependent mechanism. Nat Commun 7:12338. https://doi.org/10.1038/ncomms12338
129. Searle BC, Pino LK, Egertson JD, Ting YS, Lawrence RT, MacLean BX, Villén J, MacCoss MJ (2018) Chromatogram libraries improve peptide detection and quantification by data independent acquisition mass spectrometry. Nat Commun 9:5128. https://doi.org/10.1038/s41467-018-07454-w
130. Shechner DM, Hacisuleyman E, Younger ST, Rinn JL (2015) Multiplexable, locus-specific targeting of long RNAs with CRISPR-display. Nat Methods 12:664–670. https://doi.org/10.1038/nmeth.3433
131. Shenoy SV, Biswas D, Mandal K, Zalevsky A, Kishishita A, Verma A, Upadhay I, He Y, Sali A, Viner R, Srivastava S, Wiita AP (2025) AlphaCross-XL: a seamless tool for automated and proteome-scale mapping of crosslinked peptides onto three-dimensional protein structures
132. Shi J-M, Pei J, Liu E-Q, Zhang L (2017) Bis(sulfosuccinimidyl) suberate (BS3) crosslinking analysis of the behavior of amyloid-β peptide in solution and in phospholipid membranes. PLoS One 12:e0173871. https://doi.org/10.1371/journal.pone.0173871
133. Shiyama T, Furuya M, Yamazaki A, Terada T, Tanaka A (2004) Design and synthesis of novel hydrophilic spacers for the reduction of nonspecific binding proteins on affinity resins. Bioorg Med Chem 12:2831–2841. https://doi.org/10.1016/j.bmc.2004.03.052
134. Singh A, Thornton ER, Westheimer FH (1962) The photolysis of diazoacetylchymotrypsin. J Biol Chem 237:3006–3008
135. Sinitcyn P, Hamzeiy H, Salinas Soto F, Itzhak D, McCarthy F, Wichmann C, Steger M, Ohmayer U, Distler U, Kaspar-Schoenefeld S, Prianichnikov N, Yılmaz Ş, Rudolph JD, Tenzer S, Perez-Riverol Y, Nagaraj N, Humphrey SJ, Cox J (2021) MaxDIA enables library-based and library-free data-independent acquisition proteomics. Nat Biotechnol 39:1563–1573. https://doi.org/10.1038/s41587-021-00968-7
136. Skene PJ, Henikoff S (2017) An efficient targeted nuclease strategy for high-resolution mapping of DNA binding sites. eLife 6:e21856. https://doi.org/10.7554/eLife.21856
137. Smith E, Collins I (2015) Photoaffinity labeling in target- and binding-site identification. Future Med Chem 7:159–183. https://doi.org/10.4155/fmc.14.152
138. Smith RA, Knowles JR (1973) Letter: Aryldiazirines. Potential reagents for photolabeling of biological receptor sites. J Am Chem Soc 95:5072–5073. https://doi.org/10.1021/ja00796a062

139. Smith RJ, Milne R, Lopez VC, Wiedemar N, Dey G, Syed AJ, Patterson S, Wyllie S (2023) Chemical pulldown combined with mass spectrometry to identify the molecular targets of antimalarials in cell-free lysates. STAR Protocols 4:102002. https://doi.org/10.1016/j.xpro.2022.102002
140. Strickland EC, Geer MA, Tran DT, Adhikari J, West GM, DeArmond PD, Xu Y, Fitzgerald MC (2013) Thermodynamic analysis of protein-ligand binding interactions in complex biological mixtures using the stability of proteins from rates of oxidation. Nat Protoc 8:148–161. https://doi.org/10.1038/nprot.2012.146
141. Ströbaek J, Tang D, Gueto-Tettay C, Gomez Toledo A, Olofsson B, Hartman E, Heusel M, Malmström J, Malmström L (2025) Epitope mapping with sidewinder: an XL-MS and structural modeling approach. IJMS 26:1488. https://doi.org/10.3390/ijms26041488
142. Suorsa M, Rantala M, Mamedov F, Lespinasse M, Trotta A, Grieco M, Vuorio E, Tikkanen M, Järvi S, Aro E (2015) Light acclimation involves dynamic re-organization of the pigment–protein megacomplexes in non-appressed thylakoid domains. Plant J 84:360–373. https://doi.org/10.1111/tpj.13004
143. Swati B, Jakub S, Aleš M, Petra J, Martin H (2025) DSSBU: a novel mass spectrometry-cleavable analogue of the BS3 cross-linker. J Proteome 310:105330. https://doi.org/10.1016/j.jprot.2024.105330
144. Tabana Y, Babu D, Fahlman R, Siraki AG, Barakat K (2023) Target identification of small molecules: an overview of the current applications in drug discovery. BMC Biotechnol 23:44. https://doi.org/10.1186/s12896-023-00815-4
145. Tacheny A, Dieu M, Arnould T, Renard P (2013) Mass spectrometry-based identification of proteins interacting with nucleic acids. J Proteome 94:89–109. https://doi.org/10.1016/j.jprot.2013.09.011
146. Tang G, Lim Y-J, Wang W, Ye Z, Zhang C-J, Yao SQ (2023) Orthogonal strategies for profiling potential cellular targets of anandamide and cannabidiol. Chemistry 29:e202300682. https://doi.org/10.1002/chem.202300682
147. Tracy TE, Madero-Pérez J, Swaney DL, Chang TS, Moritz M, Konrad C, Ward ME, Stevenson E, Hüttenhain R, Kauwe G, Mercedes M, Sweetland-Martin L, Chen X, Mok S-A, Wong MY, Telpoukhovskaia M, Min S-W, Wang C, Sohn PD, Martin J, Zhou Y, Luo W, Trojanowski JQ, Lee VMY, Gong S, Manfredi G, Coppola G, Krogan NJ, Geschwind DH, Gan L (2022) Tau interactome maps synaptic and mitochondrial processes associated with neurodegeneration. Cell 185:712–728.e14. https://doi.org/10.1016/j.cell.2021.12.041
148. Tsai SQ, Joung JK (2016) Defining and improving the genome-wide specificities of CRISPR–Cas9 nucleases. Nat Rev Genet 17:300–312. https://doi.org/10.1038/nrg.2016.28
149. Tucholski T, Knott SJ, Chen B, Pistono P, Lin Z, Ge Y (2019) A top-down proteomics platform coupling serial size exclusion chromatography and Fourier transform ion cyclotron resonance mass spectrometry. Anal Chem 91:3835–3844. https://doi.org/10.1021/acs.analchem.8b04082
150. Tyanova S, Temu T, Cox J (2016) The MaxQuant computational platform for mass spectrometry-based shotgun proteomics. Nat Protoc 11:2301–2319. https://doi.org/10.1038/nprot.2016.136
151. Ule J, Jensen K, Mele A, Darnell RB (2005) CLIP: a method for identifying protein–RNA interaction sites in living cells. Methods 37:376–386. https://doi.org/10.1016/j.ymeth.2005.07.018
152. van der Zouwen AJ, Witte MD (2021) Modular approaches to synthesize activity- and affinity-based chemical probes. Front Chem 9:644811. https://doi.org/10.3389/fchem.2021.644811
153. Van Mierlo G, Vermeulen M (2021) Chromatin proteomics to study epigenetics—challenges and opportunities. Mol Cell Proteomics 20:100056. https://doi.org/10.1074/mcp.R120.002208
154. Van Nostrand EL, Pratt GA, Shishkin AA, Gelboin-Burkhart C, Fang MY, Sundararaman B, Blue SM, Nguyen TB, Surka C, Elkins K, Stanton R, Rigo F, Guttman M, Yeo GW (2016) Robust transcriptome-wide discovery of RNA-binding protein binding sites with enhanced CLIP (eCLIP). Nat Methods 13:508–514. https://doi.org/10.1038/nmeth.3810

155. Vanderboom PM, Dasari S, Ruegsegger GN, Pataky MW, Lucien F, Heppelmann CJ, Lanza IR, Nair KS (2021) A size-exclusion-based approach for purifying extracellular vesicles from human plasma. Cell Reports Methods 1:100055. https://doi.org/10.1016/j.crmeth.2021.100055
156. Villamil M, Xiao W, Yu C, Huang L, Xu P, Kaiser P (2022) The ubiquitin interacting motif-like domain of Met4 selectively binds K48 polyubiquitin chains. Mol Cell Proteomics 21:100175. https://doi.org/10.1016/j.mcpro.2021.100175
157. Vo T, Paul A, Kumar A, Boykin DW, Wilson WD (2019) Biosensor-surface plasmon resonance: a strategy to help establish a new generation RNA-specific small molecules. Methods 167:15–27. https://doi.org/10.1016/j.ymeth.2019.05.005
158. Wang J, Cui W, Ren CL, Ho EA (2024) Ready-to-use microwave sensor modified by antibody-AuNPs nanoconjugate for highly sensitive and selective detection of the SARS-CoV-2 virus. ACS Meas Sci Au 4:651–658. https://doi.org/10.1021/acsmeasuresciau.4c00032
159. West GM, Tang L, Fitzgerald MC (2008) Thermodynamic analysis of protein stability and ligand binding using a chemical modification- and mass spectrometry-based strategy. Anal Chem 80:4175–4185. https://doi.org/10.1021/ac702610a
160. West GM, Tucker CL, Xu T, Park SK, Han X, Yates JR, Fitzgerald MC (2010) Quantitative proteomics approach for identifying protein-drug interactions in complex mixtures using protein stability measurements. Proc Natl Acad Sci USA 107:9078–9082. https://doi.org/10.1073/pnas.1000148107
161. Wienen-Schmidt B, Oebbeke M, Ngo K, Heine A, Klebe G (2021) Two methods, one goal: structural differences between cocrystallization and crystal soaking to discover ligand binding poses. ChemMedChem 16:292–300. https://doi.org/10.1002/cmdc.202000565
162. Wierer M, Mann M (2016) Proteomics to study DNA-bound and chromatin-associated gene regulatory complexes. Hum Mol Genet 25:R106–R114. https://doi.org/10.1093/hmg/ddw208
163. Wöhlbrand L, Ruppersberg HS, Feenders C, Blasius B, Braun H, Rabus R (2016) Analysis of membrane–protein complexes of the marine sulfate reducer *Desulfobacula toluolica* Tol2 by 1D blue native-PAGE complexome profiling and 2D blue native-/SDS-PAGE. Proteomics 16:973–988. https://doi.org/10.1002/pmic.201500360
164. Woo AJ, Dods JS, Susanto E, Ulgiati D, Abraham LJ (2002) A proteomics approach for the identification of DNA binding activities observed in the electrophoretic mobility shift assay. Mol Cell Proteomics 1:472–478. https://doi.org/10.1074/mcp.T200003-MCP200
165. Wright MH, Sieber SA (2016) Chemical proteomics approaches for identifying the cellular targets of natural products. Nat Prod Rep 33:681–708. https://doi.org/10.1039/c6np00001k
166. Xu H, Wang H, Zhao Y, Zhan W, Chen S, Wang P (2024) Detection of small molecule compounds bound to proteins. Chin J Anal Chem 52:100398. https://doi.org/10.1016/j.cjac.2024.100398
167. Yang J, Shamji A, Matchacheep S, Schreiber SL (2007) Identification of a small-molecule inhibitor of class Ia PI3Ks with cell-based screening. Chem Biol 14:371–377. https://doi.org/10.1016/j.chembiol.2007.02.004
168. Yang L, Zhang H, Bruce JE (2009) Optimizing the detergent concentration conditions for immunoprecipitation (IP) coupled with LC-MS/MS identification of interacting proteins. Analyst 134:755. https://doi.org/10.1039/b813335b
169. Yang P, Mathieu C, Kolaitis R-M, Zhang P, Messing J, Yurtsever U, Yang Z, Wu J, Li Y, Pan Q, Yu J, Martin EW, Mittag T, Kim HJ, Taylor JP (2020) G3BP1 is a tunable switch that triggers phase separation to assemble stress granules. Cell 181:325–345.e28. https://doi.org/10.1016/j.cell.2020.03.046
170. Ye Y, Li K, Ma Y, Zhang X, Li Y, Yu T, Wang Y, Ye M (2023) The introduction of detergents in thermal proteome profiling requires lowering the applied temperatures for efficient target protein identification. Molecules 28:4859. https://doi.org/10.3390/molecules28124859
171. Yin J, Chu J-W, Ricci MS, Brems DN, Wang DIC, Trout BL (2004) Effects of antioxidants on the hydrogen peroxide-mediated oxidation of methionine residues in granulocyte

colony-stimulating factor and human parathyroid hormone fragment 13–34. Pharm Res 21:2377–2383. https://doi.org/10.1007/s11095-004-7692-4
172. Yoshikawa H, Sundaramoorthy R, Mariyappa D, Jiang H, Lamond A (2021) Efficient and rapid analysis of polysomes and ribosomal subunits in cells and tissues using Ribo mega-SEC. BIO-PROTOCOL 11. https://doi.org/10.21769/BioProtoc.4106
173. Yu F, Teo GC, Kong AT, Fröhlich K, Li GX, Demichev V, Nesvizhskii AI (2023) Analysis of DIA proteomics data using MSFragger-DIA and FragPipe computational platform. Nat Commun 14:4154. https://doi.org/10.1038/s41467-023-39869-5
174. Zhang L, Wang Y, Zheng C, Zhou Z, Chen Z (2024) Cellular thermal shift assay: an approach to identify and assess protein target engagement. Expert Rev Proteomics 21:387–400. https://doi.org/10.1080/14789450.2024.2406785
175. Zhang Q, Major MB, Takanashi S, Camp ND, Nishiya N, Peters EC, Ginsberg MH, Jian X, Randazzo PA, Schultz PG, Moon RT, Ding S (2007) Small-molecule synergist of the Wnt/beta-catenin signaling pathway. Proc Natl Acad Sci USA 104:7444–7448. https://doi.org/10.1073/pnas.0702136104
176. Zhang X, Ruan C, Zhu H, Li K, Zhang W, Wang K, Hu L, Ye M (2020) A simplified thermal proteome profiling approach to screen protein targets of a ligand. Proteomics 20:e1900372. https://doi.org/10.1002/pmic.201900372
177. Zhang Y, Zhang J, Li M, Qiao Y, Wang W, Ma L, Liu K (2023) Target discovery of bioactive natural products with native-compound-coupled CNBr-activated Sepharose 4B beads (NCCB): applications, mechanisms and outlooks. Bioorg Med Chem 96:117483. https://doi.org/10.1016/j.bmc.2023.117483
178. Zhu H, Wang S (2023) TurboID coupled with APEX2: application prospects for deciphering proteome trafficking and interactions in neuroscience. Brain-X 1:e42. https://doi.org/10.1002/brx2.42
179. Zhu S, Wurdak H, Wang J, Lyssiotis CA, Peters EC, Cho CY, Wu X, Schultz PG (2009) A small molecule primes embryonic stem cells for differentiation. Cell Stem Cell 4:416–426. https://doi.org/10.1016/j.stem.2009.04.001
180. Ziegler S, Pries V, Hedberg C, Waldmann H (2013) Target identification for small bioactive molecules: finding the needle in the haystack. Angew Chem Int Ed 52:2744–2792. https://doi.org/10.1002/anie.201208749
181. Zinn N, Werner T, Doce C, Mathieson T, Boecker C, Sweetman G, Fufezan C, Bantscheff M (2021) Improved proteomics-based drug mechanism-of-action studies using 16-Plex isobaric mass tags. J Proteome Res 20:1792–1801. https://doi.org/10.1021/acs.jproteome.0c00900

Peptidomics in the Spotlight: Advanced Sample Treatment Techniques and Analytical Insights

Alessio Lenzi, Mariano De Cristofaro, Denise Biagini, Silvia Ghimenti, Fabio Di Francesco, and Tommaso Lomonaco

Abstract In recent years, the analysis of intact peptides (peptidomics) has remained a relatively under-explored field, where its potential as a bridge between proteomics and metabolomics has only recently been recognized. Currently, immunoassay techniques are the primary method for quantifying endogenous peptides in biological samples, mainly due to their high-throughput and minimal sample manipulation. However, antibody-based strategies face important limitations, particularly in terms of specificity.

Liquid chromatography–tandem mass spectrometry has emerged as a powerful and reliable alternative for the analysis of native peptides in biological fluids. While conventional sample preparation approaches, such as protein precipitation, enable rapid and broad peptide extraction, they are often inadequate for quantifying low-abundance peptides, leading to matrix effects and ion suppression.

This chapter explores the critical aspects of antibody-free sample preparation in peptidomics, emphasizing advanced and innovative approaches. We also discuss methods for peptide isolation and purification, spanning from traditional approaches such as protein precipitation and solid-phase extraction to more recent microextraction techniques. Special attention is given to microextraction by packed sorbent, ultrafiltration, and electromembrane extraction, which offer significant advantages in efficiency, sensitivity, and scalability. By minimizing both solvent and sample requirements, these approaches provide new opportunities for improving analytical performance in peptidomics.

Keywords Clinical peptidomics · Sample treatment · Intact peptides · LC-MS/MS · Microextraction techniques

---

A. Lenzi (✉)
Department of Veterinary Sciences, University of Pisa, Pisa, Italy
e-mail: alessio.lenzi@unipi.it

M. De Cristofaro · D. Biagini · S. Ghimenti · F. Di Francesco · T. Lomonaco
Department of Chemistry and Industrial Chemistry, University of Pisa, Pisa, Italy

J.-L. Capelo-Martínez (ed.), *Emerging Sample Treatments in Proteomics*,
Advances in Experimental Medicine and Biology 1510,
https://doi.org/10.1007/978-3-032-21638-0_9

1 Introduction to Clinical Peptidomics

In recent decades, the landscape of molecular diagnostics has been profoundly transformed by the rapid expansion of omics sciences [1]. Among these disciplines, clinical peptidomics has emerged as a powerful approach for the comprehensive study of endogenous peptides in biological systems. Advances in instrumental technologies and sample preparation strategies allow the detection of peptides at very low concentration levels (in the pg/mL range). These developments have opened new perspectives for early disease detection, patient stratification, and the monitoring of physiological and pathological responses.

For the purpose of this discussion, peptides will be defined as biomolecules composed of a chain of amino acids with a molecular mass lower than 15 kDa.

1.1 Peptidomics: Defining an Emerging Omics Frontier

In contemporary biomedical research, omics-based approaches have become fundamental tools for the identification and quantification of biomolecules in clinical settings. First introduced in the early 2000s [2], peptidomics refers to a conceptual bridge between proteomics, which focuses on proteins and their post-translational modifications (PTMs), and metabolomics, which addresses small molecules [3–5]. Peptidomics is increasingly recognized for its unique ability to capture the interplay between protein synthesis, processing, and degradation, providing a unique and dynamic snapshot of physiological responses [6]. Despite its growing relevance in the field, the definition of "peptidomics" in the literature remains somewhat arbitrary [2].

Peptides below approximately 15 kDa are often bioactive and capable of diffusing through membranes, circulate systemically, and participate in rapid signaling processes [7]. Endogenous peptides are ubiquitous across biological fluids and tissues, and variations in their concentration levels have been linked to key biological processes, such as immune regulation [8], cell proliferation [9], and inflammatory responses [10]. Due to their small size and relatively simpler structures, peptides may reflect dynamic states of protein synthesis, maturation, and degradation, making them particularly suited for monitoring disease progression and therapeutic response [11, 12]. Moreover, peptides can passively diffuse across endothelial barriers, enabling their detection in peripheral biological fluids such as saliva and sweat [7]. This property enhances their physiological relevance and supports their use in noninvasive diagnostic and monitoring approaches [12].

From a biochemical perspective, peptides in the human body originate through three main mechanisms: (i) proteolytic degradation of precursor proteins by enzymes such as proteases and peptidases, (ii) direct synthesis from gene-encoded transcripts, and (iii) gene-independent enzymatic formation in vivo, involving amino acid anabolism [13]. Figure 1 shows these three mechanisms of peptide biosynthesis.

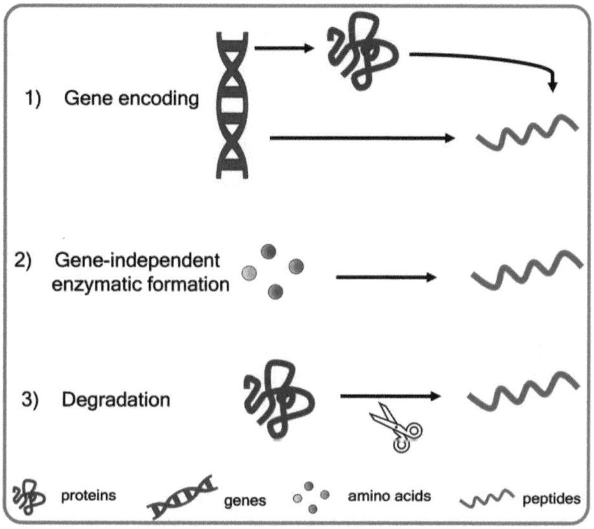

Fig. 1 In vivo generation of peptides. Endogenous peptides primarily originate through (1) gene encoding, (2) gene-independent enzymatic formation, and (3) protein degradation

Based on their origin and function, human peptides are generally classified into two categories. Bioactive peptides, such as hormones, exhibit intrinsic biological activity and function as signaling molecules [14]. In contrast, peptide fragments generated through in vivo enzymatic cleavage of larger proteins often act as by-products or intermediates of protein catabolism [15]. These fragments frequently reflect an individual's metabolic state and provide complementary information relevant to health monitoring and disease assessment.

The inherent complexity of the peptidome underscores the importance of distinguishing between several forms that share portions of the same amino acid sequence. Thus, accurate and robust analytical protocols are essential for profiling disease-associated changes in the peptide landscape [16].

1.2 Analytical Strategies in Peptidomics

Common approaches to determine peptides in biological samples are based on immunoassay techniques due to their speed and ease of use [17]. However, immunoassays are limited by their specificity, as cross-reactivity with structurally similar peptides or degraded forms of the target analyte can compromise accuracy [18]. Additionally, the lack of harmonization across platforms further complicates inter-laboratory comparisons and multi-center pilot studies [15, 17, 19].

In recent years, LC-MS/MS has emerged as a robust alternative platform for peptide analysis, due to its specificity, sensitivity, and multiplexing capabilities (Fig. 2) [18]. Although LC-MS instruments require a higher initial investment, they become cost-effective by enabling the simultaneous detection of multiple peptides in a single analysis. Advancements in instrumentation, sample handling, and

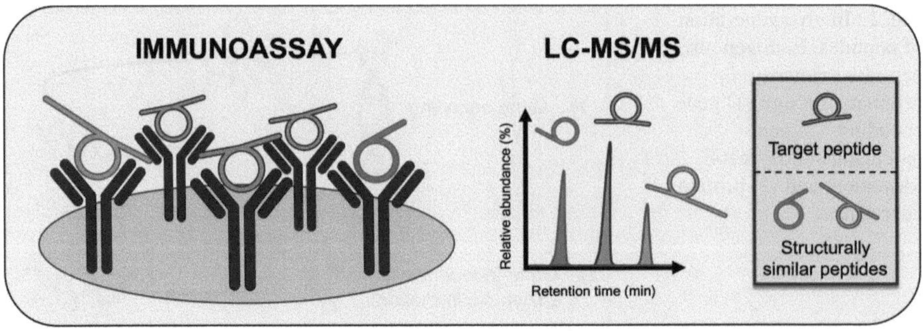

Fig. 2 Comparison of immunoassay and LC-MS/MS in peptidomics. Immunoassays exhibit cross-reactivity between peptides that share the same amino acid sequence, leading to reduced specificity. LC-MS/MS allows for accurate quantification of peptides, providing more reliable results

preparation have further enhanced the high-throughput and reliability of LC-MS/MS for profiling peptides in biological fluids [20].

Peptide analysis by LC-MS/MS is typically performed using either top-down or bottom-up approaches. In bottom-up peptidomics, target compounds are enzymatically digested into smaller peptide fragments (e.g., using trypsin and Lys-C enzymes). A surrogate fragment of the native peptide is then monitored to confirm its presence and quantify its concentration [20, 21]. The bottom-up approach, however, has some limitations, including incomplete and/or inconsistent digestion and the potential lack of specificity of the surrogate fragment, which can introduce analytical bias [20, 22, 23]. Top-down approaches analyze intact peptides, enabling the identification of peptide isoforms and providing information about PTMs, which are often lost in bottom-up workflows [24–27]. While top-down methods may be challenging due to the wide diversity of endogenous peptides in terms of size, charge, and amino acid sequence, they are increasingly employed in peptidomics studies, providing a more comprehensive and accurate snapshot of the peptidome.

The internal standards (i.e., stable isotope-labeled intact peptides, surrogate fragments of proteins, or analog peptides with slight amino acid modifications) are widely employed in both methodologies. These compounds help to reduce the variability introduced by sample manipulation, matrix effects, or instrumental fluctuations, although they can be expensive and may not be available for the analytes of interest [25, 28–30].

1.3 *Pre-analytical Variability: A Central Challenge in Intact Peptides Analysis for Clinical Applications*

Native peptides typically exhibit very short half-lives, ranging from a few seconds up to several minutes, due to their susceptibility to enzymatic degradation [11]. Enzymes such as serine, cysteine, aspartic, and metalloproteases remain active in many biological samples, and some retain proteolytic activity even after sample collection [31]. Thus, inadequate sample collection, handling, or storage may result in substantial loss of target peptides, sometimes occurring before the sample reaches the laboratory. These pre-analytical factors profoundly affect the quality, reliability, and robustness of peptidomic analyses, making careful optimization of all sample-handling steps essential [32].

Biological and physiological factors
Physiological factors, such as body temperature, stress levels, nutritional intake, circadian rhythms, and medication, significantly influence peptide secretion and degradation [33]. Accurate documentation of patient conditions is therefore essential for the correct interpretation of peptidomic data in any well-designed clinical study. For example, Manconi et al. demonstrated that the use of denture adhesive can alter the composition of salivary peptides [34]. Similar variations have been observed in response to exercise, circadian cycles, fasting, and the menstrual cycle, which can either generate false-positive biomarker signals or mask relevant biological changes [35]. These findings underline the critical need for rigorous standardization of sample collection procedures during patient recruitment [12, 36].

Sample collection and handling
Pre-analytical variability is also influenced by procedural factors. Improper sample handling, delayed centrifugation, sample mislabeling, or inadequate storage temperatures can compromise peptide integrity [7, 33]. In high-throughput clinical environments, even minor procedural errors can result in coagulation, hemolysis, or protease activation, emphasizing the importance of robust quality assurance and adequate personnel training.

Among biological matrices, blood is one of the most informative and complex specimens used in peptidomics [37]. It contains a wide array of endogenous peptides derived from endocrine secretion, immune activation, enzymatic cleavage, and cellular turnover. Since blood is rich in proteolytic enzymes (i.e., thrombin, kallikrein, plasmin, and aminopeptidases), rapid processing and appropriate inhibition of enzymatic activity are fundamental for preserving endogenous peptide integrity [25, 31].

A key consideration in blood-based peptidomics is the choice between plasma and serum. Plasma is generally preferred, as it avoids coagulation, which can enhance peptide proteolysis and compromise analytical reliability [1]. Figure 3 shows the general degradation of bioactive peptides mediated by a protease, resulting in different fragments that may lack biological activity.

Blood collection tubes require anticoagulants (e.g., EDTA, heparin, citrate), which can significantly influence peptide composition [38, 39]. Careful selection of anticoagulants based on the physicochemical properties of the peptides of interest is critical.

The addition of protease inhibitor "cocktails" before or immediately after blood collection is standard practice. These cocktails typically target serine, cysteine, aspartic, and metalloproteases, forming a broad defense against enzymatic

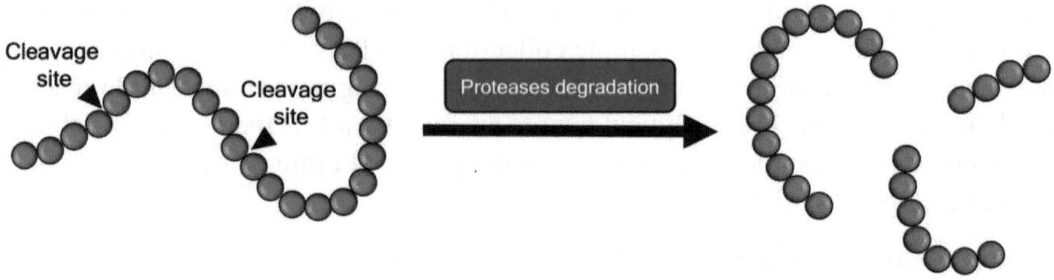

Fig. 3 Generic proteolytic degradation of bioactive peptides in blood leads to biologically inactive fragments, augmenting the complexity of the sample

degradation. Common inhibitors include PMSF, leupeptin, E-64, and EDTA, while some inhibitors are themselves peptides (e.g., aprotinin and pepstatin A) and must be considered in LC-MS analyses to avoid interference or signal suppression [7]. The protective effect of these inhibitors is well-documented; for example, BNP can degrade by over 50% within 30 minutes of collection without inhibitors but remains stable for several hours when inhibitors are present [25, 31].

To further mitigate pre-analytical variability, commercial blood collection tubes pre-loaded with stabilizing reagents have been introduced. These specialized tubes combine anticoagulants and inhibitors, enabling rapid sample stabilization and reducing pre-analytical variability [7]. However, they remain relatively costly, raising concerns about scalability and reproducibility across different laboratories. Additionally, the stabilizer–peptide interactions are peptide-specific, and experimental validation is required for each analyte.

A non-exhaustive list of the main commercially available blood tubes, along with their inhibitor composition and suppliers, is reported in Table 1.

An often underappreciated but critical issue in blood-based peptidomics is hemolysis, which refers to the rupture of red blood cells and the consequent release of intracellular components (hemoglobin) into plasma [40]. Hemolysis can result from factors such as improper needle gauge, rough handling, excessive centrifugation time, or exposure to extreme temperatures. Even minor hemolysis can dramatically affect analysis; for example, hemoglobin may produce ion suppression during electrospray ionization [41]. Standard operating procedures should include systematic assessment of hemolysis, either visually using colorimetric methods or, ideally, through quantitative measurement of hemoglobin. Furthermore, documentation of hemolysis status should accompany every peptidomic dataset [19, 42].

Proteases retain activity even at refrigeration temperatures (+4 °C), making rapid cooling and ultralow-temperature storage essential. It is generally recommended that samples should be flash-frozen in liquid nitrogen or stored immediately at −80 °C to minimize enzymatic degradation. Storage at −20 °C, while slowing peptide degradation, still allows for continued loss of sensitive peptides [12, 43]. Repeated freeze–thaw cycles further exacerbate degradation by promoting aggregation, hydrolysis, and oxidative stress. Therefore, aliquoting samples into single-use containers is widely adopted to preserve peptide integrity during long-term storage [7, 33, 40].

Table 1 A non-exhaustive list of the main commercially available blood tubes with their inhibitor mixture composition, target proteases, anticoagulant, and supplier

Product name	Inhibitor(s)	Target protease(s)	Anticoagulant	Supplier
BD™ P800 Blood Collection System	Proprietary Cocktail of Protease, Esterase, and DPP-IV Inhibitors	Dipeptidyl Peptidase IV, Esterases	K_2EDTA	BD Biosciences
BD™ P100 Blood Collection System	Proprietary Cocktail of Protease	n/a	K_2EDTA	BD Biosciences
VACUETTE® TUBE 4 ml EDTA Aprotinin	Aprotinin	Serine proteases	K_3EDTA	Greiner Bio-One
Sample collection tubes 9-SCAT-I-5	Proprietary Cocktail of Protease	n/a	EDTA	Cryopep
Sample collection tubes 9-SCAT-II-3	Phe-Pro-Arg-chloromethylketone	Thrombin	Sodium citrate	Cryopep
Sample collection tubes 9-SCAT-27-1.8/5	Corn trypsin inhibitor	Serine proteases	Citric acid	Cryopep
BD Vacutainer™ Glass Aprotinin K3EDTA Tube	Aprotinin	Serine proteases	K_3EDTA	Fisher Scientific
Sample collection tubes HT-SCAT-I	Phe-Pro-Arg-chloromethylketone, aprotinin	Thrombin, serine proteases	EDTA	Haematologic Technologies

Chemical modification of peptides, particularly oxidation, can also occur during storage and analysis. Amino acid residues such as methionine, cysteine, histidine, and tryptophan are especially susceptible [44]. Oxidation of amino acid residue alters peptide mass (e.g., +16 Da) and fragmentation patterns, potentially hindering MS identification and inactivating biological function. This process is promoted by prolonged exposure to air, light, or trace metal ions [25, 45]. Preventive measures include storing samples in amber containers, purging containers with inert gases, and adding antioxidants or metal chelators such as methionine, ascorbic acid, or EDTA [46].

Nonspecific adsorption to material surfaces

In recent years, nonspecific adsorption of peptides to material surfaces has emerged as a key concern in the development of peptidomics platforms, as it can impact the accuracy of the assay [47]. Adsorption occurs when peptides form a film on the surface of the material, leading to a progressive reduction in MS signal intensity over time until a plateau is reached [47–49]. This effect is particularly pronounced at low peptide concentration levels, where the number of available surface binding sites exceeds the number of analyte molecules. Maes and coworkers [36] identified three physicochemical parameters that primarily influence peptide solubility and adsorption: (i) the total number of amino acid residues, (ii) the number of non-polar residues, and (iii) the overall charge and net charge of the molecule. Positively charged basic amino acids may establish electrostatic interactions with negatively

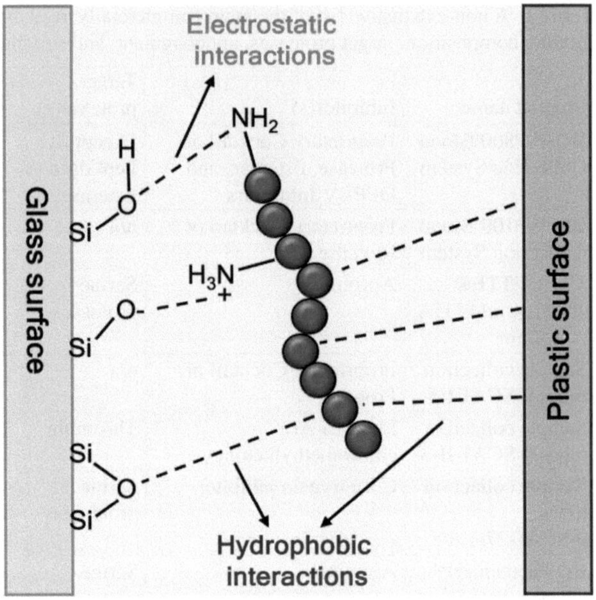

Fig. 4 Schematic nonspecific interaction mechanisms between peptide in solution and material surfaces

charged silanol groups on glass surfaces, while non-polar amino acids are more prone to adsorb onto the hydrophobic surfaces of plastic materials [36].

Figure 4 shows the typical mechanisms of nonspecific peptide adsorption on glass and plastic surfaces.

Adding small concentrations of surfactants, proteins (e.g., bovine serum albumin, BSA), or organic solvents such as acetonitrile to storage containers can help to mitigate peptide adsorption. However, this phenomenon can also occur within the LC-MS system itself. Adsorption to tubing, connectors, injectors, and column frits may cause analyte carryover or alter signal intensity [25, 48, 49].

Therefore, when developing new peptidomics procedures, careful consideration of analyte-dependent adsorption is essential to ensure accurate and reliable results.

1.4 Methodological Challenges in LC-MS/MS Analysis

LC-MS/MS offers efficient separation of complex peptide mixtures and satisfactory sensitivity for their detection and quantification [44, 50].

One of the main issues in reversed-phase liquid chromatography (RP-LC) of peptides is peak tailing, which adversely affects chromatographic peak shape and resolution. This effect is largely driven by secondary interactions between basic amino acid residues (e.g., lysine, arginine, and histidine) and residual silanol groups on silica-based stationary phase [50]. Strategies to mitigate peak tailing include the use of an end-capped column or the addition of ion-pairing reagents such as trifluoroacetic acid (TFA) to the mobile phase. However, TFA is well-known to cause ion suppression in ESI, reducing MS sensitivity. To balance chromatographic and MS performance, alternatives such as difluoroacetic acid [51] or mixed acid systems (e.g., formic acid) combined with TFA have been explored, although each introduces its own trade-offs.

A further limitation of RP-LC lies in the poor retention of highly polar or hydrophophilic peptides, which interact weakly with hydrophobic stationary phase and elute early, resulting in suboptimal separation. In such cases, hydrophilic interaction liquid chromatography (HILIC) has emerged as a viable alternative, offering improved retention and signal intensity for polar analytes. Nonetheless, HILIC may introduce higher matrix effects, especially when analyzing complex biological samples, potentially affecting quantification reliability and robustness [52, 53].

In summary, RP-LC-MS/MS remains the cornerstone analytical technique in peptidomics, but its application is hindered by challenges such as peak tailing, ion suppression, poor retention of hydrophilic peptides, and matrix effects. Addressing these limitations requires careful and meticulous method optimization, the adoption of emerging techniques, and ongoing innovation in chromatographic and mass spectrometric technologies.

At the same time, a successful clinical peptidomics platform relies on rigorous standardized pre-analytical protocols. From patient enrollment and blood collection tube selection to temperature control, inhibitor usage, and sample storage, every step must be consistent, validated, and thoroughly documented. As this field matures from experimental research toward routine clinical application, establishing robust methodologies is pivotal to harnessing its full diagnostic potential.

2 Traditional Peptide Extraction Methods for LC-MS/MS Analysis

Sample preparation is widely recognized as a pivotal and time-consuming procedure and a major source of variability in analytical workflows. This is particularly evident in peptidomics, where peptides exhibit physicochemical properties similar to metabolites and proteins, creating unique analytical challenges [54, 55].

The development of a reliable LC-MS/MS-based platform for intact peptide analysis depends on robust sample preparation strategies aimed at preserving and enriching intact peptides. These biomolecules are typically present at very low concentrations within complex biological matrices that also contain highly abundant proteins, lipids, and salts; all these species can hinder peptide detection and quantification in LC-MS/MS analyses. To address these challenges, the analysis of peptides may adapt sample preparation strategies already implemented in metabolomics (e.g., protein precipitation) and proteomics (e.g., desalting), while ensuring the integrity of the peptide [7, 12].

Figure 5 shows the typical steps employed in a traditional peptidomics sample preparation workflow for biological samples.

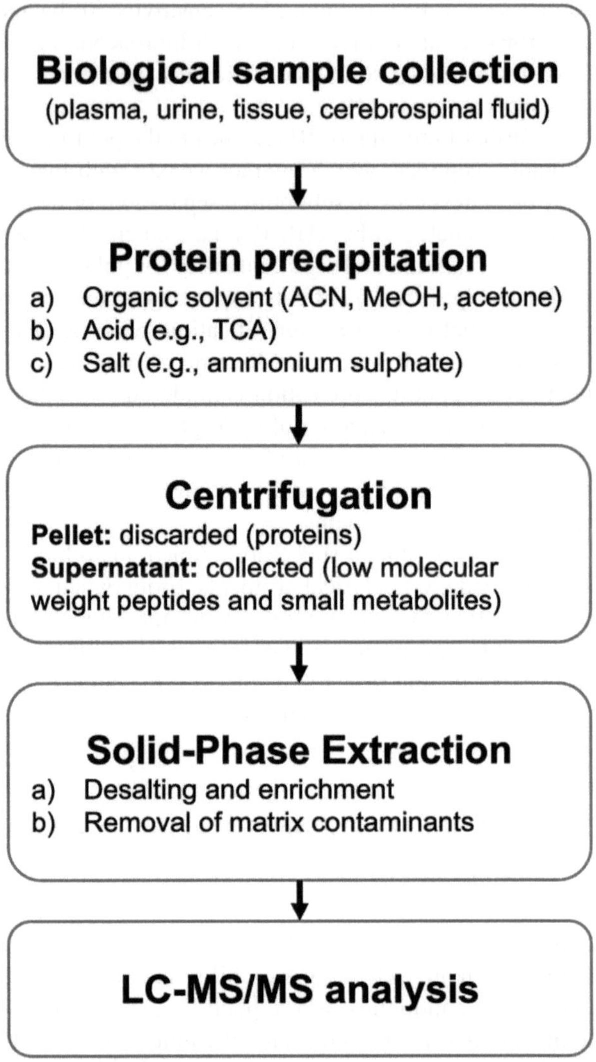

Fig. 5 Overview of a sample preparation workflow employed in traditional peptidomics platforms

Enriching the peptidome: protein precipitation and solid-phase extraction procedures

Protein precipitation is one of the most commonly employed methods in peptidomics workflows for biological sample preparation [56, 57]. Common agents added to the biological samples are organic solvents (such as acetonitrile (ACN) [57, 58], methanol (MeOH) [59], ethanol [60], and acetone [61], acids (e.g., trichloroacetic acid [62]), or salts (e.g., ammonium sulfate [63]). The main mechanism of these solvents is to disrupt large protein hydrophobic interactions and hydrogen bonds, leading to their denaturation and subsequent aggregation in the liquid medium. Within these conditions, the dielectric constant decreases, making protein–protein interactions favorable by displacing the water layer from the protein surface, triggering their precipitation [64]. Resulting protein pellets can be removed by centrifugation, leaving low-molecular-weight peptides and small metabolites in solution [25, 57, 65]. Despite its widespread use, organic solvent extraction presents some limitations. First, the potential loss of peptides can either bind to proteins or co-precipitate under certain solvent conditions. Moreover, it often lacks selectivity during the peptide extraction process. For example, most hydrophobic peptides may partition into the protein pellet rather than remaining in the supernatant, reducing their recovery [56]. For these reasons, during method development, the composition of the precipitating solvent, as well as its pH, should be carefully evaluated to ensure optimal peptide recovery from the matrix. Incubation time and temperature can also impact peptide extraction yield, with colder conditions (e.g., −20 °C) improving protein precipitation and phase separation.

Trichloroacetic acid is frequently employed for protein precipitation thanks to its capability to disrupt the hydration layer surrounding proteins [62]. As a result, proteins start to aggregate and become no longer soluble and can be eliminated by centrifugation, leaving more soluble peptides in the supernatant [62, 64, 66]. Despite its effectiveness, TCA precipitation presents two main limitations: (i) it creates an acidic environment (pH close to 1), affecting some peptides, leading to hydrolysis; (ii) its efficiency strongly depends on the sample protein content and the target peptides, making it less universally applicable [62].

Salting-out-based approaches using high concentrations of salts (e.g., ammonium sulfate) are an alternative procedure. Ammonium sulfate is a kosmotropic reagent in the Hofmeister series that promotes salting-out by reducing free water molecules available to solvate protein hydrophobic surface regions [63], resulting in their aggregation and precipitation [63]. Ammonium sulfate is typically considered a milder precipitant than organic solvents or TCA, making it suitable for preserving protein and peptide integrity during precipitation as well as for serving as a preliminary concentration step in their purification workflows [67]. However, the high amount of salt required for achieving a complete protein precipitation is incompatible with mass spectrometry instrumentation, as it causes significant ion suppression during electrospray ionization. In this case, a desalting step is mandatory before LC-MS/MS analysis [25].

Precipitation-based methods remain the common approach in peptidomic biological sample preparation thanks to their rapidity, versatility, and affordability. These characteristics make them highly attractive as first-line strategies for depleting abundant proteins while enriching low-molecular-weight peptides.

In recent years, combinations of precipitation agents have been progressively employed to improve the effectiveness of precipitation protocols, enabling the simultaneous recovery of both high-molecular-weight proteins and small peptides [61, 68, 69]. These strategies typically promote protein aggregation, thereby enhancing protein and peptide overall precipitation efficiency when supplementing organic solvents with salts. Such approaches address a limitation of traditional organic solvent precipitation, which shows poor recoveries of soluble low-molecular-weight peptides. For example, Nickerson et al. [68] developed an acetone–NaCl mixture that enabled almost quantitative recovery of the proteome and low-molecular-weight peptidome down to 10 kDa. Extending this approach, Baghalabadi et al. [61] demonstrated that supplementing acetone with 100 mM $ZnSO_4$ can induce aggregation and precipitation of even smaller tryptic peptides with molecular weights down to 1 kDa. These findings highlight the importance of the synergistic effects between organic solvents and ionic strength determined by salts in modulating protein and peptide precipitation. Lenzi et al. [25] reported similar findings for a panel of natriuretic peptides (2–3 kDa). In this work, saturated ammonium sulfate was added directly to plasma samples in a 1:9 v/v ratio, resulting in the nearly quantitative precipitation of natriuretic peptides as illustrated in Fig. 6.

These studies highlight the potential to tune precipitation agents to provide the flexibility required for optimizing protocols according to the fitness-for-purpose of the analysis. Tailoring precipitation strategies offers a powerful tool to both extend the coverage of the peptidome or to maximize the recovery of low-molecular-weight peptides.

Solid-phase extraction (SPE), either alone or combined with protein precipitation (see Fig. 5), is one of the most widely used methods for peptides purification and enrichment in mass spectrometry-based assays [26, 70–73]. Unlike protein precipitation, SPE offers a more selective and cleaner peptide extract from complex biological samples, while removing interferents such as salts, detergents, and lipids [74, 75]. The working principle of SPE relies on the reversible retention of peptides under controlled solvent conditions, followed by sequential washing and selective elution. SPE is typically performed using cartridges, tips, or 96-well plate formats that are packed with sorbent phases. The choice of sorbent material determines which peptides are retained and how effectively contaminants are removed. Among the most widely used sorbents in the analysis of multi-charged intact peptides are mixed-mode sorbents that combine hydrophobic and ionic retention mechanisms or

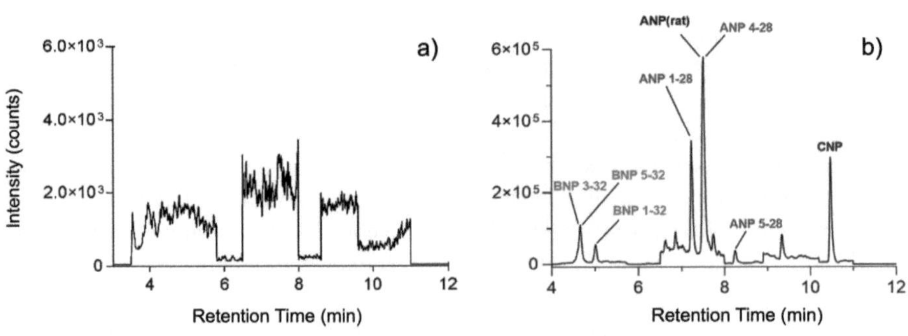

Fig. 6 LC-MS/MS chromatogram obtained from the analysis of (**a**) supernatant and (**b**) protein pellet collected after salting-out with saturated ammonium sulfate of a pooled blank plasma sample (1:9 v/v ratio) spiked with a panel of natriuretic peptides at 0.5 ng/mL. (Adapted with modifications from Lenzi et al. [25] *Talanta*, 2024, https://doi.org/10.1016/j.talanta.2024.126077. Licensed under CC BY 4.0). BNP B-type natriuretic peptide, ANP atrial natriuretic peptide, CNP C-type natriuretic peptide

hydrophilic–lipophilic balanced (HLB) polymeric sorbents [26, 70–72]. Other sorbents include reversed-phase C18 materials, which retain peptides primarily through hydrophobic interactions [73]. The general workflow involves four steps:

- Conditioning of the sorbent with organic solvent and equilibration with an aqueous solution;
- Sample loading during which peptides are retained in the stationary phase;
- Washing poorly retained contaminants such as salts and hydrophilic compounds;
- Elution of peptides using an organic solvent-rich mixture (ACN or MeOH) with a small amount of acid, such as formic acid or trifluoroacetic acid.

Figure 7 illustrates a general SPE procedure for peptidomics applications, highlighting each stage from sorbent conditioning to the final recovery of purified peptides.

The extraction efficiency depends on several factors, including sorbent type, solvent composition, pH, and the physicochemical properties of the peptides. For instance, maintaining a slightly acidic environment during sample loading enhances peptide retention on C18 sorbents by reducing ionization of acidic side chains, thereby increasing hydrophobic interactions. Conversely, basic peptides may exhibit

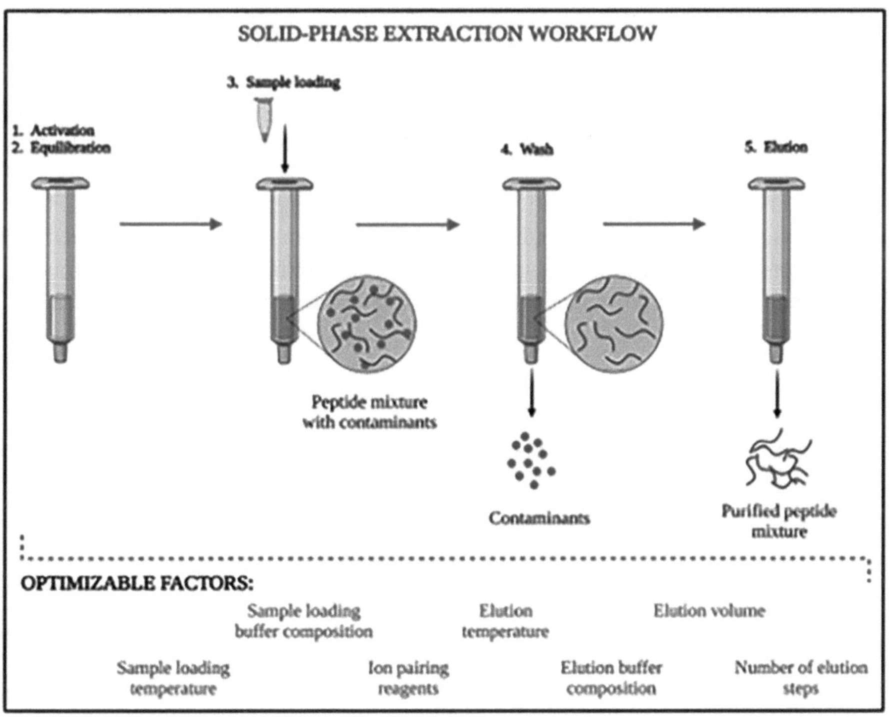

Fig. 7 General SPE workflow employed in peptidomics sample treatment. (Adapted from Bugyi et al. [76], *Journal of Mass Spectrometry*, 2023, https://doi.org/10.1002/jms.4965. Licensed under CC BY 4.0)

stronger binding and require careful elution conditions to ensure recovery [76, 77]. For a more detailed explanation of the parameters to be optimized in peptide SPE procedures and troubleshooting, see [76]. It is worth noting that it mainly focuses on small tryptic peptides.

Despite its benefits, SPE is not without limitations. One of the most significant challenges is the risk of biased peptide recovery due to differences in peptide hydrophobicity, charge, and size. Hydrophilic peptides may be lost during the washing step, while strongly hydrophobic peptides may require harsher elution conditions that risk peptide modification or carryover of contaminants [76].

Variability in cartridge performance, batch-to-batch differences in sorbent properties, and the manual nature of traditional SPE workflows can also affect reproducibility. Furthermore, for studies involving small sample volumes or peptides of very low abundance, nonspecific adsorption to tube surfaces or incomplete elution can significantly impact peptide yield. Moreover, it necessitates large volumes of organic solvents, posing both environmental and safety concerns [74, 76].

The integration of protein precipitation with solid-phase extraction (SPE) provides a synergistic workflow that enhances peptide enrichment and sample cleanliness for LC-MS analysis. By first removing bulk proteins and then eliminating salts while concentrating peptides, this two-step process ensures that endogenous peptides are detectable even at low abundance levels [78]. This complementary approach, combining protein precipitation and SPE, has become a standard in clinical peptidomics workflows, particularly in biomarker discovery studies where reproducibility and selectivity are critical. Although further refinements are needed to reduce biases and improve reproducibility, conventional cartridge- and plate-based SPE remains a benchmark method due to its versatility, adaptability to automation, and compatibility with LC-MS/MS. Emerging techniques such as microextraction, on-line SPE, and microfluidic platforms are being developed, yet the combined use of protein precipitation and SPE continues to serve as a foundational technique in peptidomics and is likely to remain central to future clinical and translational applications.

3 Advanced Sample Treatment Techniques in Peptidomics

3.1 Microextraction by Packed Sorbent (MEPS)

In recent years, increasing emphasis on the principles of green analytical chemistry has spurred the development of microextraction techniques as viable alternatives to conventional sample preparation methods, like liquid–liquid extraction (LLE) and SPE [79]. These techniques were able to fulfill many of the requirements of green analytical methods, including miniaturization, automation, high-throughput performance, online coupling with analytical instruments, and low-cost operation with little or no solvent consumption. Among the miniaturized solid-based approaches, microextraction by packed sorbent (MEPS) was adopted as a simple, fast, and online sample preparation technique after its introduction in 2004 by Abdel-Rehim [80].

MEPS is a miniaturized form of the conventional SPE technique that combines different steps of a sample pretreatment procedure, i.e., extraction, enrichment, and clean-up, in a single device composed of a syringe (typically volume of 100–500 μL) and a cartridge that is directly inserted into the syringe, as shown in Fig. 8.

MEPS cartridge, also known as Barrel Insert and Needle (BIN), contains the sorbent packed bed (1–4 mg) where analytes can be retained based on their physicochemical properties when the sample is drawn through the syringe [81]. As a general consideration, MEPS and SPE follow the same principles, making it easy to transfer a method from traditional SPE to MEPS. Compared to SPE, in MEPS procedure, the solution flows in two directions (up and down), allowing better performance. Moreover, SPE cartridges are disposable, whereas MEPS BIN can be reused several times (up to one hundred cycles) without any loss of extraction efficiency, depending on the nature of the sample [82]. In fact, MEPS works well with liquid samples; however, complex biological matrices like plasma and urine may require preliminary additional steps of sample pretreatment to avoid sorbent clogging, thus prolonging BIN life. Similar to SPE, various types of sorbents are available. Generally, MEPS sorbents are based on silica particles or polymeric phases, which are modified to obtain different chemical properties, resulting in a large portfolio of specificities for target analytes. The most employed are silica-based sorbents (C2, C8, C18), but also strong and weak anion and cation exchange (SAX and SCX), mixed sorbents, restricted access material (RAM), carbon, polystyrene–divinylbenzene copolymer, or molecular imprinted polymers (MIPs) can be used [83]. MEPS

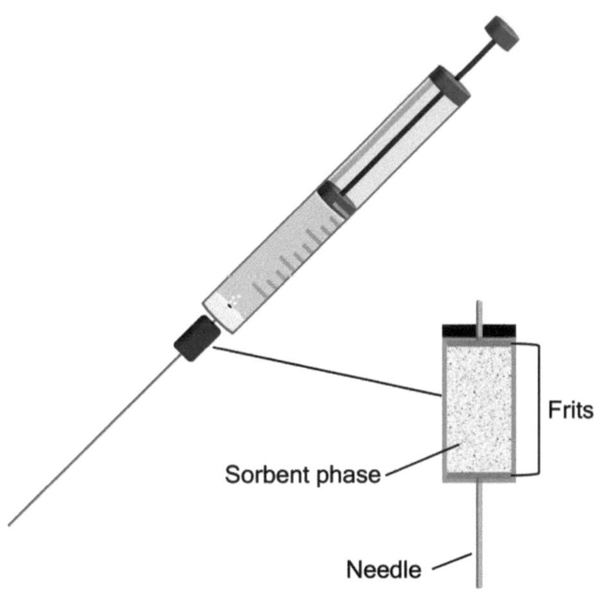

Fig. 8 Overview of the microextraction by packed sorbent (MEPS) syringe with barrel insert and needle (BIN) containing the sorbent packed bed for analyte retention

is commercially available in offline and online formats. The first commercially available format consists of manual MEPS, where the MEPS syringe is manually operated via a Hamilton-type syringe. This format has been recently replaced by the semi-automatic MEPS version (eVol-MEPS) introduced by SGE and Thermo Fisher Scientific. The eVol couples a digitally controlled and programmable electronic drive and a XCHANGE® analytical syringe to provide customizable and precise flow rate and volume, allowing for repeatable extraction of target analyte. The third form of MEPS is the fully automatic MEPS that can be easily achieved using the same MEPS XCHANGE® syringes on autosamplers (e.g., CTC Analytics, Zwingen, Switzerland). This coupling allows the direct interface of MEPS with both LC-MS and GC-MS systems, thus providing a completely automated MEPS-LC-MS or MEPS-GC-MS platform [84]. Regardless of the format, extraction is performed by loading sample and solvents up-down through the sorbent once or more (draw–eject in the same vial or draw and eject into waste). The entire MEPS procedure consists of four sequential steps as conventional SPE, i.e., conditioning, sample loading, washing, and elution (Fig. 9).

Briefly, after sorbent conditioning with appropriate solvents, usually methanol followed by water, the sample is drawn through the needle into the syringe once or several times (depending on experimental optimization), allowing the retention of

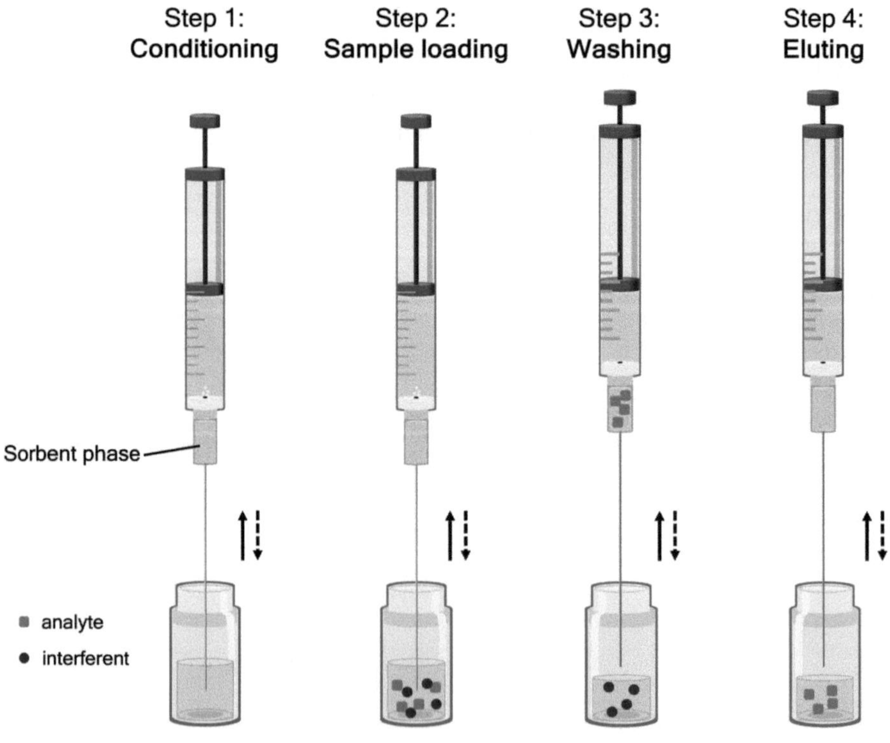

Fig. 9 Schematic steps of the microextraction by packed sorbent (MEPS) procedure

the target analytes in the sorbent. This is followed by a washing step to remove matrix interferences, most often using acidic water containing a small amount of organic solvent. Finally, the target analytes are eluted using an appropriate elution solvent, usually an organic solution (e.g., methanol, acetonitrile) into a vial or directly into the LC or GC injector without the need for concentration steps (e.g., evaporation under nitrogen stream). To reuse an MEPS cartridge, the sorbent is washed 3–4 times with water and 4–5 times with elution solvent to eliminate carryover.

Despite its simplicity, MEPS procedure needs to be carefully optimized to achieve a suitable extraction efficiency. For instance, the selection of a proper sorbent is important to accomplish satisfactory clean-up and obtain good analyte recovery. The sample volume used should be optimized to obtain the best equilibrium between a good analytical performance and a good extraction methodology. Therefore, ideally, the proper concentration factor for the target analytes should be obtained in the minimum sample volume of loading cycles, but it is not unusual to observe MEPS extraction protocols using up to ten or more extraction cycles. This creates another optimization opportunity by loading the sample several times, both discarding and reloading the aliquot at each sampling cycle. Beyond the sample loading, the washing step is useful to remove matrix interferents that are retained from the sorbent along with target analytes. Washing step is typically performed with a small amount of organic solvent for an efficient removal of weakly retained undesired compounds leading to an increase in extraction efficiency. However, the amount of organic solvent in the washing solution should be carefully evaluated since it can lead to a partial loss of the target analytes prior to elution. Elution of the analytes from the sorbent is performed using the smallest possible volume of organic solvent (possibly added with acid or base). MEPS has the advantage of eluting in smaller volumes compared to the initial amount of sample, thus allowing a considerable enrichment factor of the analytes prior to injection into an LC or a GC system.

MEPS has been successfully used to extract a wide range of analytes (metabolites and drugs) in different biological matrices, such as urine, plasma, serum, saliva, whole blood, and dried blood spots [85–87]. Moreover, MEPS applications have been extended to the environmental and forensic fields, particularly for the analysis of pesticides, polycyclic aromatic hydrocarbons, and other organic pollutants in environmental water samples and for the screening of drugs of abuse [88]. To the best of our knowledge, MEPS has been used very few times in the field of peptidomics. Ashri et al. in 2012 have used MEPS for the determination of peptide bovine adrenal medulla, BAM8–22, and its antagonist, BAM22–8, in human plasma [89]. Comparing MEPS with other extraction techniques (e.g., UF with a 20 kDa cutoff and protein precipitation with ACN), the authors obtained clearer samples and higher recoveries (around 90% for MEPS compared to approximately 60% for the other tested approaches). Moreover, the use of MEPS resulted in reduced preparation time for plasma samples, which is crucial given the poor stability of BAM peptides in plasma. More recently, in 2024, Lenzi et al. employed MEPS technique for the determination of natriuretic peptides (i.e., ANP1–28, ANP4–28, ANP5–28, BNP1–32, BNP3–32, BNP5–32, and CNP), in human plasma samples with satisfactory performances [25]. MEPS is a promising technique for peptide sample preparation because it is fast and easy to use; it can be fully automated for online procedures; it reduces the solvent and sample volume, as well as the waste produced. Overall, these MEPS properties create a high-throughput and efficient sample extraction technique, being much more environmentally friendly and less expensive than any commercially available SPE or LLE approaches.

3.2 Ultrafiltration

Ultrafiltration (UF) is a separation process that relies on semipermeable membranes to discriminate between solutes of different sizes, typically retaining high-molecular-weight compounds while allowing smaller molecules and solvents to pass through [90, 91]. Its efficiency is defined by the molecular weight cutoff (MWCO), which determines the size threshold of separation. Commercial UF devices are often designed as centrifugal units composed of two chambers divided by a semipermeable membrane. Under centrifugal force, molecules smaller than the MWCO pass through, while larger ones are retained. This straightforward principle makes UF attractive, but conventional pressure-driven systems also face limitations, including fouling and restricted selectivity when solutes are of similar molecular size. To address these challenges, innovations have incorporated additional driving forces such as electrical fields and concentration gradients (see Paragraph 1.5).

Additionally, ultrasound-assisted UF has emerged as a method to counteract flux decline due to fouling and concentration polarization. Ultrasonic waves generate cavitation and acoustic streaming, which prevent deposition on the membrane surface and enhance separation efficiency [92]. Despite the improvements in selectivity and efficiency offered by these new technologies, pressure-driven systems remain the most widely used at present due to their simplicity and speed of operation.

Pressure-driven UF devices are commercially available from a variety of suppliers (www.sartorius.com; www.millipore.com; www.corning.com; www.pall.com). Systems like Vivaspin®, Centricon®, Nanosep®, and protein concentrators use centrifugal force to concentrate biomolecules, as well as membranes with defined MWCOs for rapid, efficient sample concentration (see Table 2).

In general, UF systems use specially designed centrifugal tubes ranging from small volumes (≈1.5 mL or less) up to 70 mL, depending on the supplier. Tubes are centrifuged using either fixed-angle or swing-out rotors, with small-volume tubes typically requiring fixed-angle rotors and larger volumes often using swing-out rotors. Tubes are constructed from materials compatible with a wide pH range (pH 1–9) and solvent type and are designed to retain minimal liquid to prevent sample drying. Vertical or angular membrane configurations reduce concentration polarization and fouling, enabling high flow rates even with protein-rich solutions. Specialized membrane treatments, such as Hydrosart® (Sartorius) and Omega™ or Bio-Inert® (Pall), are proposed to minimize nonspecific adsorption onto the membrane.

Table 2 List of commercially available pressure-driven UF devices

Device	Supplier	Sample volume (mL)	MWCO range (kDa)	Notes
Vivaspin®	Sartorius	0.5–15	2–1000	Efficient protein concentration, Hydrosart® low-binding membranes
Centricon®	Merck Millipore	0.5–15	3–1000	High recovery, vertical membrane design
Nanosep®	Pall	0.05–0.5	3–300	Compact, Omega™ and Bio-Inert® low-binding membranes
Vivacon®	Sartorius	15–100	5–1000	Larger sample volumes, clinical/diagnostic use
Amicon® Ultra	Merck Millipore	0.5–15	3–1000	Rapid concentration and purification
Microcon®	Merck Millipore	0.5	3–100	Small volumes, high recovery
Spin-X®	Corning	0.2–2	3–100	Variety of membranes for different applications

Rotor type, centrifugation speed, and membrane pore size must be selected carefully according to the sample and device specifications [93]. For example, a 100 kDa cutoff membrane will preferentially retain immunoglobulins (~156 kDa) while allowing most serum albumin (~67 kDa) to pass, depending on membrane properties. This approach is widely used to separate low-molecular-weight from high-molecular-weight protein fractions in proteomic analyses [94]. Greening et al. have shown that the filter material used by different manufacturers significantly affects the efficiency of isolating small proteins (<25 kDa) from human plasma, and batch-to-batch variability can occur even for filters made from the same material [91]. Sartorius membranes (Vivaspin® 20 kDa and Centrisart® 20 kDa) effectively excluded proteins >50 kDa while recovering low-MW components, identifying 44 proteins from 266 unique peptides using SDS-PAGE and nanoLC-MS/MS. Conversely, Millipore devices (Microcon® 30 kDa and Amicon Ultra® 30 kDa) showed slightly lower overall recovery (87–93%) and less selective fractionation, highlighting the impact of membrane chemistry and design on plasma peptidome enrichment [91]. The suboptimal performance of Amicon Ultra® systems in peptide isolation has also been highlighted in a recent study by Lenzi et al., which focused on the fractionation of natriuretic peptides in plasma [25].

Typically, MWCO membranes ranging from 3 to 50 kDa and centrifugation speeds between 3000 × g and 4000 × g are employed [91]. To select an effective membrane, the "rule of 1.5–2" is commonly used: the membrane cutoff should be roughly half the molecular weight of the target protein in the retentate. For example, a 30 kDa MWCO membrane effectively removes proteins ≥65 kDa, achieving >90–95% recovery of smaller proteins in the filtrate. Additional factors influencing performance include flow rate (flux), solute concentration and composition, and temperature [12]. While wide-pore UF membranes separate molecules primarily through a sieving mechanism, narrower pores impose stricter molecular cutoffs, making separation dependent on solute solubility, membrane hydrophobicity, and polarity. For tight UF membranes (1–2 kDa) and nanofiltration (NF), electrostatic interactions between the membrane surface and solutes can further enhance selectivity if the chemical environment is properly controlled [95]. This emphasizes the concept that both pore size and chemical properties influence protein recovery and filtrate purity.

Figure 10 illustrates an ultrafiltration device applied in peptidomics, in which the membrane retains large proteins, allowing smaller peptides to pass into the permeate.

Membranes and adsorption phenomena

Traditional UF membranes are typically fabricated from cellulose derivatives or polysulfone [96]. Cellulose membranes are highly hydrophilic, reducing nonspecific adsorption and fouling, but they lack mechanical and chemical robustness [97]. Polysulfone membranes, in contrast, offer greater rigidity and stability but are more prone to fouling due to their hydrophobic character [95]. Polyethersulfone, a less hydrophobic variant, combines improved water interactions with resistance to protein adsorption, while cellulose triacetate and regenerated cellulose (including Hydrosart®) are valued for minimal nonspecific binding and high protein recovery [91]. Other synthetic polymers, including polyvinylidene fluoride (PVDF), polyacrylonitrile (PAN), and cellulose acetate (CA), are also extensively commercialized thanks to their permeability, flux, and relatively low cost, making polymeric membranes the mainstream choice in both laboratory and industrial UF [96].

Regardless of the type of membrane, the key challenge is fouling, which reduces permeation flux and lowers productivity over time. The formation of a fouling layer induced by peptides at the membrane surface is mainly dependent on peptide MW, charge, and hydrophobicity. The physicochemical parameters of the membrane

Fig. 10 Schematic representation of ultrafiltration device in peptidomics: complex matrices (retentate) are separated through a proper MWCO membrane, allowing peptides and small molecules (permeate) to pass through while retaining larger proteins and macromolecules

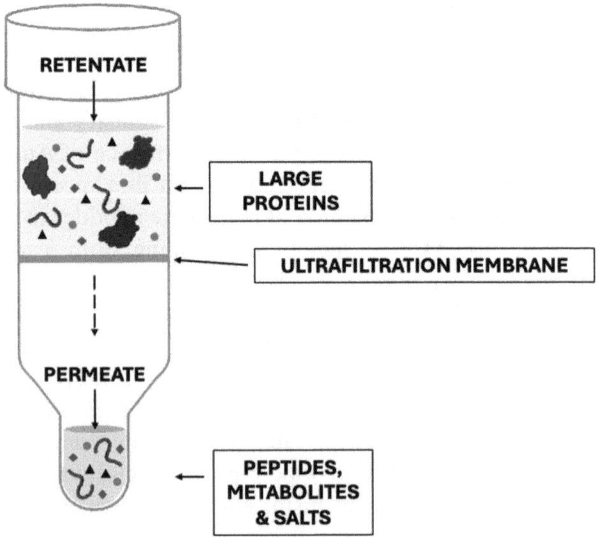

material also have a significant impact on peptide adsorption [98]. Although regular cleaning can address fouling, it is often difficult and time-consuming. Fouling can be mitigated by using hydrophilic membranes, designing open-channel modules, or operating below the critical flux to prevent flux decline. Achieving high permeation flux also requires careful membrane design, considering factors like porosity, pore size distribution, asymmetric layer thickness, and surface hydrophilicity [99]. To overcome the limitations of fouling and poor selectivity, advances in membrane design have included chemical surface modifications, copolymerization, and hybrid organic–inorganic composites that improve hydrophilicity and antifouling properties. Liu et al. prepared zwitterionic organic–inorganic composite ultrafiltration membranes by zwitterionic chitosan (ZICS), polyvinyl alcohol (PVA), and tetraethyl orthosilicate (TEOS) [100]. Beyond synthetic polymers, nanoporous materials such as mesoporous silica and metal–organic frameworks (MOFs) have gained attention in proteomics. Their highly ordered pore structures provide precise size-exclusion capabilities and tunable surface chemistries, enabling selective enrichment of endogenous peptides, though high cost and complex fabrication remain barriers to widespread use [101]. For example, boronic acid–functionalized mesoporous graphene–silica composites (GO@mSiO_2) have proven suitable for glycopeptide analysis in human serum [102], while Fe_3O_4@mSiO_2 microspheres have been employed to capture endogenous peptides from the serum of gastric cancer patients [103]. Many nanocomposites are costly and require complex synthetic routes, which significantly limit their practical applications. Therefore, there is an urgent need to develop simpler and more cost-effective materials for peptidome analysis. Recently, natural polymeric membranes have emerged as sustainable alternatives, combining biodegradability with competitive peptide separation performance (i.e., chitosan, cellulose, silk fibroin, alginate, keratin, fibrin, and gelatin) [104]. Silk fibroin membranes have been successfully employed for separating peptides with molecular weights ranging from 1 to 10 kDa [105], whereas chitosan membranes have been demonstrated to have excellent separation properties for molecules with molecular weights ranging from 0.2 to 100 kDa [106]. Their properties can be further enhanced by blending with nanoparticles or synthetic polymers, improving strength, hydrophilicity, and antimicrobial activity. While issues of stability and scalability remain, natural polymer membranes represent a promising, eco-friendly direction for next-generation UF technologies [107].

Despite advances in ultrafiltration materials, a considerable fraction of proteins can still adhere nonspecifically to ultracentrifugation membranes. Highly hydrophobic or membrane-associated small proteins are prone to loss during cell lysis or may bind tightly to devices designed to enrich small proteins, such as MWCO filters. Nonspecific adsorption of proteins and peptides to UF containers and membranes remains a critical source of analytical bias despite the development of low-bind plastics and advanced membrane chemistries [33]. The dominant mechanisms are electrostatic interactions, such as the affinity of basic residues (Lys, Arg) for negatively charged silanol groups on glass, and hydrophobic interactions between nonpolar side chains and polyolefin surfaces such as polypropylene or polyethylene [33]. Adsorptive losses are strongly concentration-dependent, disproportionately affecting analytes in the low nanomolar to picomolar range, and are further enhanced near the isoelectric point where peptides exhibit minimal net charge and reduced electrostatic repulsion [108]. Adsorption is rapid, reaching equilibrium within minutes, and increases with handling time. Several strategies have been proposed to mitigate these effects. Adjustment of solvent composition and pH can improve peptide solubility and decrease surface affinity; for example, addition of 5–30% acetonitrile or methanol, combined with weak acids such as formic or acetic acid, reduces both hydrophobic and electrostatic binding, while alkaline buffers such as ammonium bicarbonate can repel acidic peptides through charge effects [108]. Solvents with the ability to solubilize both polar and nonpolar compounds, such as dimethyl sulfoxide and dimethylformamide, might also be good alternatives, but they are not suitable for every peptide because they may induce uncontrolled oxidation of the peptides [109]. Surfactants provide an alternative approach. Non-ionic detergents such as Tween-20 or Triton X-100, or MS-compatible acid-labile agents including RapiGest, PPS, or MaSDeS, block surface adsorption sites, though concentrations must be carefully optimized to minimize ion suppression in LC-MS [110]. Surface pre-blocking with proteins, notably BSA (0.5% *m/v* overnight) or Prionex, has also proven effective in occupying nonspecific binding sites and stabilizing recovery of labile peptides such as parathyroid hormone fragments [33]. Ultrafiltration introduces further challenges: hydrophobic and membrane-associated peptides are frequently retained on MWCO membranes, with transmission governed by peptide charge state, ionic strength, centrifugation speed, and membrane design [108]. Pretreatment of membranes with water [91], ammonium acetate [74], or chaotropes such as urea 6 M [111] has been shown to reduce aggregation and minimize irreversible binding. Lower centrifugation speed and longer centrifugation times than those recommended by the manufacturer have been shown to improved recovery of low-molecular-weight proteins and peptides [112]. Collectively, these findings emphasize the need for empirical optimization of buffer composition, container and membrane materials, and blocking strategies to ensure reproducible recovery of low-molecular-weight peptides in proteomic and bioanalytical workflows.

UF advantages and case studies

UF offers notable advantages such as simplicity, relatively low cost, availability of membranes with diverse solvent compatibilities, and the ability to concentrate analytes while simultaneously removing larger proteins [93]. Compared to precipitation with organic solvents, UF is faster and less labor-intensive and is easily scalable from small laboratory samples to larger volumes [113]. However, its performance is frequently hindered by nonspecific adsorption to membranes, fouling, and concentration polarization, which may compromise reproducibility and reduce peptide recovery [114]. Losses are particularly critical for hydrophobic peptides, analytes near their isoelectric point, or peptides bound to abundant carriers such as albumin, which may fail to pass through the membrane unless denaturing conditions or organic modifiers (e.g., 20–25% acetonitrile) are used to disrupt protein–protein interactions [90]. Moreover, UF devices are single-use, prone to clogging, and require optimization of centrifugal force and buffer composition to prevent either incomplete depletion of high-molecular-weight proteins or inadvertent passage of albumin through compromised membranes [115].

While UF remains attractive for its speed, accessibility, and versatility, it is rarely sufficient as a stand-alone approach and often needs to be integrated with additional enrichment techniques to achieve comprehensive coverage of the peptidome [12]. Aristoteli et al. employed a 10 kDa MWCO filter centrifuged at 3000 × g for 15–30 minutes, achieving optimal recovery for MALDI analysis in the 800–4000 m/z range. While SPE yielded higher peptide diversity, UF effectively removed high-molecular-weight proteins, particularly those above 10 kDa [116]. Parker et al. used a 10 kDa MWCO ultrafiltration device combined with multiplexed LC-MS/MS to track exercise-induced changes in the plasma peptidome, identifying hundreds of low-abundance peptides from stress, inflammatory, and metabolic proteins that would otherwise be masked by high-molecular-weight proteins [66]. Williams et al. compared UF with organic precipitation and selective extraction methods, showing that precipitation followed by aqueous-organic extraction could capture peptides from proteins, such as transferrin and fibrinogens, which UF alone missed [117]. Similarly, Zheng et al. used 10 kDa Vivaspin® filters with 10% acetonitrile to disrupt peptide–protein interactions in human serum, achieving efficient depletion of high-molecular-weight proteins while preserving low-molecular-weight peptides, although some losses of albumin-bound peptides were observed [118]. Lenzi et al. provided a detailed analysis of UF efficiency and peptide distribution in plasma using a 10 kDa Amicon Ultra® device. UHPLC-MS/MS analysis revealed differential distribution of natriuretic peptides (MW <4 kDa): ANPs were roughly equally distributed among ultrafiltrate, retentate, and surfaces, while others (e.g., CNP) were predominantly retained on the membrane despite being below the MWCO, thus emphasizing the importance of nonspecific adsorption phenomena [25]. These studies collectively demonstrate that UF is highly effective for preparing serum and plasma peptides but benefits from combination with complementary extraction methods for maximal peptidome coverage. Finoulst et al. applied UF to cerebrospinal fluid, followed by nano-LC-MS/MS, identifying 563 neuropeptides derived from 91 precursors, many of which were not detected in parallel proteome profiling, emphasizing the importance of separating LMW peptides from larger proteins [119]. Hartman et al. extracted peptides from infected and uninfected pig and human wound fluids using Microcon 30 centrifugal filters with 6 M urea in 10 mM Tris buffer, followed by SPE [111]. This workflow enabled detailed temporal profiling of wound peptides under different infection conditions, showcasing the utility of UF in dynamic clinical studies and infection biomarker discovery. Finally, salivary peptide analysis has been advanced by UF in combination with various pretreatments. Vitorino et al. compared UF, organic solvent precipitation, and acid treatment to extract peptides from saliva. Saliva pretreatment with guanidine, acetonitrile, or acid improved recovery of peptides, especially those bound to high-molecular-weight glycoproteins such as MUC5B and MUC7. The combination of organic solvent extraction with ultrafiltration yielded the best recovery, identifying 2497 peptides, including clusters from histatin 1, aPRP, bPRP2, and SMR3B [120]. Across multiple biofluids, centrifugal UF demonstrates robust capability for isolating low-molecular-weight peptides, crucial for clinical peptidomics and biomarker discovery. Key considerations include MWCO selection, pretreatment with denaturants or organic solvents, mitigation of nonspecific adsorption, and, when needed, combination with complementary methods such as SPE or precipitation. Together, these strategies maximize peptide recovery as well as enhance downstream MS analysis.

3.3 Electromembrane Extraction

Electromembrane extraction (EME) was first introduced in 2006 by Pedersen-Bjergaard and Rasmussen as an innovative method for the extraction of charged analytes [121]. Initially developed for biological fluids, environmental samples, and food products, EME has gained recognition for its high selectivity, low solvent consumption, and efficient cleanup [122–124]. Its operating principle combines both electrokinetic migration (electrophoresis) and diffusive migration (liquid–liquid extraction), offering a remarkable alternative to conventional extraction methods, such as SPE and LLE [125].

EME relies on the migration of charged analytes from a donor solution (an aqueous sample) through a supported liquid membrane (SLM) into an acceptor solution, under an applied external electrical field [121]. Two platinum electrodes, connected to a direct current power supply, are immersed in both the sample and the acceptor solutions to generate the electric field.

The fundamentals of EME are shown in Fig. 11.

To enable analyte migration across the SLM, sample pH must be adjusted based on the compound's physicochemical properties. For instance, basic analytes require a pH that guarantees the presence of a net positive charge throughout the process [126]. Therefore, the cathode is placed in the acceptor solution, whereas the anode is placed in the sample. Once the electric field is generated by the external power supply, it drives positively charged analytes through the SLM and traps them in the acceptor solution. Finally, the acceptor solution can be collected and further processed or analyzed by LC-MS. For the extraction of acidic analytes, the conditions of pH and polarity should be reversed accordingly.

The extraction time optimization is needed to maximize analyte recovery while avoiding a prolonged extraction cycle that may lead to poor recovery and unstable current [127]. Additionally, the agitation of the system should be carefully evaluated to enhance mass transfer from the sample to the liquid membrane and improve the extraction efficiency [127]. Since the sample volume is typically larger than the acceptor solution, EME enables valuable preconcentration and clean-up steps, making it suitable for high-throughput workflows [128].

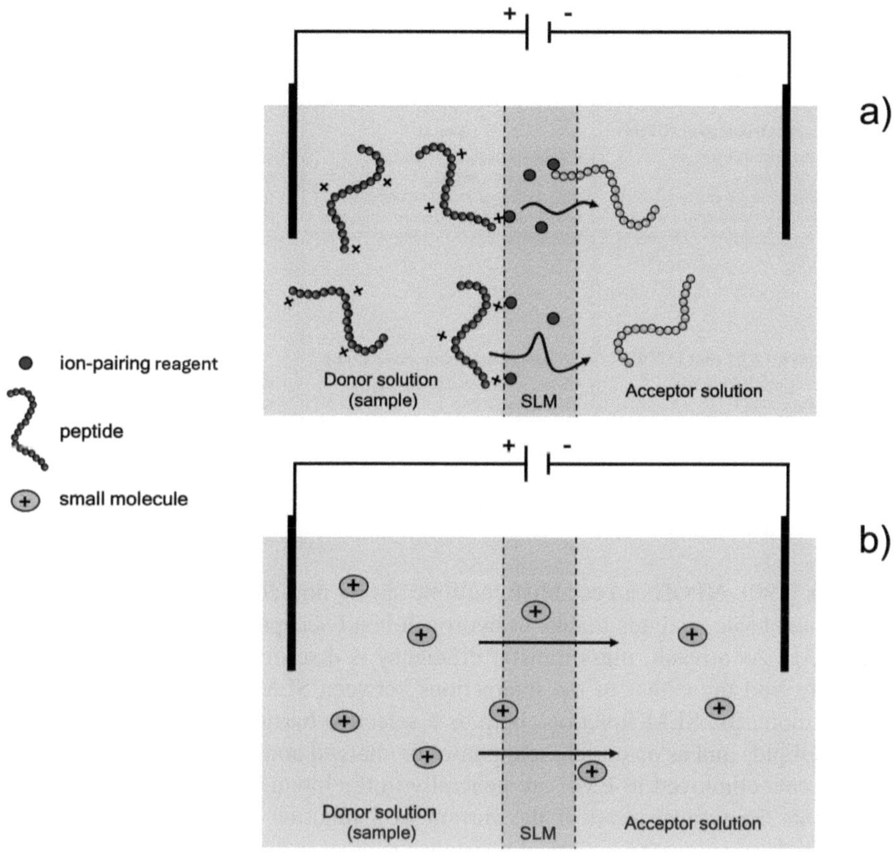

Fig. 11 Schematic representation of EME fundamentals employed for (**a**) intact peptides and (**b**) small molecules. Red circles represent ion-pairing reagents supplemented in the supported liquid membrane (SLM)

Importance of the supported liquid membrane

SLM plays a pivotal role in determining the selectivity and efficiency of the procedure. Its chemical composition influences analyte mass transfer, system stability, and extraction performance, determining the compounds that can successfully migrate across the interface [129].

Typically, SLM is prepared by immobilizing a microliter-range volume of a water-immiscible organic solvent in the pores of a solid porous polymeric membrane using capillary forces [128, 129]. Ideally, the liquid membrane's water solubility should be less than 0.5 mg/mL to prevent leakage into the sample and acceptor and ensure overall system [128]. Most common solvents include nitroaromatic ethers, alcohols, and ketones, as shown in Fig. 12.

Among these solvents, 2-nitrophenyl octyl ether (NPOE, logP = 4.9) is extensively used to extract non-polar basic analytes (logP >2) efficiently from aqueous

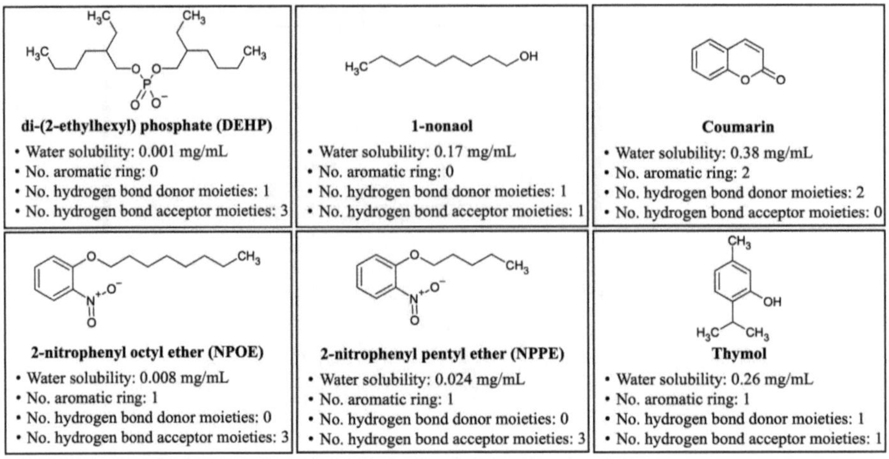

Fig. 12 Molecular structure and physicochemical properties of some of the most used SLMs. (Adapted from Ref. [131])

samples [130]. NPOE can establish multiple strong non-covalent interactions with protonated basic analytes thanks to hydrogen bond acceptor sites and an aromatic ring [131]. As a result, mass transfer efficiency is determined by solvent polarity, viscosity, and the extent of the interactions between SLM and the target analyte. Furthermore, the SLM functions both as a selective barrier to salts, proteins, and phospholipids and as an enrichment phase for charged analytes [131].

Currents employed in EME are generally in the low μA range. Figure 13 (left) represents the typical trend of the current as a function of time for an unstable SLM. In these cases, the electrical resistance decreases due to solvent leakage and to the diffusion of water into the hydrophobic liquid membrane, leading to a rise in current. On the other hand, Fig. 13 (right) illustrates the current response in a stable SLM. Experts recommend constant current monitoring in the circuit during EME and to operate with systems below <50 μA to ensure reproducibility and reliability of the extraction [131].

It has been demonstrated that experimental outcomes can deviate from theoretical behavior, with drops in recovery at longer extraction times. This can be primarily attributed to pH changes in both acceptor and donor solutions caused by electrolysis [132, 133], according to the following equations:

$$\text{Cathode}: 2H^+ + 2e^- \rightarrow H_2 \tag{1}$$

$$\text{Anode}: H_2O \rightarrow 2H^+ + 1/2O_2 + 2e^- \tag{2}$$

Specifically, from Eqs. (1) and (2), it can be observed that pH increases in the sample and decreases in the acceptor solution [133].

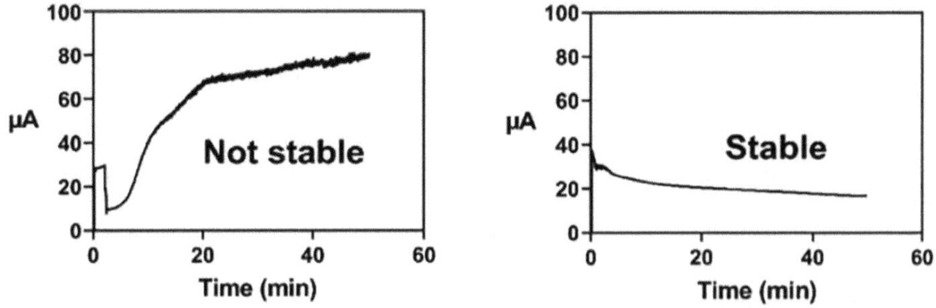

Fig. 13 Plots showing the relationship between current and extraction time with (left) an unstable and (right) a stable SLM. A rising current indicates instability due to membrane leakage. (Adapted from Hansen et al. [131], *Advances in Sample Preparation*, 2022, https://doi.org/10.1016/j.sampre.2022.100020. Licensed under CC BY 4.0). SLM supported liquid membrane

EME case studies

Despite EME's extensive use in small-molecule extraction protocols, intact peptide extraction poses additional challenges that should be carefully addressed [134–139].

Peptide migration across the SLM is inherently more complex due to their intrinsic hydrophilicity, their variable net charge and size, and a structurally intricate nature. As a result, optimizing the chemical composition of the SLM is crucial for promoting favorable interactions and achieving satisfactory mass transfer of the peptide under the electric field [131].

The molecular structure of the SLM, particularly its hydrogen bond donor (HBD) and acceptor (HBA) capabilities and aromatic properties, plays a crucial role in optimizing peptide migration across the membrane [128]. For polar peptides (logP <0), the combination of hydrogen bond and cation-π may have a limited effect on their solubility in hydrophobic SLM, and the addition of an ion-pair reagent, such as di-(2-ethylhexyl) phosphate (DEHP), is required to enhance peptide mass transfer in the EME system (Fig. 11a) [140].

In the past decade, the importance of membrane chemical composition, ion-pairing strategies, and non-covalent interactions for peptide EME has been highlighted. The following case studies emphasized the implementation of EME in a range of peptidomic workflows, from model systems to real biological matrices.

In their 2008 innovative study, Balchen et al. demonstrated for the first time the suitability of EME for extracting a panel of eight polar peptides [141]. The selected model peptides covered a wide range of physicochemical properties, including size, polarity, and net charge, as summarized in Table 3.

The extraction was carried out in 5 min with an SLM composed of 1-octanol containing 15% (w/w) of DHEP as an ion-pairing reagent. The results revealed a high variability in the recovery rates (22–56%), reflecting a strong dependence on the physicochemical properties of each peptide. A 20-fold preconcentration factor was also achieved. Notably, peptides with a higher net positive charge and moderate hydrophobicity (i.e., Ile-Pro-Ile, angiotensin II) showed the highest extraction

Table 3 Physicochemical properties of the peptide involved in the study. Adapted from [141]

Peptide	Amino acid sequence	No. of hydrophobic residues	Basic/acidic moieties	Net charge at pH 3
Angiotensin I	NRVYIHPFHL	3	4/2	3.1
Angiotensin II	NRVYIHPF	2	3/2	2.5
Angiotensin III	RVYIHPF	2	3/1	2.3
Glu-Glu-Leu	EEL	1	1/2	0.08
Ile-Pro-Ile	IPI	2	1/1	0.16
Neurotensin	pyroELYENKPRRPYIL	3	3/2	1.95
Neurotensin 1–6	pyroELYENK	1	1/2	0
Neurotensin 1–8	pyroELYENKPR	1	2/2	1

efficiency. Conversely, more hydrophilic peptides with low or neutral net charge (i.e., neurotensin 1–6) exhibited poor recovery, with over 50% of the peptide retained in the sample.

Overall, this study established EME as a promising approach in the selective extraction of peptides, demonstrating both its technical feasibility and analytical potential within peptidomic platforms.

In a follow-up study [142], EME was employed for the simultaneous determination of three low-abundant neuropeptides (angiotensin II, Leu-enkephalin, and endomorphin I) in real plasma samples. Due to their low abundance and the degradation operated by proteases, a reliable analysis of these peptides in plasma is challenging. In this setting, EME may offer selective enrichment while reducing matrix interferences. To ensure membrane stability and promote mass transfer, a ternary mixture comprising 1-octanol, di-isobutyl ketone, and DHEP was designed and optimized. Plasma was acidified to a pH of 3 with 50 mM HCl to ensure peptide protonation and migration under the electrical field (Fig. 14). The extraction lasted 5 min at 20 V, and the enriched acceptor solution was injected into the LC-MS/MS system, resulting in signal-to-noise ratios in the range of 24–60 pg/mL. This study demonstrated the reliability and feasibility of EME in analyzing real biological specimens by enriching low-abundant bioactive peptides.

In 2024, Rye et al. conducted the first large-scale, systematic assessment of EME's performance across a diverse library of nearly 6000 peptides spanning sequence, hydrophilicity, charge, and size [135]. While previous studies demonstrated the feasibility of EME in the extraction of small sets of bioactive peptides, this work provides a comprehensive evaluation of its applicability in modern peptidomics workflows. EME experimental conditions involved the use of an SLM composed of a 1:1 mixture of 2-nitrophenyl octyl ether and carvacrol, supplemented with 2%DEHP, and a voltage of 10 V over 45 min at pH 3 and 1.8 in donor and acceptor solutions, respectively. More than half of the peptides exhibited a higher MS signal in the acceptor solution than in the sample, demonstrating the suitability of the technique. The use of a machine learning approach permitted the study of

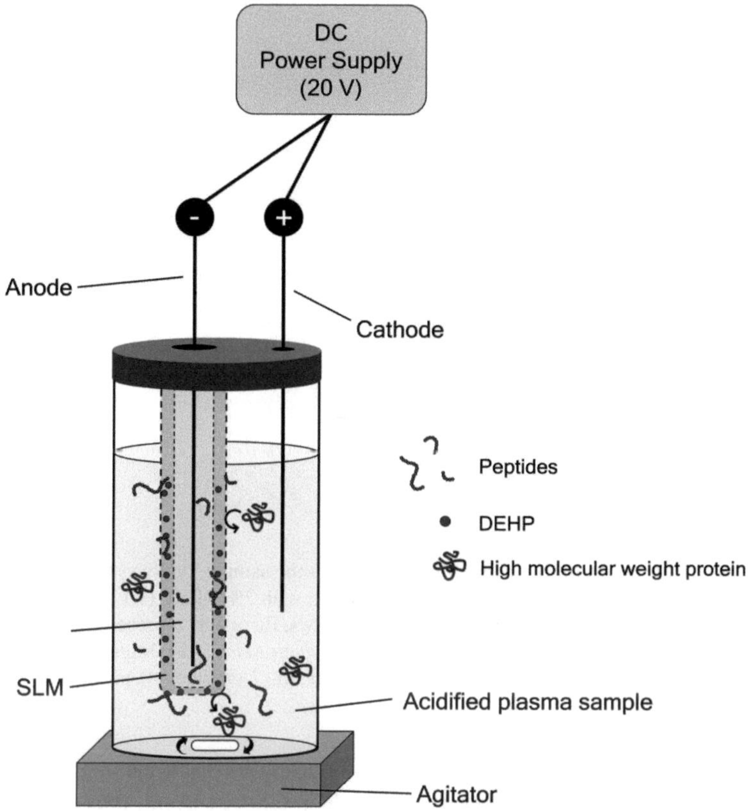

Fig. 14 Schematic illustration of the EME setup employed in the case study [142]

mass transfer dependence on the balance between net charge and hydrophobicity. They found out that peptides with a net positive charge (≥ +2) and a Grand Average of Hydropathicity (GRAVY) index >0 were most efficiently extracted. On the other hand, those with low or no charge at acidic pH or strong hydrophilicity showed poor mass transfer. This phenomenon can be rationalized by the influence of both electrokinetic migration and **non-covalent interactions** (including **hydrogen bonding**, **π-cation interactions**, and **ion pairing** with DEHP) on transport modulation.

While this systematic understanding provides a rationale for the observed mass transfer behavior, the study presents certain limitations. Extractions were performed using buffered aqueous solutions rather than complex biological matrices, whereas experimental conditions may not have been optimal for every peptide. However, the insights provided by Rye et al. offer critical implications for advancing EME in future peptidomics.

Deep eutectic solvents (DESs) have emerged as a greener alternative to traditional organic solvents, offering a tunable and hydrophilic environment for improving polar peptide mass transfer in EME approaches [140]. Typically, DESs are composed of a mixture of two (or more) solid components, one acting as HBD and

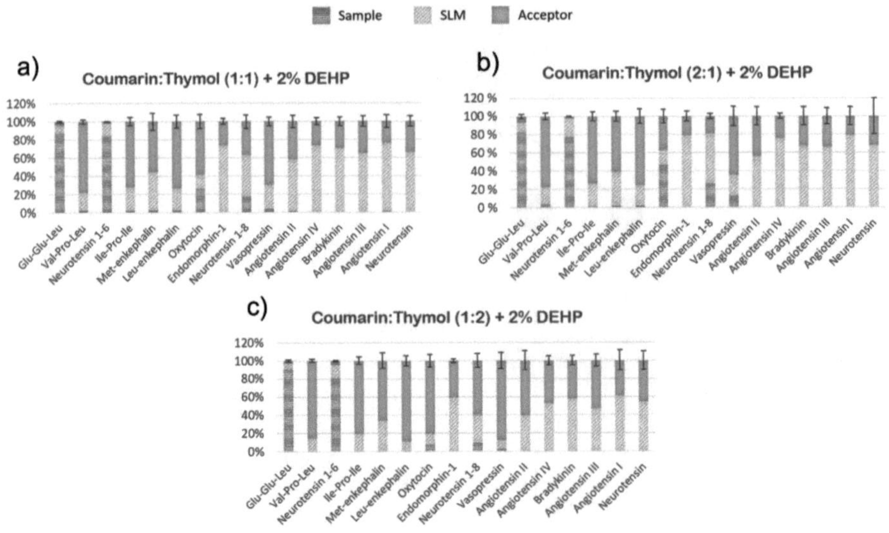

Fig. 15 Relative amounts of peptides (blue) retained in the sample, (yellow) trapped in the SLM, and (green) migrated in the acceptor solution after EME with 2% DEHP (v/v) added to coumarin-to-thymol in a ratio of (**a**) 1:1, (**b**) 2:1, and (**c**) 1:2 as SLMs. Error bars represent standard deviation ($n = 4$). (Adapted from Rye et al. [140], *Analytica Chimica Acta*, 2024, https://doi.org/10.1016/j.aca.2021.338717. Licensed under CC BY 4.0). DEHP di-(2-ethylhexyl) phosphate, SLM supported liquid membrane

the other as HBA. Upon mixing, hydrogen bonding interactions result in a significantly reduced melting point, known as the deep eutectic point, thus forming a stable liquid at room temperature. Compared to ionic liquids, DESs are generally considered more biodegradable, less toxic, and cheaper, encouraging their use in green analytical chemistry [143].

In a pivotal study, Rye et al. demonstrated the viability of using DES-based SLMs in the extraction of a panel of 16 peptides ranging from 3 to 13 amino acids [140]. They evaluated the performance of a DES system composed of coumarin and thymol in various molar ratios (1:1, 2:2, and 1:2), supplemented with 2% DEHP. The extent of peptide retained in the donor solution following extraction is shown in Fig. 15, highlighting how DES composition may influence extraction performance.

The results indicated that an increased thymol content (1:2 coumarin-to-thymol ratio) boosted extraction efficiency, likely due to the hydrogen bonding between thymol (HBD) and peptide residues (HBA), improving peptide–SLM interactions. As illustrated in Fig. 15c, optimized conditions permitted the achievement of (i) recoveries above 80% for five peptides; (ii) recoveries between 40% and 80% for nine peptides; and (iii) poor recoveries (<5%) for two acidic peptides (i.e., neurotensin 1–6 and Glu-Glu-Leu) due to electrostatic repulsion and insufficient ion pairing. Conversely, the method presented great performance for basic peptides. Notably, prolonging extraction times improved the average peptide recovery from 55% at 15 min to 83% at 45 min, outperforming traditional hydrophobic SLMs.

This study highlighted the importance of accounting not only for peptide physicochemical properties but also for peptide-specific sequence, given their role in impacting peptide–membrane interactions.

Table 4 summarizes an overview of the main research regarding applications of EME in peptidomics.

Challenges and future perspectives
Electromembrane extraction represents a promising and innovative approach in the development of peptidomics workflows. It offers advantages, including high selectivity, effective sample cleanup, and minimal sample handling, which enhance peptide detection and mitigate degradation risks.

However, the successful incorporation of EME into peptidomics workflows still faces challenges. One of the major limitations is the compatibility of ion-pairing reagents (e.g., DEHP) with biological samples. These ionic carriers, while playing a pivotal role in peptide mass transfer through SLM, can destabilize the system, triggering leakage and elevated current flow. For these reasons, balancing the concentration of the carrier in SLM is crucial, since higher concentrations increase peptide recovery, but can simultaneously compromise membrane integrity. Moreover, method recovery is highly peptide-dependent, reflecting the complex interplay between peptide physicochemical properties, amino acid sequence, and SLM characteristics. Alternative ion-pairing reagents should also be evaluated to improve acidic peptide extraction efficiency. In this context, the incorporation of DESs in liquid membranes represents a remarkable improvement, since they can provide a more sustainable and tunable platform tailored to peptide-specific properties.

In summary, EME offers promising opportunities for both targeted and untargeted peptidomics. However, its feasibility in such workflows still requires validation by widening the range of peptides to be extracted. Currently, the upper size limit for efficient peptide extraction by EME remains insufficiently characterized, highlighting the need for further investigations to extend its utility for larger peptides [128]. Overall, while EME has already demonstrated great potential in peptide analysis, future studies are needed for optimizing membrane compositions, ion-pairing strategies, and system integration for high-throughput workflows. These improvements will be pivotal to elevate EME from a specialized technique to a reliable tool in clinical and proteomic research.

4 Summary

In clinical peptidomics, sample preparation is fundamental for the accurate analysis of intact endogenous peptides. Unlike traditional proteomics, peptidomics workflows normally avoid reduction, alkylation, and enzymatic digestion to preserve native peptide structures and modifications. This makes sample treatment even more challenging, as peptides are often present at very low concentrations (pg/mL

Table 4 Overview of EME applications in the determination of peptides in biological fluids

EME design	Sample type	SLM	Acceptor solution	Extraction time and voltage	Target peptides	Recovery (%)	Limit of detection (ng mL^{-1})[a]	References
Hollow fiber	500 µL human plasma acidified with HCl to pH 3	1-octanol + 8% DEHP	25 µL 50 mM HCl	10 min at 15 V	Angiotensin I Angiotensin II Angiotensin III	36 25 43	nd 0.24 nd	[144]
Hollow fiber	2 mL plasma acidified with HCl to pH 3	1-octanol: di-isobutyl ketone: DHEP (55:35:10 w/w/w)	25 µL 50 mM HCl (pH 1.3)	5 min at 20 V	Angiotensin I Leu-enkephalin Endomorphin I	3 4 10	0.06 0.024 0.024	[142]
Hollow fiber	5 mL human plasma diluted with water (pH 7)	1 -octanol + 5% DEHP	30 µL 100 mM HCl, (pH 1)	15 min at 20 V	Alarelin Leuprolide Triptorelin Buserelin	49 71 56 58	0.6 0.6 0.6 0.6	[145]
Hollow fiber	4 mL rabbit plasma acidified with HCl to pH 4.5	1-octanol: 2-ethylhexanol (1:1) + 10% DEHP	25 µL (pH 1)	5 min at 5 V	Leuprolide Triptorelin	80.3 75.5	0.15 0.15	[139]
Conductive vial	300 µL human plasma diluted with phosphate-citrate buffer to pH 3	Tris(pentyl) phosphate	300 µL 50 mM phosphoric acid (pH 1.8)	15 min at 20 V	Met-enkephalin Leu-enkephalin Endomorphin	40 59 66	0.27 0.03 0.08	[134]

[a]Calculated as signal-to-noise ratio of 3
nd not determined

range) and are prone to degradation, chemical modifications, and nonspecific adsorption to material surfaces.

To address these issues, in recent years, advanced techniques such as microextraction by packed sorbent (MEPS), electromembrane extraction, and ultrafiltration have been employed to enrich and purify peptides while minimizing loss and alteration. MEPS is attractive due to its automation potential and environmentally friendly use of minimal sample and solvent volumes. Ultrafiltration provides simple and effective size-based separation, while electromembrane extraction offers high selectivity and clean isolation, both enhancing peptidomics workflows.

In the future, the integration and optimization of these techniques may pave the way to further improve sensitivity, selectivity, and throughput in clinical peptidomics research.

References

1. Melby JA, Roberts DS, Larson EJ, Brown KA, Bayne EF, Jin S, Ge Y (2021) Novel strategies to address the challenges in top-down proteomics. J Am Soc Mass Spectrom 32:1278–1294. https://doi.org/10.1021/jasms.1c00099
2. Schulz-Knappe P, Zucht H-D, Heine G, Jürgens M, Hess R, Schrader M (2001) Peptidomics: the comprehensive analysis of peptides in complex biological mixtures. Comb Chem High Throughput Screen 4(2):207–217
3. Messner CB, Demichev V, Wang Z, Hartl J, Kustatscher G, Mülleder M, Ralser M (2023) Mass spectrometry-based high-throughput proteomics and its role in biomedical studies and systems biology. Proteomics 23. https://doi.org/10.1002/pmic.202200013
4. Wishart DS (2019) Metabolomics for investigating physiological and pathophysiological processes. Physiol Rev 99:1819–1875. https://doi.org/10.1152/physrev.00035.2018.-Metabolomics
5. Schrader M, Schulz-Knappe P, Fricker LD (2014) Historical perspective of peptidomics. EuPA Open Proteom 3:171–182. https://doi.org/10.1016/j.euprot.2014.02.014
6. Schulte I, Tammen H, Selle H, Zucht HD, Schulz-Knappe P (2005) Chapter 7 Clinical peptidomics: peptide-biomarker discovery in blood. Compr Anal Chem 46:385–409. https://doi.org/10.1016/S0166-526X(05)46007-9
7. De Cristofaro M, Lenzi A, Ghimenti S, Biagini D, Bertazzo G, Vivaldi FM, Armenia S, Pugliese NR, Masi S, Di Francesco F, Lomonaco T (2024) Decoding the challenges: navigating intact peptide mass spectrometry-based analysis for biological applications. Crit Rev Anal Chem. https://doi.org/10.1080/10408347.2024.2427140
8. Serrao S, Firinu D, Olianas A, Deidda M, Contini C, Iavarone F, Sanna MT, Boroumand M, Amado F, Castagnola M, Messana I, Del Giacco S, Manconi B, Cabras T (2020) Top-down proteomics of human saliva discloses significant variations of the protein profile in patients with Mastocytosis. J Proteome Res 19:3238–3253. https://doi.org/10.1021/acs.jproteome.0c00207
9. Torrini F, Scarano S, Palladino P, Minunni M (2023) Advances and perspectives in the analytical technology for small peptide hormones analysis: a glimpse to gonadorelin. J Pharm Biomed Anal 228. https://doi.org/10.1016/j.jpba.2023.115312
10. Schrader M (2018) Origins, technological development, and applications of peptidomics. In: Methods in molecular biology. Humana Press Inc., pp 3–39. https://doi.org/10.1007/978-1-4939-7537-2_1

11. Vasan RS (2006) Biomarkers of cardiovascular disease: molecular basis and practical considerations. Circulation 113:2335–2362. https://doi.org/10.1161/CIRCULATIONAHA.104.482570
12. Foreman RE, George AL, Reimann F, Gribble FM, Kay RG (2021) Peptidomics: a review of clinical applications and methodologies. J Proteome Res 20:3782–3797. https://doi.org/10.1021/acs.jproteome.1c00295
13. Peng J, Zhang H, Niu H, Wu R (2020) Peptidomic analyses: the progress in enrichment and identification of endogenous peptides. TrAC Trends Anal Chem 125. https://doi.org/10.1016/j.trac.2020.115835
14. Cai T, Yang F (2017) Strategies for characterization of low-abundant intact or truncated low-molecular-weight proteins from human plasma. In: Enzymes (Essen). Academic Press, pp 105–123. https://doi.org/10.1016/bs.enz.2017.08.004
15. Heurtault B, Reix N, Meyer N, Gasser F, Wendling MJ, Ratomponirina C, Jeandidier N, Sapin R, Agin A (2014) Extensive study of human insulin immunoassays: promises and pitfalls for insulin analogue detection and quantification. Clin Chem Lab Med 52:355–362. https://doi.org/10.1515/cclm-2013-0427
16. Pauletti GM, Gangwar S, Knipp GT, Nerurkar MM, Okumu FW, Tamura K, Siahaan TJ, Borchardt RT (1996) Structural requirements for intestinal absorption of peptide drugs. 41:3
17. Wirobski G, Schaebs FS, Range F, Marshall-Pescini S, Deschner T (2021) Analytical and physiological validation of an enzyme immunoassay to measure oxytocin in dog, wolf, and human urine samples. Sci Rep 11. https://doi.org/10.1038/s41598-021-92356-z
18. Holst JJ, Albrechtsen NJW (2019) Methods and guidelines for measurement of glucagon in plasma. Int J Mol Sci 20. https://doi.org/10.3390/ijms20215416
19. Saenger AK, Rodriguez-Fraga O, Ler R, Ordonez-Llanos J, Jaffe AS, Goetze JP, Apple FS (2017) Specificity of B-type natriuretic peptide assays: cross-reactivity with different BNP, NT-proBNP, and proBNP peptides. Clin Chem 63:351–358. https://doi.org/10.1373/clinchem.2016.263749
20. Hu L, Ye M, Jiang X, Feng S, Zou H (2007) Advances in hyphenated analytical techniques for shotgun proteome and peptidome analysis-a review. Anal Chim Acta 598:193–204. https://doi.org/10.1016/j.aca.2007.07.046
21. Jiang Y, Rex DAB, Schuster D, Neely BA, Rosano GL, Volkmar N, Momenzadeh A, Peters-Clarke TM, Egbert SB, Kreimer S, Doud EH, Crook OM, Yadav AK, Vanuopadath M, Hegeman AD, Mayta ML, Duboff AG, Riley NM, Moritz RL, Meyer JG (2024) Comprehensive overview of bottom-up proteomics using mass spectrometry. ACS Meas Sci Au 4:338–417. https://doi.org/10.1021/acsmeasuresciau.3c00068
22. Wang EH, Combe PC, Schug KA (2016) Multiple reaction monitoring for direct quantitation of intact proteins using a triple quadrupole mass spectrometer. J Am Soc Mass Spectrom 27:886–896. https://doi.org/10.1007/s13361-016-1368-2
23. Van Den Broek I, Van Dongen WD (2015) LC-MS-based quantification of intact proteins: perspective for clinical and bioanalytical applications. Bioanalysis 7:1943–1958. https://doi.org/10.4155/bio.15.113
24. Trenchevska O, Koska J, Sinari S, Yassine H, Reaven PD, Billheimer DD, Nelson RW, Nedelkov D (2016) Association of cystatin C proteoforms with estimated glomerular filtration rate. Clin Mass Spectrom 1:27–31. https://doi.org/10.1016/j.clinms.2016.11.001
25. Lenzi A, De Cristofaro M, Biagini D, Ghimenti S, Armenia S, Pugliese NR, Masi S, Di Francesco F, Lomonaco T (2024) Development of a high-throughput liquid chromatography-tandem mass spectrometry platform for the determination of intact natriuretic peptides in human plasma. Talanta 275. https://doi.org/10.1016/j.talanta.2024.126077
26. Chambers EE, Legido-Quigley C, Smith N, Fountain KJ (2013) Development of a fast method for direct analysis of intact synthetic insulins in human plasma: the large peptide challenge. Bioanalysis 5:65–81. https://doi.org/10.4155/bio.12.290
27. Feickert M, Burckhardt BB (2020) A design of experiments concept for the minimization of nonspecific peptide adsorption in the mass spectrometric determination of substance P and related hemokinin-1. J Sep Sci 43:818–828. https://doi.org/10.1002/jssc.201901038

28. Maurer J, de Groot A, Martin L, Grouzmann E, Wuerzner G, Eugster PJ (2024) Quantification of endogenous Angiotensin 1-10, 1-9, 1-8, 1-7, and 1-5 in human plasma using micro-UHPLC-MS/MS: outlining the importance of the pre-analytics for reliable results. J Pharm Biomed Anal 243. https://doi.org/10.1016/j.jpba.2024.116101
29. Stokvis E, Rosing H, Beijnen JH (2005) Stable isotopically labeled internal standards in quantitative bioanalysis using liquid chromatography/mass spectrometry: necessity or not? Rapid Commun Mass Spectrom 19:401–407. https://doi.org/10.1002/rcm.1790
30. Bults P, Spanov B, Olaleye O, van de Merbel NC, Bischoff R (2019) Intact protein bioanalysis by liquid chromatography—high-resolution mass spectrometry. J Chromatogr B Analyt Technol Biomed Life Sci 1110–1111:155–167. https://doi.org/10.1016/j.jchromb.2019.01.032
31. Niederkofler EE, Kiernan UA, O'Rear J, Menon S, Saghir S, Protter AA, Nelson RW, Schellenberger U (2008) Detection of endogenous B-type natriuretic peptide at very low concentrations in patients with heart failure. Circ Heart Fail 1:258–264
32. Rai AJ, Vitzthum F (2006) Effects of preanalytical variables on peptide and protein measurements in human serum and plasma: implications for clinical proteomics. Expert Rev Proteomics 3:409–426. https://doi.org/10.1586/14789450.3.4.409
33. Maurer J, Grouzmann E, Eugster PJ (2023) Tutorial review for peptide assays: an ounce of pre-analytics is worth a pound of cure. J Chromatogr B Analyt Technol Biomed Life Sci 1229. https://doi.org/10.1016/j.jchromb.2023.123904
34. Manconi B, Liori B, Cabras T, Iavarone F, Manni A, Messana I, Castagnola M, Olianas A (2017) Top-down HPLC-ESI–MS proteomic analysis of saliva of edentulous subjects evidenced high levels of cystatin A, cystatin B and SPRR3. Arch Oral Biol 77:68–74. https://doi.org/10.1016/j.archoralbio.2017.01.021
35. Haupt S, Eckstein ML, Wolf A, Zimmer RT, Wachsmuth NB, Moser O (2021) Eat, train, sleep—retreat? Hormonal interactions of intermittent fasting, exercise and circadian rhythm. Biomolecules 11. https://doi.org/10.3390/biom11040516
36. Maes E, Oeyen E, Boonen K, Schildermans K, Mertens I, Pauwels P, Valkenborg D, Baggerman G (2019) The challenges of peptidomics in complementing proteomics in a clinical context. Mass Spectrom Rev 38:253–264. https://doi.org/10.1002/mas.21581
37. Li L, Wu J, Lyon CJ, Jiang L, Hu TY (2023) Clinical peptidomics: advances in instrumentation, analyses, and applications. BME Front 4. https://doi.org/10.34133/bmef.0019
38. Hanon EA, Sturgeon CM, Lamb EJ (2013) Sampling and storage conditions influencing the measurement of parathyroid hormone in blood samples: a systematic review. Clin Chem Lab Med 51:1925–1941. https://doi.org/10.1515/cclm-2013-0315
39. Hartog H, van der Graaf WTA, Wesseling J, van der Veer E, Boezen HM (2008) Measurement of insulin-like growth factor-1 and insulin-like growth factor binding protein-3 after delayed separation of whole blood samples. Clin Biochem 41:636–639. https://doi.org/10.1016/j.clinbiochem.2008.02.001
40. Cemin R, Daves M (2015) Pre-analytic variability in cardiovascular biomarker testing. J Thorac Dis 7:E395–E401. https://doi.org/10.3978/j.issn.2072-1439.2015.10.03
41. Forgrave LM, Van Der Gugten JG, Nguyen Q, Demarco ML (2022) Establishing pre-analytical requirements and maximizing peptide recovery in the analytical phase for mass spectrometric quantification of amyloid-β peptides 1-42 and 1-40 in CSF. Clin Chem Lab Med 60:198–206. https://doi.org/10.1515/cclm-2021-0549
42. Grankvist K, Gomez R, Nybo M, Lima-Oliveira G, Von Meyer A (2019) Preanalytical aspects on short- and long-term storage of serum and plasma. Diagnosi 6:51–56. https://doi.org/10.1515/dx-2018-0037
43. Dufresne J, Florentinus-Mefailoski A, Ajambo J, Ferwa A, Bowden P, Marshall J (2017) Random and independent sampling of endogenous tryptic peptides from normal human EDTA plasma by liquid chromatography micro electrospray ionization and tandem mass spectrometry. Clin Proteom 14:1–15. https://doi.org/10.1186/s12014-017-9176-7
44. Carr D (n.d.) The handbook of analysis and purification of peptides and proteins by reversed-phase HPLC. www.harvaardapparatus.com

45. Wang Y, Chen H, Zhao M, Feng L, Liu Z, Zeng Q, Shi W, Zhu W, Song L, Zhu J, Lu H (2023) Oxidation and reduction analysis of therapeutic recombinant human interleukin-15 by HPLC and LC-MS. Appl Microbiol Biotechnol 107:3217–3227. https://doi.org/10.1007/s00253-023-12508-1
46. Li S, Schöneich C, Borchardt RT (1995) Chemical instability of protein pharmaceuticals: mechanisms of oxidation and strategies for stabilization. Biotechnol Bioeng 48:490–500. https://doi.org/10.1002/bit.260480511
47. Aubry AF (2011) LC-MS/MS bioanalytical challenge: ultra-high sensitivity assays. Bioanalysis 3:1819–1825. https://doi.org/10.4155/bio.11.166
48. Bongaerts J, De Bundel D, Smolders I, Mangelings D, Vander Heyden Y, Van Eeckhaut A (2023) Improving the LC-MS/MS analysis of neuromedin U-8 and neuromedin S by minimizing their adsorption behavior and optimizing UHPLC and MS parameters. J Pharm Biomed Anal 228. https://doi.org/10.1016/j.jpba.2023.115306
49. Goebel-Stengel M, Stengel A, Taché Y, Reeve JR (2011) The importance of using the optimal plasticware and glassware in studies involving peptides. Anal Biochem 414:38–46. https://doi.org/10.1016/j.ab.2011.02.009
50. Fekete S, Veuthey JL, Guillarme D (2012) New trends in reversed-phase liquid chromatographic separations of therapeutic peptides and proteins: theory and applications. J Pharm Biomed Anal 69:9–27. https://doi.org/10.1016/j.jpba.2012.03.024
51. Kellet J, Birdsall RE, Qing Yu Y (n.d.) Application of difluoroacetic acid to improve optical and MS performance in peptide LC-UV/MS
52. Simon R, Enjalbert Q, Biarc J, Lemoine J, Salvador A (2012) Evaluation of hydrophilic interaction chromatography (HILIC) versus C18 reversed-phase chromatography for targeted quantification of peptides by mass spectrometry. J Chromatogr A 1264:31–39. https://doi.org/10.1016/j.chroma.2012.09.059
53. Wang J, Kong S, Yan J, Jin G, Guo Z, Shen A, Xu J, Zhang X, Zou L, Liang X (2014) Hydrophilic interaction liquid chromatography-solid phase extraction directly combined with protein precipitation for the determination of triptorelin in plasma. J Chromatogr B Analyt Technol Biomed Life Sci 960:214–221. https://doi.org/10.1016/j.jchromb.2014.04.036
54. Hellinger R, Sigurdsson A, Wu W, Romanova EV, Li L, Sweedler JV, Süssmuth RD, Gruber CW (2023) Peptidomics. Nat Rev Methods Primers 3. https://doi.org/10.1038/s43586-023-00205-2
55. Descamps A, De Spiegeleer B, Wynendaele E, Naumovski P, De Spiegeleer A (2025) Sample preparation in analytical peptidomics: it's all about timing. Talanta Open 11. https://doi.org/10.1016/j.talo.2025.100450
56. Polson C, Sarkar P, Incledon B, Raguvaran V, Grant R (2003. www.elsevier.com/locate/chromb) Optimization of protein precipitation based upon effectiveness of protein removal and ionization effect in liquid chromatography-tandem mass spectrometry. 785:263
57. Chertov O, Biragyn A, Kwak LW, Simpson JT, Boronina T, Hoang VM, Prieto DRA, Conrads TP, Veenstra TD, Fisher RJ (2004) Organic solvent extraction of proteins and peptides from serum as an effective sample preparation for detection and identification of biomarkers by mass spectrometry. Proteomics 4:1195–1203. https://doi.org/10.1002/pmic.200300677
58. Kay R, Barton C, Ratcliffe L, Matharoo-Ball B, Brown P, Roberts J, Teale P, Creaser C (2008) Enrichment of low molecular weight serum proteins using acetonitrile precipitation for mass spectrometry based proteomic analysis. Rapid Commun Mass Spectrom 22:3255–3260. https://doi.org/10.1002/rcm.3729
59. Yang Y, Bao X, Ning J, Huang R, Liang Y, Yan Z, Chen H, Ding L, Shu C (2024) A sensitive and specific LC–MS/MS method for determination of a novel antihypertensive peptide FR-6 in rat plasma and pharmacokinetic study. Heliyon 10. https://doi.org/10.1016/j.heliyon.2024.e26209
60. Yoshikawa H, Hirano A, Arakawa T, Shiraki K (2012) Mechanistic insights into protein precipitation by alcohol. Int J Biol Macromol 50:865–871. https://doi.org/10.1016/j.ijbiomac.2011.11.005

61. Baghalabadi V, Doucette AA (2020) Mass spectrometry profiling of low molecular weight proteins and peptides isolated by acetone precipitation. Anal Chim Acta 1138:38–48. https://doi.org/10.1016/j.aca.2020.08.057
62. Koontz L (2014) TCA precipitation. In: Methods enzymol. Academic Press Inc., pp 3–10. https://doi.org/10.1016/B978-0-12-420119-4.00001-X
63. Wingfield P (1998) Protein precipitation using ammonium sulfate. Curr Protoc Protein Sci 13. https://doi.org/10.1002/0471140864.psa03fs13
64. Kramer RM, Shende VR, Motl N, Pace CN, Scholtz JM (2012) Toward a molecular understanding of protein solubility: increased negative surface charge correlates with increased solubility. Biophys J 102:1907–1915. https://doi.org/10.1016/j.bpj.2012.01.060
65. Periasamy P, Rajandran S, Ziegman R, Casey M, Nakamura K, Kore H, Datta K, Gowda H (2021) A simple organic solvent precipitation method to improve detection of low molecular weight proteins. Proteomics 21. https://doi.org/10.1002/pmic.202100152
66. Parker BL, Burchfield JG, Clayton D, Geddes TA, Payne RJ, Kiens B, Wojtaszewski JFP, Richter EA, James DE (2017) Multiplexed temporal quantification of the exercise-regulated plasma peptidome. Mol Cell Proteomics 16:2055–2068. https://doi.org/10.1074/mcp.RA117.000020
67. Pringels L, Broeckx V, Boonen K, Landuyt B, Schoofs L (2018) Abundant plasma protein depletion using ammonium sulfate precipitation and Protein A affinity chromatography. J Chromatogr B Analyt Technol Biomed Life Sci 1089:43–59. https://doi.org/10.1016/j.jchromb.2018.04.045
68. Nickerson JL, Doucette AA (2020) Rapid and quantitative protein precipitation for proteome analysis by mass spectrometry. J Proteome Res 19:2035–2042. https://doi.org/10.1021/acs.jproteome.9b00867
69. Niu L, Zhang H, Wu Z, Wang Y, Liu H, Wu X, Wang W (2018) Modified TCA/acetone precipitation of plant proteins for proteomic analysis. PLoS One 13. https://doi.org/10.1371/journal.pone.0202238
70. Gangnus T, Burckhardt BB (n.d.) Targeted LC-MS/MS platform for the comprehensive determination of peptides in the kallikrein-kinin system. https://doi.org/10.1007/s00216-021-03231-9
71. Wang Y, Qu Y, Bellows CL, Ahn JS, Burkey JL, Taylor SW (2012) Simultaneous quantification of davalintide, a novel amylin-mimetic peptide, and its active metabolite in beagle and rat plasma by online SPE and LC-MS/MS. Bioanalysis 4:2141–2152. https://doi.org/10.4155/bio.12.184
72. Demeuse J, Huyghebaert L, Determe W, Schoumacher M, Grifnée E, Massonnet P, Dubrowski T, Rechchad M, Segura JF, Peeters S, Cavalier E, Le Goff C (2024) Development and validation of an LC-MS/MS method for the simultaneous quantitation of angiotensin (1–7), (1–8), (1–9) and (1–10) in human plasma. J Chromatogr B Analyt Technol Biomed Life Sci 1232. https://doi.org/10.1016/j.jchromb.2023.123943
73. Cui L, Nithipatikom K, Campbell WB (2007) Simultaneous analysis of angiotensin peptides by LC-MS and LC-MS/MS: metabolism by bovine adrenal endothelial cells. Anal Biochem 369:27–33. https://doi.org/10.1016/j.ab.2007.06.045
74. Maráková K, Renner BJ, Thomas SL, Opetová M, Tomašovský R, Rai AJ, Schug KA (2023) Solid phase extraction as sample pretreatment method for top-down quantitative analysis of low molecular weight proteins from biological samples using liquid chromatography—triple quadrupole mass spectrometry. Anal Chim Acta 1243. https://doi.org/10.1016/j.aca.2023.340801
75. Thomas SL, Thacker JB, Schug KA, Maráková K (2021) Sample preparation and fractionation techniques for intact proteins for mass spectrometric analysis. J Sep Sci 44:211–246. https://doi.org/10.1002/jssc.202000936
76. Bugyi F, Turiák L, Drahos L, Tóth G (2023) Optimization of reversed-phase solid-phase extraction for shotgun proteomics analysis. J Mass Spectrom 58. https://doi.org/10.1002/jms.4965

77. Herraiz T, Casal V (1995) Evaluation of solid-phase extraction procedures in peptide analysis. J Chromatogr A 708:209
78. Badawy MEI, El-Nouby MAM, Kimani PK, Lim LW, Rabea EI (2022) A review of the modern principles and applications of solid-phase extraction techniques in chromatographic analysis. Anal Sci 38:1457–1487. https://doi.org/10.1007/s44211-022-00190-8
79. Filippou O, Bitas D, Samanidou V (2017) Green approaches in sample preparation of bioanalytical samples prior to chromatographic analysis. J Chromatogr B Analyt Technol Biomed Life Sci 1043:44–62. https://doi.org/10.1016/j.jchromb.2016.08.040
80. Altun Z, Abdel-Rehim M, Blomberg LG (2004) New trends in sample preparation: on-line microextraction in packed syringe (MEPS) for LC and GC applications: part III: determination and validation of local anaesthetics in human plasma samples using a cation-exchange sorbent, and MEPS-LC-MS-MS. J Chromatogr B Analyt Technol Biomed Life Sci 813:129–135. https://doi.org/10.1016/j.jchromb.2004.09.020
81. Abdel-Rehim M (2011) Microextraction by packed sorbent (MEPS): a tutorial. Anal Chim Acta 701:119–128. https://doi.org/10.1016/j.aca.2011.05.037
82. Ashri NY, Abdel-Rehim M (2011) Sample treatment based on extraction techniques in biological matrices. Bioanalysis 3:2003–2018. https://doi.org/10.4155/bio.11.201
83. Yang L, Said R, Abdel-Rehim M (2017) Sorbent, device, matrix and application in microextraction by packed sorbent (MEPS): a review. J Chromatogr B Analyt Technol Biomed Life Sci 1043:33–43. https://doi.org/10.1016/j.jchromb.2016.10.044
84. Pereira J, Câmara JS, Colmsjö A, Abdel-Rehim M (2014) Microextraction by packed sorbent: an emerging, selective and high-throughput extraction technique in bioanalysis. Biomed Chromatogr 28:839–847. https://doi.org/10.1002/bmc.3156
85. Biagini D, Bertazzo G, Ghimenti S, Lenzi A, Oger C, Galano JM, Balas L, Durand T, Pugliese NR, Armenia S, Masi S, Di Francesco F, Lomonaco T (2025) Dried blood spot analysis of long-chain polyunsaturated fatty acids and oxylipins for monitoring heart failure. Free Radic Biol Med 233:13–23. https://doi.org/10.1016/j.freeradbiomed.2025.03.020
86. Biagini D, Ghimenti S, Lenzi A, Bonini A, Vivaldi F, Oger C, Galano JM, Balas L, Durand T, Salvo P, Di Francesco F, Lomonaco T (2023) Salivary lipid mediators: key indexes of inflammation regulation in heart failure disease. Free Radic Biol Med 201:55–65. https://doi.org/10.1016/j.freeradbiomed.2023.03.015
87. Biagini D, Mrakic-Sposta S, Bondi D, Ghimenti S, Lenzi A, Vivaldi F, Santangelo C, Verratti V, Pietrangelo T, Vezzoli A, Giardini G, Oger C, Galano JM, Balas L, Durand T, D'Angelo G, Lomonaco T, Di Francesco F (2024) A MEPS-UHPLC-MS/MS analytical platform to detect isoprostanoids and specialized pro-resolving mediators in the urinary extracellular vesicles of mountain ultramarathon runners. Talanta 279. https://doi.org/10.1016/j.talanta.2024.126619
88. Pereira JAM, Gonçalves J, Porto-Figueira P, Figueira JA, Alves V, Perestrelo R, Medina S, Câmara JS (2019) Current trends on microextraction by packed sorbent-fundamentals, application fields, innovative improvements and future applications. Analyst 144:5048–5074. https://doi.org/10.1039/c8an02464b
89. Ashri NY, Daryanavard M, Abdel-Rehim M (2013) Microextraction by packed sorbent and liquid chromatography-tandem mass spectrometry as a tool for quantification of peptides in plasma samples: determination of sensory neuron-specific receptors agonist BAM8-22 and antagonist BAM22-8 in plasma samples. Biomed Chromatogr 27:396–403. https://doi.org/10.1002/bmc.2806
90. Luque-Garcia JL, Neubert TA (2007) Sample preparation for serum/plasma profiling and biomarker identification by mass spectrometry. J Chromatogr A 1153:259–276. https://doi.org/10.1016/j.chroma.2006.11.054
91. Greening DW, Simpson RJ (2011) Low-molecular weight plasma proteome analysis using centrifugal ultrafiltration. In: Methods in molecular biology. Humana Press Inc., pp 109–124. https://doi.org/10.1007/978-1-61779-068-3_6

92. Saxena A, Tripathi BP, Kumar M, Shahi VK (2009) Membrane-based techniques for the separation and purification of proteins: an overview. Adv Colloid Interf Sci 145:1–22. https://doi.org/10.1016/j.cis.2008.07.004
93. Goldring JPD (2015) Methods to concentrate proteins for protein isolation, proteomic, and peptidomic evaluation. In: Detection of blotted proteins: methods and protocols. Springer, pp 5–18. https://doi.org/10.1007/978-1-4939-2718-0_2
94. Wan Y, Ghosh R, Cui Z (2004) Separation of human serum albumin and human immunoglobulins using carrier phase ultrafiltration. Biotechnol Prog 20:1103–1112. https://doi.org/10.1021/bp0343716
95. Galanakis CM (2015) Separation of functional macromolecules and micromolecules: from ultrafiltration to the border of nanofiltration. Trends Food Sci Technol 42:44–63. https://doi.org/10.1016/j.tifs.2014.11.005
96. Martínez-Maqueda D, Hernández-Ledesma B, Amigo L, Miralles B, Gómez-Ruiz JÁ (2013) Extraction/fractionation techniques for proteins and peptides and protein digestion. In: Proteomics in foods: principles and applications. Springer, pp 21–50. https://doi.org/10.1007/978-1-4614-5626-1_2
97. Doyen A, Beaulieu L, Saucier L, Pouliot Y, Bazinet L (2011) Impact of ultrafiltration membrane material on peptide separation from a snow crab byproduct hydrolysate by electrodialysis with ultrafiltration membranes. J Agric Food Chem 59:1784–1792. https://doi.org/10.1021/jf103739m
98. Boukil A, Suwal S, Chamberland J, Pouliot Y, Doyen A (2018) Ultrafiltration performance and recovery of bioactive peptides after fractionation of tryptic hydrolysate generated from pressure-treated Β-lactoglobulin. J Memb Sci 556:42–53. https://doi.org/10.1016/j.memsci.2018.03.079
99. Kammakakam I, Lai Z (2023) Next-generation ultrafiltration membranes: a review of material design, properties, recent progress, and challenges. Chemosphere 316. https://doi.org/10.1016/j.chemosphere.2022.137669
100. Liu S, Li Z, Yu B, Wang S, Shen Y, Cong H (2020) Recent advances on protein separation and purification methods. Adv Colloid Interf Sci 284. https://doi.org/10.1016/j.cis.2020.102254
101. Li J, He X, Deng Y, Yang C (2019) An update on isolation methods for proteomic studies of extracellular vesicles in biofluids. Molecules 24. https://doi.org/10.3390/molecules24193516
102. Kong S, Zhang Q, Yang L, Huang Y, Liu M, Yan G, Zhao H, Wu M, Zhang X, Yang P, Cao W (2021) Effective enrichment strategy using boronic acid-functionalized mesoporous graphene-silica composites for intact N-and O-linked glycopeptide analysis in human serum. Anal Chem 93:6682–6691. https://doi.org/10.1021/acs.analchem.0c05482
103. Meng C, Zhen-Xian Z, Peng C, Kun W, Gang-Tian Z, Yu-Qi F, Shi-Hai Y, Cheng-Cai H, Zhi-Rong G, Xiao-Dong M, Ning-Wei Z, Chao L (2016) Monodisperse magnetic mesoporous silica microspheres facilitate the studies of gastric cancer-specific peptides in sera. RSC Adv 6:39963–39971. https://doi.org/10.1039/c5ra27378a
104. Kalita V, Bhattacharyya J, Bora SJ, Kalita R, Dutta PP (2025) Fractionation and separation of bioactive peptides using natural polymeric membrane. Braz J Pharm Sci 61. https://doi.org/10.1590/s2175-97902025e24081
105. Jiang F, Liu K, Zhao M, Tao X, Hu X, Lu S, Lu S (2020) Tunable high-molecular-weight silk fibroin polypeptide materials: fabrication and self-assembly mechanism. ACS Appl Bio Mater 3:3248–3259. https://doi.org/10.1021/acsabm.0c00231
106. Román-Doval R, Torres-Arellanes SP, Tenorio-Barajas AY, Gómez-Sánchez A, Valencia-Lazcano AA (2023) Chitosan: properties and its application in agriculture in context of molecular weight. Polymers (Basel) 15. https://doi.org/10.3390/polym15132867
107. Thimmiah BR, Chien BTC, Fui KS, Yon LS, Nallathambi G, Jeevanandam J, Danquah MK (2022) Nanoformulation of peptides for pharmaceutical applications: in vitro and in vivo perspectives. Appl Sci 12. https://doi.org/10.3390/app122412777

108. Arrutia F, Rubio R, Riera FA (2016) Production and membrane fractionation of bioactive peptides from a whey protein concentrate. J Food Eng 184:1–9. https://doi.org/10.1016/j.jfoodeng.2016.03.010
109. Van Midwoud PM, Rieux L, Bischoff R, Verpoorte E, Niederländer HAG (2007) Improvement of recovery and repeatability in liquid chromatography-mass spectrometry analysis of peptides. J Proteome Res 6:781–791. https://doi.org/10.1021/pr0604099
110. Verbeke F, Bracke N, Debunne N, Wynendaele E, De Spiegeleer B (2020) LC-MS compatible antiadsorption diluent for peptide analysis. Anal Chem 92:1712–1719. https://doi.org/10.1021/acs.analchem.9b01840
111. Hartman E, Forsberg F, Kjellström S, Petrlova J, Luo C, Scott A, Puthia M, Malmström J, Schmidtchen A (2024) Peptide clustering enhances large-scale analyses and reveals proteolytic signatures in mass spectrometry data. Nat Commun 15. https://doi.org/10.1038/s41467-024-51589-y
112. Harper RG, Workman SR, Schuetzner S, Timperman AT, Sutton JN (2004) Low-molecular-weight human serum proteome using ultrafiltration, isoelectric focusing, and mass spectrometry. Electrophoresis 25:1299–1306. https://doi.org/10.1002/elps.200405864
113. Dallas DC, Guerrero A, Parker EA, Robinson RC, Gan J, German JB, Barile D, Lebrilla CB (2015) Current peptidomics: applications, purification, identification, quantification, and functional analysis. Proteomics 15:1026–1038. https://doi.org/10.1002/pmic.201400310
114. Ratnaningsih E, Reynard R, Khoiruddin K, Wenten IG, Boopathy R (2021) Recent advancements of UF-based separation for selective enrichment of proteins and bioactive peptides—a review. Appl Sci (Switz) 11:1–36. https://doi.org/10.3390/app11031078
115. Georgiou HM, Rice GE, Baker MS (2001) Proteomic analysis of human plasma: failure of centrifugal ultrafiltration to remove albumin and other high molecular weight proteins. Proteomics 1:1503–1506. https://doi.org/10.1002/1615-9861(200111)1:12<1503::AID-PROT1503>3.0.CO;2-M
116. Aristoteli LP, Molloy MP, Baker MS (2007) Evaluation of endogenous plasma peptide extraction methods for mass spectrometric biomarker discovery. J Proteome Res 6:571–581. https://doi.org/10.1021/pr0602996
117. Williams D, Ackloo S, Zhu P, Bowden P, Evans KR, Addison CL, Lock C, Marshall JG (2010) Precipitation and selective extraction of human serum endogenous peptides with analysis by quadrupole time-of-flight mass spectrometry reveals posttranslational modifications and low-abundance peptides. Anal Bioanal Chem 396:1223–1247. https://doi.org/10.1007/s00216-009-3345-0
118. Zheng X, Baker H, Hancock WS (2006) Analysis of the low molecular weight serum peptidome using ultrafiltration and a hybrid ion trap-Fourier transform mass spectrometer. J Chromatogr A 1120:173–184. https://doi.org/10.1016/j.chroma.2006.01.098
119. Finoulst I, Pinkse M, Van Dongen W, Verhaert P (2011) Sample preparation techniques for the untargeted LC-MS-based discovery of peptides in complex biological matrices. J Biomed Biotechnol 2011. https://doi.org/10.1155/2011/245291
120. Vitorino R, Barros AS, Caseiro A, Ferreira R, Amado F (2012) Evaluation of different extraction procedures for salivary peptide analysis. Talanta 94:209–215. https://doi.org/10.1016/j.talanta.2012.03.023
121. Pedersen-Bjergaard S, Rasmussen KE (2006) Electrokinetic migration across artificial liquid membranes: new concept for rapid sample preparation of biological fluids. J Chromatogr A:183–190. https://doi.org/10.1016/j.chroma.2006.01.025
122. Wan L, Gao H, Gao H, Du R, Wang F, Wang Y, Chen M (2022) Selective extraction and determination of steroidal glycoalkaloids in potato tissues by electromembrane extraction combined with LC-MS/MS. Food Chem 367. https://doi.org/10.1016/j.foodchem.2021.130724
123. Pedersen-Bjergaard S, Huang C, Gjelstad A (2017) Electromembrane extraction–recent trends and where to go. J Pharm Anal 7:141–147. https://doi.org/10.1016/j.jpha.2017.04.002

124. Hay AO, Hansen FA (2024) Exploring electromembrane extraction coupled to fast LC-MS/MS as a high-throughput platform for determination of 12 polar endogenous metabolites in human plasma. J Chromatogr A 1737. https://doi.org/10.1016/j.chroma.2024.465451
125. Seip KF, Jensen H, Sønsteby MH, Gjelstad A, Pedersen-Bjergaard S (2013) Electromembrane extraction: distribution or electrophoresis? Electrophoresis 34:792–799. https://doi.org/10.1002/elps.201200587
126. Stalmach A, Albalat A, Mullen W, Mischak H (2013) Recent advances in capillary electrophoresis coupled to mass spectrometry for clinical proteomic applications. Electrophoresis 34:1452–1464. https://doi.org/10.1002/elps.201200708
127. Dowlatshah S, Hay AO, Noreng L, Hansen FA (2025) A practical tutorial for optimizing electromembrane extraction methods by response surface methodology. Adv Sample Prep 14. https://doi.org/10.1016/j.sampre.2025.100170
128. Rye TK, Dowlatshah S, Hansen FA, Halvorsen TG, Pedersen-Bjergaard S (2024) Electromembrane extraction of peptides—looking retrospectively into published studies. Adv Sample Prep 9. https://doi.org/10.1016/j.sampre.2024.100106
129. Martins RO, de Araújo GL, Simas RC, Chaves AR (2023) Electromembrane extraction (EME): fundamentals and applications. Talanta Open 7. https://doi.org/10.1016/j.talo.2023.100200
130. Zhou C, Dowlatshah S, Hay AO, Schüller M, Pedersen-Bjergaard S, Hansen FA (2023) Generic liquid membranes for electromembrane extraction of bases with low or moderate hydrophilicity. Anal Chem 95:8982–8989. https://doi.org/10.1021/acs.analchem.3c01052
131. Hansen FA, Pedersen-Bjergaard S (2022) Electromembrane extraction—looking closer into the liquid membrane. Adv Sample Prep 2. https://doi.org/10.1016/j.sampre.2022.100020
132. Kubáň P, Boček P (2015) The effects of electrolysis on operational solutions in electromembrane extraction: the role of acceptor solution. J Chromatogr A 1398:11–19. https://doi.org/10.1016/j.chroma.2015.04.024
133. Huang C, Jensen H, Seip KF, Gjelstad A, Pedersen-Bjergaard S (2016) Mass transfer in electromembrane extraction - the link between theory and experiments. J Sep Sci 39:188–197. https://doi.org/10.1002/jssc.201500905
134. Dowlatshah S, Hansen FA, Zhou C, Ramos-Payán M, Halvorsen TG, Pedersen-Bjergaard S (2023) Electromembrane extraction of peptides based on hydrogen bond interactions. Anal Chim Acta 1275. https://doi.org/10.1016/j.aca.2023.341610
135. Rye TK, Lee CY, Zellner A, Moen SH, Dowlatshah S, Grønhaug Halvorsen T, Pedersen-Bjergaard S, Hansen FA (2024) Electromembrane extraction of peptides based on charge, hydrophobicity, and size—a large-scale fundamental study of the extraction window. J Sep Sci 47. https://doi.org/10.1002/jssc.202400292
136. Pourahadi A, Nojavan S, Hosseiny Davarani SS (2020) Gel-electromembrane extraction of peptides: determination of five hypothalamic agents in human plasma samples. Talanta 217. https://doi.org/10.1016/j.talanta.2020.121025
137. Huang C, Gjelstad A, Pedersen-Bjergaard S (2015) Exhaustive extraction of peptides by electromembrane extraction. Anal Chim Acta 853:328–334. https://doi.org/10.1016/j.aca.2014.10.017
138. Rye TK, Hansen FA, Halvorsen TG, Pedersen-Bjergaard S (2024) Electromembrane extraction of human chorionic gonadotropin—a case study on mass transfer of intact protein versus signature peptide. Adv Sample Prep 11. https://doi.org/10.1016/j.sampre.2024.100126
139. Hosseiny Davarani SS, Pourahadi A, Ghasemzadeh P (2019) Quantification of controlled release leuprolide and triptorelin in rabbit plasma using electromembrane extraction coupled with HPLC–UV. Electrophoresis 40:1074–1081. https://doi.org/10.1002/elps.201800481
140. Rye TK, Martinovic G, Eie LV, Hansen FA, Halvorsen TG, Pedersen-Bjergaard S (2021) Electromembrane extraction of peptides using deep eutectic solvents as liquid membrane. Anal Chim Acta 1175. https://doi.org/10.1016/j.aca.2021.338717
141. Balchen M, Reubsaet L, Pedersen-Bjergaard S (2008) Electromembrane extraction of peptides. J Chromatogr A 1194:143–149. https://doi.org/10.1016/j.chroma.2008.04.041

142. Balchen M, Lund H, Reubsaet L, Pedersen-Bjergaard S (2012) Fast, selective, and sensitive analysis of low-abundance peptides in human plasma by electromembrane extraction. Anal Chim Acta 716:16–23. https://doi.org/10.1016/j.aca.2011.02.058
143. Kudłak B, Owczarek K, Namieśnik J (2015) Selected issues related to the toxicity of ionic liquids and deep eutectic solvents—a review. Environ Sci Pollut Res 22:11975–11992. https://doi.org/10.1007/s11356-015-4794-y
144. Balchen M, Halvorsen TG, Reubsaet L, Pedersen-Bjergaard S (2009) Rapid isolation of angiotensin peptides from plasma by electromembrane extraction. J Chromatogr A 1216:6900–6905. https://doi.org/10.1016/j.chroma.2009.08.037
145. Nojavan S, Bidarmanesh T, Mohammadi A, Yaripour S (2016) Electromembrane extraction of gonadotropin-releasing hormone agonists from plasma and wastewater samples. Electrophoresis 37:826–833. https://doi.org/10.1002/elps.201500555

Index

J.-L. Capelo-Martínez (ed.), *Emerging Sample Treatments in Proteomics*, Advances in Experimental Medicine and Biology 1510,
https://doi.org/10.1007/978-3-032-21638-0